T0256195

# Modern Anti-windup Synthesis

Princeton Series in Applied Mathematics

Series Editors: Ingrid Daubechies (Princeton University); Weinan E (Princeton University); Jan Karel Lenstra (Eindhoven University); Endre Süli (University of Oxford)

The Princeton Series in Applied Mathematics publishes high quality advanced texts and monographs in all areas of applied mathematics. Books include those of a theoretical and general nature as well as those dealing with the mathematics of specific applications areas and real-world situations.

*Chaotic Transitions in Deterministic and Stochastic Dynamical Systems: Applications of Melnikov Processes in Engineering, Physics, and Neuroscience*, Emil Simiu

*Selfsimilar Processes*, Paul Embrechts and Makoto Maejima

*Self-Regularity: A New Paradigm for Primal-Dual Interior Point Algorithms*, Jiming Peng, Cornelis Roos, and Tamas Terlaky

*Analytic Theory of Global Bifurcation: An Introduction*, Boris Buffoni and John Toland

*Entropy*, Andreas Greven, Gerhard Keller, and Gerald Warnecke, editors

*Auxiliary Signal Design for Failure Detection*, Stephen L. Campbell and Ramine Nikoukhah

*Thermodynamics: A Dynamical Systems Approach*, Wassim M. Haddad, VijaySekhar Chellaboina, and Sergey G. Nersesov

*Optimization: Insights and Applications*, Jan Brinkhuis and Vladimir Tikhomirov

*Max Plus at Work, Modeling and Analysis of Synchronized Systems: A Course on Max-Plus Algebra and Its Applications*, Bernd Heidergott, Geert Jan Olsder, and Jacob van der Woude

*Impulsive and Hybrid Dynamical Systems Stability, Dissipativity, and Control*, Wassim M. Haddad, VijaySekhar Chellaboina, and Sergey G. Nersesov

*The Traveling Salesman Problem: A Computational Study*, David L. Applegate, Robert E. Bixby, Vasek Chvatal, and William J. Cook

*Positive Definite Matrices*, Rajendra Bhatia

*Genomic Signal Processing*, Ilya Shmulevich and Edward Dougherty

*Wave Scattering by Time-Dependent Perturbations: An Introduction*, G. F. Roach

*Algebraic Curves over a Finite Field*, J.W.P. Hirschfeld, G. Korchmáros, and F. Torres

*Distributed Control of Robotic Networks: A Mathematical Approach to Motion Coordination Algorithms*, Francesco Bullo, Jorge Cortés, and Sonia Martínez

*Robust Optimization*, Aharon Ben-Tal, Laurent El Ghaoui, and Arkadi Nemirovski

*Control Theoretic Splines: Optimal Control, Statistics, and Path Planning*, Magnus Egerstedt and Clyde Martin

*Matrices, Moments, and Quadrature with Applications*, Gene Golub and Gérard Meurant

*Totally Nonnegative Matrices*, Shaun M. Fallat and Charles R. Johnson

*Matrix Completions, Moments, and Sums of Hermitian Squares*, Mihály Bakonyi and Hugo J. Woerdeman

*Modern Anti-windup Synthesis: Control Augmentation for Actuator Saturation*, Luca Zaccarian and Andrew R. Teel

# Modern Anti-windup Synthesis

*Control Augmentation for Actuator Saturation*

# Luca Zaccarian and Andrew R. Teel

PRINCETON UNIVERSITY PRESS

PRINCETON AND OXFORD

Copyright © 2011 by Princeton University Press
Published by Princeton University Press, 41 William Street, Princeton, New Jersey
08540
In the United Kingdom: Princeton University Press, 6 Oxford Street, Woodstock,
Oxfordshire OX20 1TW
press.princeton.edu

Cover art: Fiore B. Zaccarian, *Lettura sotto gli alberi* (reading under the trees), oil
on canvas, 1970. Courtesy of Dr. Paolo Zaccarian.

Library of Congress Cataloging-in-Publication Data
Zaccarian, Luca.
    Modern anti-windup synthesis: control augmentation for actuator saturation /
    Luca Zaccarian, Andrew R. Teel.
        p. cm. – (Princeton series in applied mathematics)
        Includes bibliographical references and index.
        ISBN 978-0-691-14732-1 (hardback)
        1. Automatic control–Mathematical models. 2. Linear control systems. 3.
        Actuators. I. Teel, Andrew R., 1965- II. Title.
    TJ213.7.Z326 2011
    629.8'3–dc22                                                    2010044248

British Library Cataloging-in-Publication Data is available

The publisher would like to acknowledge the authors of this volume for providing
the camera-ready copy from which this book was printed.

Printed on acid-free paper. ∞

Printed in the United States of America

10 9 8 7 6 5 4 3 2 1

# Contents

# Preface

When feedback control is synthesized for a linear plant, an often neglected but important feature of the feedback loop is a saturation nonlinearity at the plant input. A saturated input typically provides a better model of reality than a linear input model does. Indeed, all real actuators have limited capabilities and these limits can have a dramatic effect on the performance of an otherwise linear feedback loop. In the face of input saturation, the control engineer must accept that the achievable large signal performance is inherently limited. Then he or she must decide how to address this fact in control synthesis.

When input saturation is expected to be a common occurrence in the plant's operation, it makes sense to look for design methodologies that account for input saturation directly. There are many useful techniques in this category that have been developed in the control literature, including the very popular "model predictive control" framework. In the case where input saturation is expected to be a less frequent occurrence and the small signal performance specifications are difficult to incorporate into a general framework like model predictive control, it makes sense to consider the "anti-windup" paradigm.

In the anti-windup synthesis problem, a controller for the saturation-free case has already been synthesized based on some performance criterion and it has been confirmed that this controller does not perform well when input saturation occurs. In this case, the task is to synthesize a controller augmentation that has no effect when saturation does not occur and that otherwise attempts to provide satisfactory performance for large signals. In this way, the small signal performance is not compromised for the sake of guaranteeing acceptable large signal performance.

This book is dedicated to the description of anti-windup synthesis algorithms. The focus is on state-space methods and synthesis algorithms that require solving linear matrix inequalities (LMIs). Many efficient software programs, such as MATLAB/Simulink,[1] are widely available for solving these LMIs. In order to follow the material in this book, the reader should be familiar with state-space descriptions and basic stability theory for linear, continuous-time systems. In order to become comfortable with the algorithms and their behaviors, the reader should be willing to invest the energy required to gain competency with an LMI solver. Several examples are provided in the text that can be used by the reader to test the understanding

---

[1]MATLAB® and Simulink® are registered trademarks of The MathWorks Inc. and are used with permission. The MathWorks does not warrant the accuracy of the text or exercises in this book. This book's use of MATLAB® and Simulink® does not constitute an endorsement or sponsorship by The MathWorks of a particular pedagogical approach or particular use of the MATLAB® and Simulink® software.

of the synthesis recipes and ability to reproduce numerically the anti-windup augmentation given in the examples. Throughout the book we emphasize algorithms over stability proofs, with the aim of making this book accessible both to industrial engineers and also graduate students and researchers interested in digesting the concepts associated with state-of-the-art anti-windup synthesis. Some advanced but still accessible ideas are provided in the text in sections marked with an asterisk. The reader may wish to skip these sections and can do so without breaking the continuity of the book's flow. Where the state of the art becomes too advanced to cover in this text, we provide references to work that can be found in the control literature.

In Part I of this book, we use examples to motivate anti-windup synthesis, we describe various anti-windup performance objectives, and we present the most common anti-windup architectures. In other words, we present the controller's structure without specifying the source of the controller's numerical values. In Part II, we provide various LMI-based anti-windup synthesis algorithms based on different large signal performance specifications. Some of these algorithms address global solutions, when possible. Otherwise, the focus is on algorithms that guarantee performance over a bounded region. We present algorithms that introduce no extra dynamic elements, in other words, "static anti-windup," but the achievable performance with such augmentation is limited. When static anti-windup is not sufficient, we provide dynamic anti-windup algorithms. We also emphasize the nonlinear nature of performance in feedback loops with saturation and provide numerical algorithms for quantifying this nonlinear performance.

In Part III, we focus on a particular architecture and goal that we call "model recovery anti-windup" (MRAW). We start with linear synthesis algorithms that can be compared to the solutions given in Part II. However, a strength of the anti-windup approach in Part III is that it lends itself nicely to the synthesis of nonlinear anti-windup augmentation, exploiting ideas that have appeared in the nonlinear control literature over the last two decades. It also permits bringing the tools of model predictive control to bear on the anti-windup synthesis problem. Finally, we conclude the book with an extensive, annotated anti-windup synthesis bibliography.

We would like to express our thanks to all of our co-authors on the topic of anti-windup synthesis, and to other scholars who helped to shape our viewpoint on this problem through fruitful discussions, at conferences, and through email. We also thank the U.S. National Science Foundation, the Air Force Office of Scientific Research, the Italian MIUR, the ENEA-Euratom Association and the International Relations Office of the University of Rome, Tor Vergata, for their support of this research over many years. We express our appreciation to Petar Kokotovic, especially for his prodding to see this book through to its completion. Finally, we thank our families for enduring the absences that were required to finish this project.

Rome, Italy                                                                 Luca Zaccarian
Santa Barbara, California, USA                                        Andrew R. Teel
Spring 2009

# Algorithms Summary

Throughout the book, several anti-windup constructions will be illustrated. They differ from each other in terms of applicability, namely, what control systems they can be applied to, stability and performance guarantees, and architecture. While most of the notation related to the concepts listed next will be discussed in Chapter 2, a general overview of all the algorithms is provided here, at the beginning of the book, as a quick reference for the different procedures and solutions available throughout.

Table 1 comparatively illustrates the applicability, architectures, and guarantees characterizing each algorithm (asterisks mean that some restrictions apply). The applicability is stated in terms of properties of the linear plant involved in the saturated control system, namely, exponentially stable (all the eigenvalues in the open left half plane), marginally stable (same as in the previous case, with possible single eigenvalues on the imaginary axis), marginally unstable (all the eigenvalues in the closed left half plane), and exponentially unstable (plants with at least one eigenvalue in the right half plane). This is not applicable for the last two algorithms because they address a special class of nonlinear plants. The architecture of the anti-windup solutions is characterized by their linear or nonlinear nature, the presence or not of dynamics in the anti-windup filter, and their interconnection properties: external or full authority. Finally, the guarantees on the compensated closed loop are distinguished as global or regional. For completeness, the page number where each algorithm appears is also listed in the second column of the table. It should be emphasized that all the direct linear anti-windup algorithms (namely, from Algorithm 1 to Algorithm 9) require linearity of the unconstrained controller to be applicable, whereas all the remaining model recovery anti-windup algorithms are applicable with any nonlinear controller and only linearity of the plant is required. Moreover, all MRAW algorithms correspond to plant-order, external augmentation.

A list of all the algorithm titles and their brief description is reported next, once again for a quick reference to the several solutions available in this book.

1. *Static full-authority global DLAW* (page 81): Simple architecture, very commonly used but not necessarily feasible for any exponentially stable plant. Global input-output gain is optimized.
2. *Static external global DLAW* (page 94): Useful when some internal states of the controller are inaccessible. Otherwise not more effective than Algorithm 1. Global input-output gain is optimized.
3. *Static full-authority regional DLAW* (page 99): Extends the applicability of

| Algor Number | Page | Applicability | | | | Architecture | | | Guar |
|:---:|:---:|:---:|:---:|:---:|:---:|:---:|:---:|:---:|:---:|
| | | Ex St | Mar St | Mar Uns | Exp Uns | Lin/Nlin | Dyn/Stat | Ext/FullAu | Glob/Reg |
| 1 | 81 | √* | | | | L | S | FA | G |
| 2 | 94 | √* | | | | L | S | E | G |
| 3 | 99 | √ | √ | √ | √ | L | S | FA | R |
| 4 | 114 | √ | | | | L | D | FA | G |
| 5 | 126 | √ | | | | L | D* | FA | G |
| 6 | 130 | √ | | | | L | D | E | G |
| 7 | 138 | √ | | | | L | D* | E | G |
| 8 | 143 | √ | √ | √ | √ | L | D | FA | R |
| 9 | 149 | √ | √ | √ | √ | L | D* | FA | R |
| 10 | 176 | √ | | | | L | D | E | G |
| 11 | 185 | √ | √ | | | L | D | E | G |
| 12 | 189 | √ | | | | L | D | E | G |
| 13 | 194 | √ | | | | L | D | E | G |
| 14 | 195 | √ | √ | √ | √ | L | D | E | R |
| 15 | 197 | √ | √ | √ | √ | L | D | E | R |
| 16 | 203 | √ | √ | √ | √ | NL | D | E | R/G |
| 17 | 206 | √ | √ | √ | √ | NL | D | E | R/G |
| 18 | 215 | √ | √ | √ | √ | NL | D | E | R/G |
| 19 | 216 | √ | √ | √ | √ | NL | D | E | R/G |
| 20 | 218 | √ | √ | √ | | NL | D | E | G |
| 21 | 221 | √ | √ | √ | | L | D | E | R |
| 22 | 221 | √ | √ | √ | | NL | D | E | G |
| 23 | 223 | √ | √ | √ | √ | NL | D | E | R |
| 24 | 248 | n/a | n/a | n/a | n/a | NL | D | E | G |
| 25 | 249 | n/a | n/a | n/a | n/a | NL | D | E | G |

Table 1  Applicability, architectures and guarantees of the algorithms illustrated in the book.

Algorithm 1 to a larger class of systems by requiring only regional properties. However, feasibility conditions still need to hold. Regional input-output gain is optimized.

4. *Dynamic plant-order full-authority global DLAW* (page 114): Dynamic anti-windup, with state dimension equal to that of the plant, overcomes the applicability limitations of Algorithm 1. Feasible for any loop containing an exponentially stable plant. Global input-output gain is optimized; it is never worse, and often better, than the input-output gain provided by Algorithm 1.

5. *Dynamic reduced-order full-authority global DLAW* (page 126): Useful when static anti-windup is infeasible or performs poorly yet a low-order anti-windup augmentation is desired. Order reduction is carried out while maintaining a

prescribed global input-output gain.

6. *Dynamic plant-order external global DLAW* (page 130): Useful when some internal states of the controller are inaccessible; otherwise not more effective than Algorithm 4. Feasible for any loop containing an exponentially stable plant. Global input-output gain is optimized; it is never worse, and often better, than the input-output gain provided by Algorithm 2.

7. *Dynamic reduced-order external global DLAW* (page 138): Useful when Algorithm 2 is not feasible or performs poorly yet a low-order external anti-windup augmentation is desired. Order reduction is carried out while maintaining a prescribed global input-output gain.

8. *Dynamic plant-order full-authority regional DLAW* (page 143): Extends the applicability of Algorithm 4 by only requiring regional properties. Applicable to any loop. Regional input-output gain is optimized.

9. *Dynamic reduced-order full-authority regional DLAW* (page 149): Useful when Algorithm 3 is not feasible or performs poorly yet a low-order anti-windup augmentation is desired. Order reduction is carried out while maintaining a prescribed global input-output gain.

10. *Stability-based MRAW for exponentially stable plants* (page 176): Special cases include IMC anti-windup and Lyapunov-based anti-windup, neither of which creates an algebraic loop. Global exponential stability is guaranteed but no performance measure is optimized.

11. *Lyapunov-based MRAW for marginally stable plants* (page 185): Does not create an algebraic loop. Global asymptotic stability is guaranteed but no performance measure is optimized.

12. *Global LQ-based MRAW* (page 189): A linear quadratic performance index related to URR is optimized subject to guaranteeing global exponential stability.

13. *Global $H_2$-based MRAW* (page 194): An $H_2$ performance index related to the unconstrained response recovery is optimized subject to guaranteeing global exponential stability.

14. *Regional LQ-based MRAW* (page 195): Extends the applicability of Algorithm 12 to any loop by requiring only regional exponential stability. Does not create an algebraic loop. A linear quadratic performance index related to small signal unconstrained response recovery is optimized.

15. *Regional $H_2$-based MRAW* (page 197): Extends the applicability of Algorithm 13 to any loop by requiring only regional exponential stability. An $H_2$ performance index related to small signal unconstrained response recovery is optimized.

16. *Switched MRAW* (page 203): Augmentation based on a family of nested ellipsoids and switching among a corresponding family of linear feedbacks. Regional unconstrained response recovery gain is optimized in each ellipsoid. Global exponential stability is guaranteed for any loop with an exponentially stable plant. Otherwise, regional exponential stability is guaranteed.

17. *Scheduled MRAW* (page 206): Augmentation based on a family of nested ellipsoids and continuous scheduling among a corresponding family of linear feedbacks. Regional unconstrained response recovery gain is optimized in

each ellipsoid. Global exponential stability is guaranteed for any loop with an exponentially stable plant. Otherwise, regional exponential stability is guaranteed.

18. *MPC-based MRAW with special terminal cost* (page 215): Sampled-data augmentation based on model predictive control for constrained, discrete-time linear systems. An appropriate terminal cost function is used to guarantee global exponential stability for any loop with an exponentially stable plant. Regional exponential stability is guaranteed for any loop.

19. *MPC-based MRAW with sufficiently long horizons* (page 216): Sampled-data augmentation based on model predictive control for constrained, discrete-time linear systems. A sufficiently long optimization horizon is used to guarantee global exponential stability for any loop with an exponentially stable plant. Regional exponential stability is guaranteed for any loop.

20. *Global MRAW using nested saturation* (page 218): Augmentation using feedback consisting of multiple nested saturation functions. Does not create an algebraic loop. Global exponential stability is guaranteed for any loop with an exponentially stable plant. For a loop with a marginally unstable plant, global asymptotic stability is guaranteed. No performance measure is optimized.

21. *Semiglobal MRAW using Riccati inequalities* (page 221): Linear augmentation providing an arbitrarily large stability region for any loop containing a marginally unstable plant. Does not create an algebraic loop. Provides a quantified characterization of the regional unconstrained response recovery gain.

22. *Global MRAW using scheduled Riccati inequalities* (page 221): Augmentation based on a family of Riccati inequalities and continuous scheduling among a corresponding family of linear feedbacks. Global exponential stability is guaranteed for any loop with an exponentially stable plant. Global asymptotic stability is guaranteed for any loop with a marginally unstable plant.

23. *MRAW with guaranteed region of attraction* (page 223): Augmentation that, in order to achieve a large operating region for a loop containing an exponentially unstable plant, uses measurements of the plant's exponentially unstable modes.

24. *MRAW for Euler-Lagrange systems with linear injection* (page 248): Augmentation for a loop containing a plant from a class of nonlinear systems. Global asymptotic stability is guaranteed for any loop containing a plant whose parameters satisfy a certain constraint.

25. *MRAW for Euler-Lagrange systems with nonlinear injection* (page 249): Extends and improves upon the performance in Algorithm 24 by replacing certain linear terms with nonlinear terms.

# PART 1
# Preparation

# Chapter One

## The Windup Phenomenon and Anti-windup Illustrated

### 1.1 INTRODUCTION

Every control system actuator has limited capabilities. A piezoelectric stack actuator cannot traverse an unlimited distance. A motor cannot deliver an unlimited force or torque. A rudder cannot deflect through an unlimited angle. An amplifier cannot produce an unlimited voltage level. A hydraulic actuator cannot change its position arbitrarily quickly. These actuator limitations can have a dramatic effect on the behavior of a feedback control system.

In this book, the term "windup" refers to the degradation in performance that occurs when a saturation nonlinearity is inserted, at the plant input, in an otherwise linear feedback control loop. Usually the term is reserved for the situation where this degradation is severe. The term has its origins in the fact that, among the simple analog control architectures that were used in the early days of electronic control, feedback loops with controllers that contained an integrator were the most likely to experience a severe performance degradation due to input saturation. Windup, as the term is use here, was said to occur because the saturation nonlinearity would slow down the response of the feedback loop and thus cause the integrator state to "wind up" to excessively large values.

"Anti-windup" refers to augmentation of a controller in a feedback loop that is prone to windup so that:

1. the closed-loop performance is unaltered when saturation never occurs, in other words, the augmentation has no effect for small signals;
2. acceptable performance is achieved, to the extent that it is possible, even when actuator saturation occurs.

Anti-windup synthesis refers to the design of such augmentation. This book provides principles, guidelines, and algorithms for anti-windup synthesis.

In order to motivate anti-windup synthesis, the rest of this chapter contains examples where windup occurs. In each of these examples, alternatives to anti-windup synthesis include investing in actuators with more capabilities, or redesigning the controller from scratch to account for input saturation directly. These strategies should be considered when the control system's actuators are continuously trying to act beyond their limits. On the other hand, suppose that hitting the actuator limits is the exception rather than the rule. In addition, suppose the operating budget or some physical constraint does not permit more capable actuators. Moreover, suppose the small signal performance is highly desirable and very difficult to reproduce with control synthesis tools that account for saturation directly. In this case, anti-windup synthesis becomes a very appealing design tool: it is uniquely qualified to address saturation with potentially dramatic performance improvement using the existing actuators without sacrificing the small signal performance for the sake of guaranteeing acceptable large signal behavior. The examples will illustrate these capabilities of anti-windup synthesis, without going into the synthesis details yet. The examples will be revisited after the anti-windup synthesis algorithms have been described.

## 1.2 ILLUSTRATIVE EXAMPLES

### 1.2.1 A SISO academic example

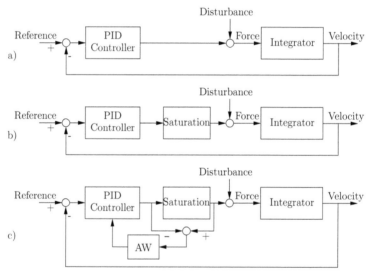

Figure 1.1  An integrator plant in negative feedback with a PID controller (a) without input saturation, (b) with input saturation, (c) with input saturation and anti-windup augmentation.

Consider the closed-loop system resulting from using a PID (proportional + integral + derivative) controller with unity gains to control a single integrator plant, as shown in Figure 1.1a. When the force applied to the object is not limited, the closed-loop system is linear and the related response to a unitary step reference corresponds to the dashed curves in Figure 1.2. During the initial transient, the applied force exhibits a large peak. Its maximum value, which exceeds the lower plot's range, is one unit. If the maximum force that the actuator can deliver is ±0.1 units, then undesirable input and output oscillations occur, as shown by the dotted curves in Figure 1.2. Although the velocity eventually converges to the desired steady-state value, the response is very sluggish: it takes approximately 60 seconds to recover the linear performance. The output oscillations consist of rising and falling ramps that correspond to large time intervals where the force sits on either

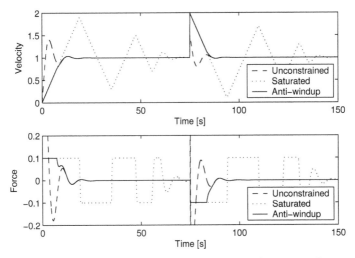

Figure 1.2  Responses of the integrator control system to a unitary step reference: unconstrained response (dashed curves), response with input saturation (dotted curves),
and response of the anti-windup augmented system according to Algorithm 11
(solid curves).

its positive or negative limit.

Although the limits on the allowable input force imposed by saturation must
cause some deviation from the ideal linear response, the large oscillations indicated
by the dotted curve in Figure 1.2 are unacceptable. Since this undesirable response
is induced by the large step reference input, in principle it could be avoided by
shaping the reference signal so that it does not feed large and sudden changes to the
control system. This solution does not address the core of the problem, however,
because similar behavior will also occur whenever large enough disturbances affect
the control system. Indeed, the response after 75 seconds in Figure 1.2 is due to an
impulsive disturbance acting at the integrator's input, as drawn in Figure 1.1a. This
impulsive disturbance resembles the action of an external element hitting the object
being controlled and remaining in contact with it for a very short time interval.
Mathematically, this is modeled as a very large pulse acting for a very short time.

The effect of the impulsive disturbance on the closed loop is essentially the same
as that of the step reference input. However, the reference can be shaped to avoid
input saturation and its undesired consequences, while the disturbance input cannot
be changed. It is therefore desirable to insert extra compensation into the control
scheme, aimed at eliminating the undesirable oscillatory behavior occurring after
the actuator reaches its magnitude limit, regardless of the reason for actuator saturation. For this example, Algorithm 11, which appears on page 185, has been used
to illustrate the capabilities of anti-windup augmentation. On page 186, Example 7.2.4 provides details of this construction for the current example. With anti-
windup augmentation, after the initial, inevitable deviation, the resulting closed-
loop velocity and force signals, corresponding to the solid curves in Figure 1.2,

converge rapidly to the unconstrained linear response after both the large reference change and the impulsive disturbance. In each case, the linear response is recovered after about 10 seconds. Thus, the PID controller's anti-windup augmentation, which has no effect for small signals, can induce a dramatic improvement for signals that cause input saturation.

### 1.2.2  A MIMO academic example

The simulations in this section are for a closed-loop system where the plant is a multi-input/multi-output (MIMO), two-state system with lightly damped modes in feedback with a MIMO PI controller. For details on the plant and controller, see Example 7.2.1 on page 178. An important feature of the plant model is that each input has a significant effect on both of the plant's states, which also correspond to the plant's outputs.

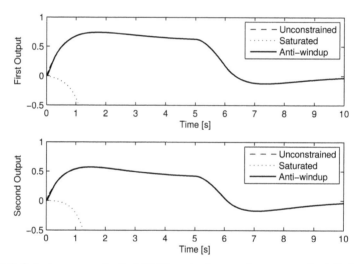

Figure 1.3  Output responses for the MIMO academic example: unconstrained (dashed), saturated (dotted), and anti-windup (solid).

Figure 1.3 shows the responses of the control system to a step reference of $[0.6, 0.4]$ for the two outputs. The dashed curves represent the response of the closed loop when no limitation is imposed on the control input. When the two inputs are limited to values between $\pm 3$ and $\pm 10$, respectively, the closed-loop response is such that the plant outputs converge to values that are far from the reference values. In turn, the driving signals to the controller's integrators approach nonzero constant values, causing the controller states to diverge, as shown in the long simulation reported in Figure 1.5. This behavior belies the fact that it is possible to almost exactly reproduce the unconstrained closed-loop response even with the given input constraints. Indeed, synthesizing anti-windup augmentation by using Algorithm 10, given on page 176, results in the anti-windup augmented closed-loop response represented by the solid line in Figure 1.3. There is very lit-

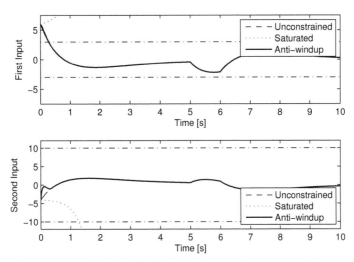

Figure 1.4  Input responses for the MIMO academic example: unconstrained (dashed), saturated (dotted), and anti-windup (solid).

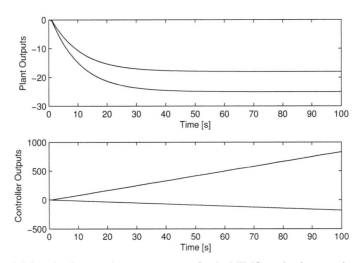

Figure 1.5  Diverging input and output responses for the MIMO academic example when the input is saturated.

tle difference between the unconstrained response and the anti-windup augmented response, while the constrained (non augmented) response is completely unacceptable. Without anti-windup augmentation, the MIMO PI controller would need to be abandoned for reference values leading to input saturation. Anti-windup design permits retaining the MIMO PI controller without modification for small reference signals and with assistance for larger reference signals.

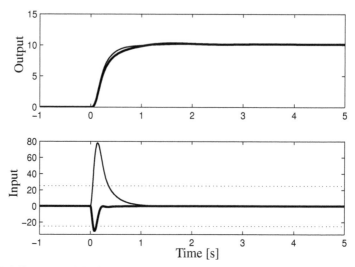

Figure 1.6  Response of F8 aircraft without input constraints. Output: pitch angle (thick) and
flight path angle (thin). Input: elevator angle (thick) and flaperon angle (thin).
Input amplitude limits NOT imposed (dotted).

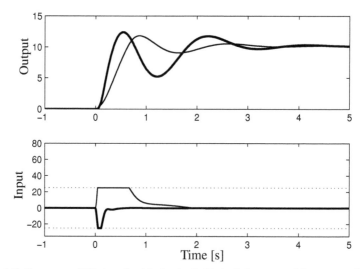

Figure 1.7  Response of F8 aircraft with physically limited elevator and flaperon angles. Output: pitch angle (thick) and flight path angle (thin). Input: elevator angle (thick)
and flaperon angle (thin). Input amplitude limits (dotted).

### 1.2.3  The longitudinal dynamics of an F8 airplane

Consider a fourth-order linear model describing the longitudinal dynamics of an
F8 aircraft with two inputs, elevator angle and flaperon angle, both measured in de-

grees, and two outputs, pitch angle and flight path angle, also measured in degrees. Additional details about this example can be found in Example 4.3.5 on page 90.

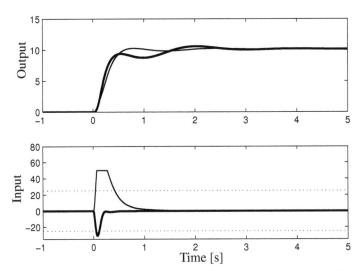

Figure 1.8  Partial performance recovery of the F8 aircraft by doubling the amplitude limits. Output: pitch angle (thick) and flight path angle (thin). Input: elevator angle (thick) and flaperon angle (thin). Previous input amplitude limits (dotted).

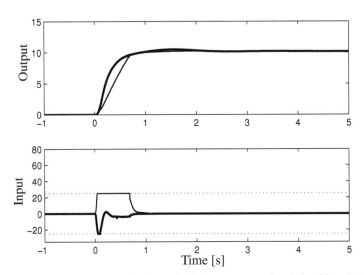

Figure 1.9  Response of F8 aircraft with anti-windup augmentation (Algorithm 4). Output: pitch angle (thick) and flight path angle (thin). Input: elevator angle (thick) and flaperon angle (thin). Input amplitude limits (dotted).

Suppose a controller has been designed following an LQG/LTR methodology,

so that, in the absence of input amplitude limits, the resulting closed loop has a
desirable response. In particular, in the absence of input constraints, the system
response to a step reference change of 10 degrees in pitch angle and flight path angle
is as shown in Figure 1.6. The input plot in Figure 1.6 shows that the controller is
attempting to use large input angles, especially flaperon angle, to effect this step
change. If elevator and flaperon angles are limited in magnitude to 25 degrees, then
the response deteriorates to what is shown in Figure 1.7. Even with the limits set to
50 degrees in magnitude, there is still some potentially undesirable oscillations in
the step response, as shown in Figure 1.8.

Nevertheless, the situation is not hopeless. The trajectory shown in Figure 1.9
corresponds to limiting the actuator angles to 25 degrees in magnitude and en-
hancing the original controller with anti-windup augmentation, synthesized using
Algorithm 4 given on page 114, so that the small signal response is not altered
and the response for large step changes is improved. In particular, the response in
Figure 1.9 shows no undesirable oscillations in the pitch and flight path angles.

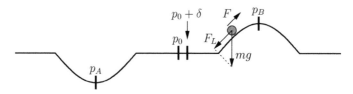

Figure 1.10  A servo-positioning system.

## 1.2.4  A servo-positioning system

Consider controlling the position of a mass, an autonomous vehicle, for example,
constrained to a one-dimensional path of variable elevation. The forces acting on
the mass are gravity and a motor force that serves as a control input. The gravity
force is state dependent, as shown in Figure 1.10, but can be modeled as an unknown external disturbance, especially when the elevation of the path is not known
ahead of time.

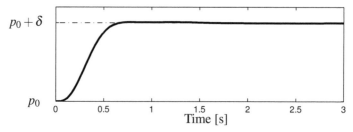

Figure 1.11  A small signal response of the servo-positioning system.

The objective is to design a control system that quickly drives the mass to a
given reference value with minimal overshoot and zero steady-state tracking error.

For small step changes, the transition from one position to another should take no more that 0.5 seconds. Moreover, this behavior should occur for all reasonable path elevation profiles. To accomplish this control objective, a third-order linear control system containing integral action and a double lead network has been designed. See Example 7.2.5 for details. A resulting step response is shown in Figure 1.11.

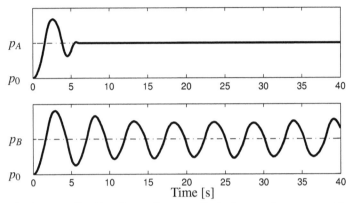

Figure 1.12 The unpredictable effects of actuator limitations on the servo-positioning system.

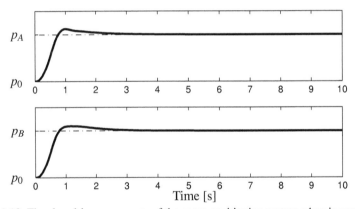

Figure 1.13 The closed-loop responses of the servo-positioning system when increasing the actuator force capability by a factor of five.

The behavior for some larger references changes is shown in Figure 1.12. The upper plot shows a step reference change from $p_0$ to $p_A$ while the lower plot shows a step reference change from $p_0$ to $p_B$. When moving to the position $p_A$, the system exhibits only a small oscillation and rapidly settles to the desired steady-state value. However, when moving to the position $p_B$, persistent oscillations occur. In both cases, the control system asks for more force than the motors can deliver. However, the effect of the force limitations is much more severe when moving toward $p_B$. The problem in moving to $p_B$ would not occur if using a motor with five times the force capability. Indeed, Figure 1.13 shows what would happen with such a

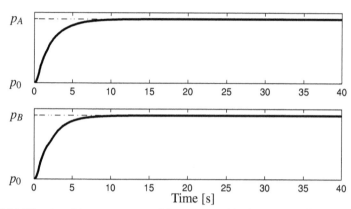

Figure 1.14  The closed-loop responses of the servo-positioning system after the augmentation of Algorithm 11.

motor. Of course, the stronger motor might still have a problem with even larger step changes. It may also be judged to be prohibitively expensive, especially when it becomes clear that, to compensate for input saturation, there is a relatively simple control software fix produced by an anti-windup augmentation algorithm.

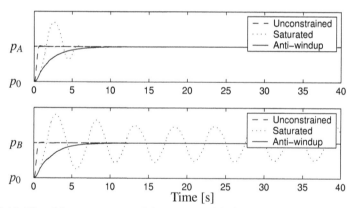

Figure 1.15  Closed-loop responses of the servo-positioning system with force limits and no additional compensation (dotted curves), without force constraints (dashed curves), and with force limits and anti-windup compensation per Algorithm 11 (solid curves).

Using the anti-windup recipe given in Algorithm 11 on page 185 results in the response shown in Figure 1.14 for the transition from $p_0$ to $p_B$. The details of the synthesis for this particular system are described in Example 7.2.5 on page 187. As usual, the augmentation is such that the response for small reference changes, like in Figure 1.11, is unchanged.

Figure 1.15 shows the transition from $p_0$ to $p_B$ for all three scenarios considered. The dotted curve in Figure 1.15 corresponds to using the original motor and the original controller without augmentation, as in the lower plot of Figure 1.12.

The dashed curve corresponds to the ideal response without input constraints, as in Figure 1.11. The solid curve corresponds to using the original motor and the original controller with anti-windup augmentation, as in Figure 1.14. Anti-windup augmentation has induced an adequate response for large reference changes without having to buy a more expensive (and heavier) motor and without compromising the response for small step changes.

### 1.2.5 The damped mass-spring system

Consider a damped mass-spring system, as shown in the diagram of Figure 1.16, where the input $u_p$ is the force exerted on the mass and the output is the mass position $q$. See Example 7.2.6 on page 190 for additional details about this example.

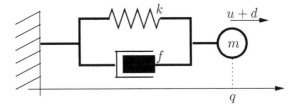

Figure 1.16  The damped mass-spring system.

Suppose a two degrees of freedom linear controller is designed such that the mass follows a reference input while rejecting constant force disturbances $d$ acting at the plant input. Without input force limits, the plant output and input responses would be as shown by the dashed curves in Figure 1.18. However, when the input is constrained the asymptotic tracking is lost completely as the system converges to a limit cycle with large amplitude shown in Figure 1.17. This response is also shown by the dotted curves in Figure 1.18. Applying Algorithm 12, which appears on page 189, the controller is enhanced with anti-windup augmentation and stability is recovered, while the tracking performance is only slightly deteriorated, as shown by the solid curves in Figure 1.18.

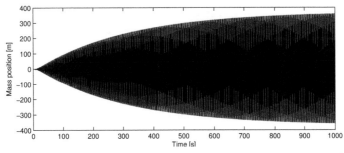

Figure 1.17  The mass-spring response with input saturation converges to a very large limit cycle.

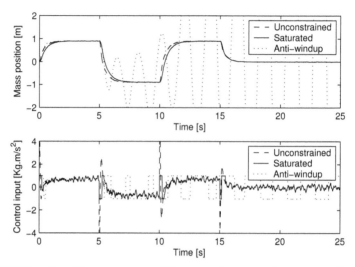

Figure 1.18  The closed-loop responses of the damped mass-spring system without input
            constraints (dashed curves), with input constraints (dotted curves), and with
            anti-windup augmentation coming from Algorithm 12 (solid curves).

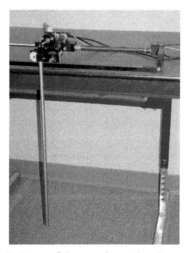

Figure 1.19  A picture of the experimental spring-gantry system.

### 1.2.6  The experimental spring-gantry system

Consider the experimental spring-gantry system shown in Figure 1.19, where a
pendulum hangs from a cart constrained to linear motion and the cart is attached to
a fixed point via a spring. The input is the voltage applied to a DC motor that drives
the cart and the outputs are the pendulum angle and cart position. A linear controller has been designed based on an LQG construction to regulate the pendulum
angle to zero quickly and to attenuate small forces. For larger forces, which lead to
a requested control input that substantially exceeds the voltage of the power supply,

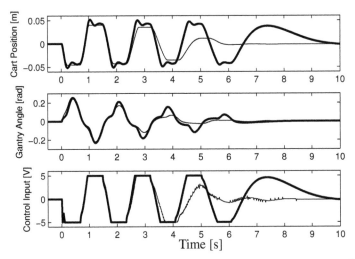

Figure 1.20 The saturated closed-loop trajectories of the spring-gantry system: simulated (thick) and experimental (thin).

the resulting output and input trajectories are highly oscillatory. This behavior is shown in Figure 1.20, both for simulations (thick curves) and for experiments (thin curves).

Using the anti-windup augmentation Algorithm 4, given on page 114, the performance of the system with input constraints can be improved without altering the response to small forces exerted on the pendulum. The result of such a modification is shown in Figure 1.21, both in simulation and in experiment.

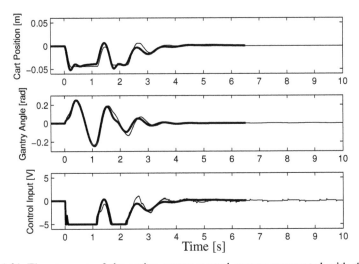

Figure 1.21 The response of the spring-gantry control system augmented with the anti-windup compensation of Algorithm 4: simulated (thick) and experimental (thin).

### 1.2.7 A robot manipulator

Consider the selective compliance assembly robot arm (SCARA). The SCARA is a common workhorse for industrial assembly tasks, typically combining components that are located on a horizontal working surface. For this task, the SCARA uses its first two rotational joints to position the tip of the robot at a desired coordinate in the horizontal plane. It uses its vertical translational joint to impose a desired tilt to the robot tip. Its last rotational joint, located at the tip, is used to impose a desired orientation angle to the robot gripper.

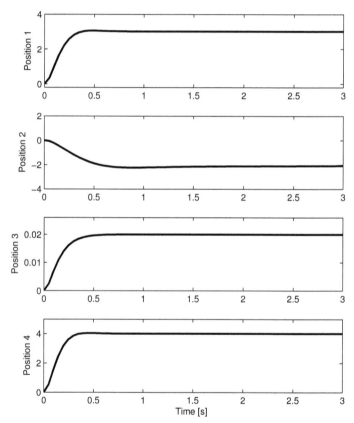

Figure 1.22 Small signal response of the "computed torque" control scheme for the SCARA robot.

The nonlinear coupling between the different joints in the SCARA is very strong. Thus it is difficult to design effective feedback controllers using linear control design techniques. However, since the SCARA has a motor on each of its joints, in principle it can be controlled effectively using the so-called "computed torque" algorithm, which is a model-based nonlinear control strategy that ignores motor torque constraints. When combined with PID feedback, the control approach can enforce a desirable linear and decoupled behavior on all the robot's joints, at least

when the commanded torques do not exceed the capabilities of the motors. By suitably selecting the PID gains of this controller (see Section 10.3.3 on page 258 for
details), the linear decoupled response of Figure 1.22 is obtained for motions that
do not approach the limits of the robot's motors. This figure represents the four
joint positions when the desired position reference step changes from (0 deg, 0 deg,
0 cm, 0 deg) to (3 deg, -2 deg, 2 cm, 2 deg).

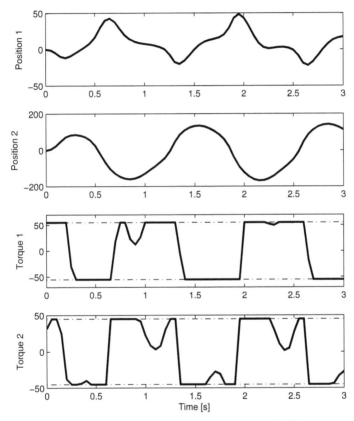

Figure 1.23  Unpredictable effects of actuator limitations on the SCARA robot control sys-
        tem.

When commanding step reference changes that are double in size, the controller
torque commands exceed the maximum torque values attainable by the motors and,
as shown in Figure 1.23, the rotational joints experience persistent oscillations that
could damage the robot. In Figure 1.23 only the first two joint responses are shown
because they are sufficient to illustrate the windup phenomenon. It is once again
evident that the effects of input limitations on an otherwise desirable control system
can be unpredictable. Indeed, there is a very small threshold over which the robot's
response changes from a decoupled linear response to a highly oscillatory nonlinear
response.

The possibilities for eliminating the undesired behavior seen in Figure 1.23 are the same as they were for the previous examples. One possibility is to increase the size and capabilities of the motors. For example, with actuators twice as big as those employed in Figure 1.23, the response would once again be desirable, linear, and decoupled.

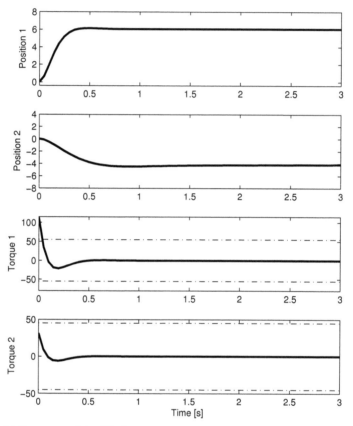

Figure 1.24 Response of the SCARA control system when increasing two times the actuators' maximum torque.

While increasing the size of the actuators may seem to be a reasonable approach to eliminate the behavior shown in Figure 1.23, this might result in overly high costs or weight of the robotic structure. Moreover, even with larger actuators, the same oscillatory behavior occurs when the reference is twice as big as the reference used in Figure 1.24. Another option for eliminating the large signal oscillations is to reduce the controller gains. However, this compromises the rate of convergence reported in Figure 1.22. Another possibility is to use anti-windup augmentation. Using Algorithm 25, the original computed torque plus PID control law is augmented with extra model-based dynamical elements that aim at preserving the small signal behavior of Figure 1.22 while eliminating the behavior shown in Figure 1.23. The resulting response for small signals is linear and decoupled and exactly matches the response in Figure 1.22. Moreover, the response to the same reference that caused

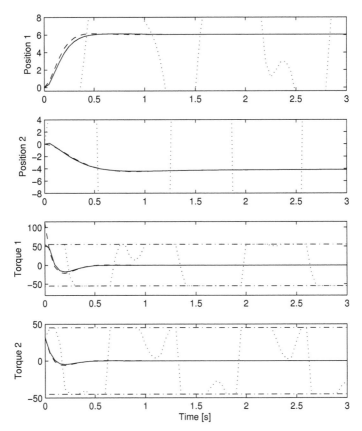

Figure 1.25  Response of the SCARA robot control system augmented with the construction
proposed in Algorithm 25 (solid) compared to the ideal linear response (dashed)
and to the undesired response of Figure 1.23 (dotted).

the undesired oscillations of Figure 1.23 is well behaved and only deviates slightly
from the ideal response of Figure 1.24, which corresponds to employing larger actuators. This can be verified in Figure 1.25, where the dotted curves, which are
largely out of range in the plot, reproduce the response of Figure 1.23. The dashed
curves represent the ideal response of Figure 1.24, and the solid curves reproduce
the response of the control system augmented according to Algorithm 25. The solid
curve is almost indistinguishable from the ideal, dashed curve. Once again, anti-
windup augmentation solves the problem of preserving the small signal response
while eliminating the deleterious behavior for large signals that results from the
interaction of motor torque limitations with the original controller.

## 1.2.8  A disturbance rejection problem

Consider a second-order plant containing an integrator subjected to both low-frequency input-matched disturbances and high-frequency output measurement noise.

A schematic block diagram of the plant is shown in Figure 1.26.

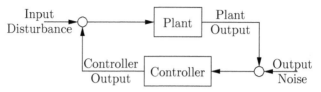

Figure 1.26 A disturbance rejection problem.

For this plant, a controller has been designed following standard loop shaping techniques (see Example 4.4.1 reported on page 102 for details about this example) so that the closed loop guarantees asymptotic rejection of constant disturbances. Moreover, the loop gain is sufficiently large at low frequencies to guarantee a $-60$ dB attenuation of disturbances below $\omega_l = 0.5\ rad/s$, and the loop gain is sufficiently small at high frequencies to make the control system insensitive to measurement noise above $\omega_h = 100\ rad/s$ acting at the plant output. See Example 4.4.1 reported on page 102 for additional information about the controller parameters.

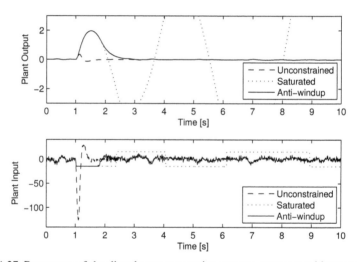

Figure 1.27 Responses of the disturbance attenuation system: response without saturation (dashed), response with saturation and no anti-windup (dotted), response with saturation and anti-windup constructed following Algorithm 3 on page 99.

The dashed line of Figure 1.27 shows the system steady-state response to specific selections of the input disturbance and output noise. From the upper plot, which shows the plant output, it is evident that disturbance attenuation is obtained. The lower plot shows that the plant input exhibits an oscillatory behavior caused, in part, by the controller reaction to the input disturbance and, in part, by the controller reaction to the output noise.

On that same figure, the dotted curves correspond to the response of the system when the plant control input is saturated between $+15$ and $-15$. In particular, to

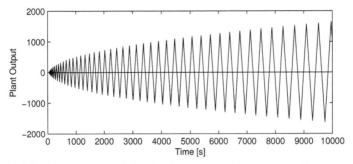

Figure 1.28  Diverging response of the disturbance attenuation system with input saturation.

simulate the closed-loop response when the saturation limits are exceeded, a pulse is added to the input disturbance at time $t = 1$. The resulting response diverges to infinity. The closed-loop system is then augmented with anti-windup compensation using Algorithm 3 on page 99. The resulting response corresponds to the solid curves in Figure 1.27, which show that stability is fully recovered and performance is only partially lost during the transient close to the pulse disturbance occurrence.

## 1.3 SUMMARY

The preceding examples illustrate the windup phenomenon, which may occur due to input saturation in a feedback loop, and the capabilities of anti-windup synthesis in mitigating the windup phenomenon. Windup can manifest itself as a sluggish response, a highly oscillatory response, or a diverging response. A sluggish response is demonstrated in Figure 1.2 on page 5 for the SISO academic example, in Figure 1.7 on page 8 for the F8 airplane, or in Figure 1.20 on page 15 for the experimental spring-gantry system. A highly oscillatory response appears in Figure 1.12 on page 11 for the servo-positioning system, in Figure 1.17 on page 13 for the damped mass-spring system, or in Figure 1.23 on page 17 for the robotic manipulator. A diverging response can be seen in Figure 1.28 on page 21 for the disturbance rejection problem, or in Figure 1.5 on page 7 for the MIMO academic example.

The windup phenomenon can occur even when the nominal control signal would need to spend a small amount of time in a saturated condition. See Figures 1.3 and 1.4 on page 6 for the MIMO academic example, Figure 1.18 on page 14 for the damped mass-spring system, or Figure 1.25 on page 19 for the robotic mainpulator.

The various manifestations of windup can be addressed with anti-windup augmentation, which does not change the small signal response but works to ensure that the large signal response is acceptable. In some instances, performance degradation is unavoidable. See the solid anti-windup response in Figure 1.2 on page 5 for the SISO academic example, in Figure 1.15 on page 12 for the servo-positioning system, or in Figure 1.27 on page 20 for the disturbance rejection problem. In other instances, especially when the nominal control signals would need to spend a very small amount of time in a saturated condition, the performance improvements gained by using anti-windup are dramatic, coming quite close to the unconstrained response. See the solid anti-windup response in Figure 1.3 on page 6 for the MIMO academic example or in Figure 1.25 on page 19 for the robotic mainpulator.

Having demonstrated the capabilities of anti-windup synthesis, it is time to be more precise about the anti-windup problem, its formal objectives, and the typical architectures used to achieve these goals. These topics are covered in the next chapter.

## 1.4  NOTES AND REFERENCES

One of the first authors to make note of the windup phenomenon in feedback loops with integrators was Lozier [PD1]. Lozier did not actually use the term "windup," which began to appear in the control literaure in the 1960's and can be found in the papers [PD2, PD6, PD7].

The SISO academic example of Section 1.2.1 was used by Åström in [CS2]. The MIMO academic example of Section 1.2.2 was introduced in [CS7] and has been used repeatedly in the anti-windup literature (e.g., see [MR3]). The model of the longitudinal dynamics of the F8 airplane first appeared in [SAT1] and was later rivisited in a large number of papers (e.g, [MI18]). In [SAT1] the example was used to illustrate a novel anti-windup technique called "error governor." That technique, which scales the controller input to avoid saturation at the controller output, inspired later work that goes under the name of *reference governor* or *command governor*. See Chapter 11 for details. The damped mass-spring system of Section 1.2.5 was used to illustrate anti-windup synthesis in [MR15] and was later used in many anti-windup papers (see, e.g., [MI23, MR23, MR27, MR28]). Anti-windup for the experimental spring-gantry system appears in [MI18], while a more detailed description of the specific experimental application can be found in [MI8]. The anti-windup results for the robot manipulator were taken from [MR19]. The model used in the disturbance rejection problem is similar to the SISO academic example of Section 1.2.1; however the second-order plant dynamics lead to an unbounded state in the presence of actuator saturation.

Several papers can be found where specific applications of anti-windup designs, rather than just simple academic examples, are illustrated. Some of them are [AP8], where anti-windup is applied to an industrial vibration isolation system; [AP10], where anti-windup is applied to the control of the gates of open water channels with specific application data from a channel in Queensland (Australia); [AP4], where anti-windup is applied to the active queue management control in TCP networks; [AP2, AP6, AP3, AP9, AP7], where anti-windup is applied to different flight control problems; [AP1], where anti-windup is applied to a turbofan engine; [AP5, AP11], where anti-windup is applied to hard-disk drive control.

# Chapter Two

## Anti-windup: Definitions, Objectives, and Architectures

### 2.1 PRELIMINARIES

#### 2.1.1 The unconstrained closed-loop system

The anti-windup problem's starting point is a plant-controller pair that, when connected in feedback without input saturation, behaves in a satisfying manner. This unconstrained feedback loop then serves as something to emulate when trying to specify and solve the anti-windup problem. Figure 2.1 shows this *unconstrained closed-loop system*. The *unconstrained plant* is represented by $\mathcal{P}$ and the *uncon-*

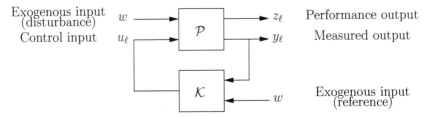

Figure 2.1 The unconstrained closed-loop system.

*strained controller* is represented by $\mathcal{K}$. For now, think of these systems as being linear systems. Eventually, nonlinear plants and controllers will be considered; most of the ensuing discussion applies equally well to that situation. The plant in the unconstrained closed loop has a control input $u_\ell$, an exogenous input $w$, a measured output $y_\ell$, and a performance output $z_\ell$. One controller input is equal to the measured plant output while the other is coming from an exogenous signal, perhaps a reference command. The exogenous inputs affecting the plant and the controller have been grouped into the common symbol $w$ for convenience. The internal state trajectory of the unconstrained closed-loop system, denoted $(x_{p\ell}, x_{c\ell})$, is called the *unconstrained response* and the signals $u_\ell$, $z_\ell$, $y_\ell$ are called, respectively, the *unconstrained control input response*, *unconstrained performance output response*, and *unconstrained measured output response*.

An example of an unconstrained closed-loop system is shown in Figure 2.2. This block diagram corresponds to the unconstrained closed-loop system for the example given in Section 1.2.1.

Among the features of the unconstrained closed-loop system that should be em-

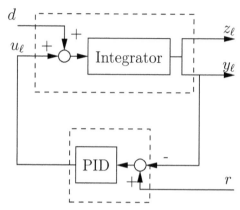

Figure 2.2 The unconstrained closed-loop system for the example described in Section 1.2.1.

ulated in the specification and solution of the anti-windup problem, one important feature is asymptotic stability. In fact, it is difficult to make sense out of an anti-windup problem when the unconstrained closed-loop system is not asymptotically stable. Hence, the origin of the unconstrained closed-loop system is assumed to be asymptotically stable when the exogenous inputs are set to zero. This automatically implies that the internal state of the plant is stabilizable through the control input.

### 2.1.2 Actuator saturation

The examples of the previous chapter illustrated some consequences of actuator limitations in linear control systems. Actuator saturation, which imposes input constraints, is described now in more detail.

When a control system gives a request or command to an actuator, the actuator typically produces an output—force, torque, displacement, or other physical quantity—within its operating range that is closest to the requested value. Values outside of the actuator's amplitude limits are mapped into the range of capabilities according to the function illustrated in Figure 2.3, which is usually referred to as the *saturation function* or *nonlinearity*.

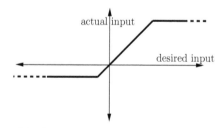

Figure 2.3 Graph of the typical saturation nonlinearity.

In Figure 2.4, the saturation nonlinearity is represented by a nonlinear block diagram resembling the shape of the saturation nonlinearity. In the same figure,

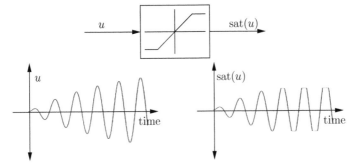

Figure 2.4  An example of a signal $u$ and its saturated version $\text{sat}(u)$.

an example of a possible desired input signal $t \mapsto u(t)$ and its saturated version $t \mapsto \text{sat}(u(t))$ are given. Note that when the signal $u(t)$ is small it coincides with its saturated version $\text{sat}(u(t))$. However, when $u(t)$ becomes too large, the amplitude of its saturated version $\text{sat}(u(t))$ is limited.

Letting $u_M$ and $u_m$ correspond to the maximal and the minimal attainable actuator value, the saturation function is described mathematically by the following equation:

$$\text{sat}(u) := \begin{cases} u_M, & \text{if } u \geq u_M \\ u, & \text{if } u_m \leq u \leq u_M \\ u_m, & \text{if } u \leq u_m. \end{cases} \tag{2.1}$$

The input $u$ is said to be in the region where the saturation nonlinearity is equal to the identity when $u_m \leq u \leq u_M$. The saturation function is called a *symmetric saturation* if $u_M = -u_m$ and it is called a *unit saturation* if $u_M = -u_m = 1$.

For the case of a multi-input control system, the notion of a vector-valued saturation function becomes relevant. In general, a vector-valued saturation function is a continuous mapping from the vector $u$ to a bounded, usually convex, region $\mathcal{U}$ with the property that the function evaluates to $u$ when $u$ belongs to $\mathcal{U}$. Perhaps the simplest vector-valued saturation function corresponds to the *decentralized saturation function*, which consists of a vector of scalar saturation functions, the $i$th function depending only on the $i$th component of the input vector. In other words, the vector-valued decentralized saturation function $u \mapsto \sigma(u)$ has the form

$$\sigma(u) = \begin{bmatrix} \text{sat}_1(u_1) \\ \text{sat}_2(u_2) \\ \vdots \\ \text{sat}_{n_u}(u_{n_u}) \end{bmatrix}, \tag{2.2}$$

where $\text{sat}_i(\cdot)$ is defined as in (2.1) for all $i$. Figure 2.5 shows a schematic interpretation of the decentralized saturation function. The vector input $u$ is said to be in the region where the saturation nonlinearity is equal to the identity when $u_m \leq u \leq u_M$, where $u_m$ and $u_M$ are vectors consisting of the minimum and maximum attainable values for each of the saturation functions and the inequality is understood in a component-wise sense. Although decentralized symmetric saturations are the most common input nonlinearities encountered in control system designs, the results reported in this book also apply to a larger class of saturation nonlinearities.

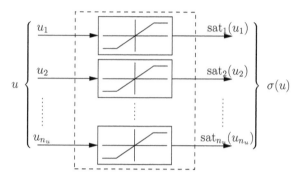

Figure 2.5 The decentralized saturation.

### 2.1.3 The saturated closed-loop system

As demonstrated in the previous chapter, the windup phenomenon can occur when plant input saturation is introduced into the unconstrained closed-loop system. The

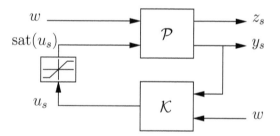

Figure 2.6 The saturated closed-loop system.

corresponding *saturated closed-loop system* is illustrated in Figure 2.6. The internal state trajectory of the saturated closed-loop system is called the *saturated response*; the trajectories $u_s$, $z_s$ and $y_s$ are called, respectively, the *saturated control input response*, *saturated performance output response*, and *saturated measured output response*. Introducing plant input saturation into the unconstrained closed-loop system of Figure 2.2 results in the saturated closed-loop system shown in Figure 2.7.

## 2.2 QUALITATIVE OBJECTIVES

In the subsections below, the qualitative objectives of anti-windup augmentation are described, in both global and regional terms. The quantitative aspects of the anti-windup problem will be discussed in a later section of this chapter.

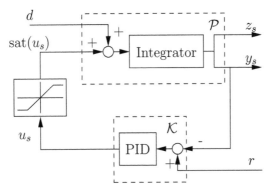

Figure 2.7  The saturated closed-loop system corresponding to Figure 2.2.

### 2.2.1 Small signal preservation

The first objective of anti-windup augmentation:

> (Small signal preservation) *To make the response of the anti-windup augmented closed-loop system match the response of the unconstrained closed-loop system whenever this is possible.*

One of the main tenets in the philosophy behind anti-windup synthesis is that the unconstrained closed loop provides a response for the anti-windup augmented closed loop to emulate. For this reason, it is logical to insist that the response of the anti-windup augmented closed-loop system match the response of the unconstrained closed-loop system whenever this is possible. This behavior will be called the *small signal preservation* property. The ability to preserve a response depends on whether or not the unconstrained control input response $u_\ell(t)$ leaves the region where the saturation nonlinearity equals the identity. If it does not leave this region, then the response is reproducible and should be preserved, at least when the initial state of the anti-windup augmentation (if it exists) is zero.

### 2.2.2 Internal stability: global vs. regional

The second objective of anti-windup augmentation:

> (Asymptotic stability) *To make, in the absence of exogenous inputs, a desired constant operating point asymptotically stable with a basin of attraction at least as big as the set of states over which the system is expected to operate.*

A baseline requirement for anti-windup augmentation for linear systems is that the origin be asymptotically stable when the exogenous inputs are identically equal to zero and zero is in the interior of the region where the saturation nonlinearity maps to the identity. Indeed, this property is presumably enforced by the unconstrained controller for the unconstrained closed-loop system. It should be maintained when using anti-windup augmentation. Asking for *local* asymptotic stability is not asking for much. It is guaranteed by not adding any anti-windup augmentation at all. More ambitious is the objective of making the domain of attraction for the origin in the absence of exogenous inputs at least as big as the set of states over which the system is expected to operate. For plants that are not exponentially unstable, it is possible to make this basin of attraction the entire state space, in other words, to induce global asymptotic stability. However, for some problems achieving global asymptotic stability might be overkill and regional asymptotic stability might be sufficient. Focusing on regional asymptotic stability can be advantageous because it may permit simpler algorithms that perform better than global algorithms over the region of interest. For exponentially unstable plants, global asymptotic stability is impossible. An example that illustrates this fact is given in the next subsection. Fortunately, many times anti-windup augmentation algorithms with regional guarantees are effective on loops with exponentially unstable plants.

### 2.2.3 Input-output stability: global vs. regional

The third objective of anti-windup augmentation:

> (Input-output stability) *To induce a bounded response for initial states and exogenous inputs that are expected during operation.*

First, it must be understood that a bounded response will be impossible for some exogenous inputs and initial states when the plant to be controlled is not exponentially stable. This will be the case regardless of the controller used. To illustrate this, consider a first-order, exponentially unstable plant modeled in state-space form as

$$\begin{aligned} \dot{x} &= x + \text{sat}(u) + w \\ y &= x, \end{aligned}$$

where "sat" denotes the symmetric unit saturation nonlinearity. Suppose the system starts at rest, i.e., $x(0) = 0$, and that $w(t) = 2$ for two seconds, after which it resets to zero. The differential equation can be used to verify that, no matter what $u(t)$ is, $x(2) \geq 2$. After this point in time the signal $t \mapsto x(t)$ will diverge exponentially no matter what $u(t)$ is. Since this behavior occurs no matter what $u(t)$ is, there is nothing that anti-windup augmentation can do to prevent this phenomenon. On the other hand, for smaller disturbances very simple anti-windup augmentation may be able to help significantly.

In general, for any constrained linear plant with exponentially unstable modes, large enough initial conditions for these modes or large enough exogenous disturbances that affect these modes lead to trajectories that diverge exponentially, no matter what the control algorithm is. Thus, for these plants, it will not be reasonable to express the anti-windup augmentation objectives in global terms.

It will be reasonable to express the anti-windup augmentation objectives in global terms for constrained linear plants that are exponentially stable. The objectives will go beyond just guaranteeing internal and external stability. Otherwise, the anti-windup task would be trivial for these systems. For marginally stable plants and marginally unstable plants, achieving certain global anti-windup properties will also be possible, but will require more sophisticated algorithms. There are times when the extra effort required to achieve global anti-windup properties is not worth the effort. For example, if the designer has a good handle on the size of states and disturbances to expect as the control system operates then it is enough to guarantee anti-windup properties in the face of such conditions. In this case, regional anti-windup properties can be pursued. This often can lead to better anti-windup performance during normal operation since the anti-windup augmentation is not designed to guard against unrealistically large states or disturbances. In any case, anti-windup augmentation should strive to guarantee boundedness of signals in the face of reasonable initial conditions and disturbances. In fact, it should strive for even more: it should also strive for unconstrained response recovery, which is described next.

### 2.2.4 Unconstrained response recovery (URR)

The fourth objective of anti-windup augmentation:

> *To recover the unconstrained closed-loop response asymptotically whenever this is possible.*

As mentioned in the subsection on small signal preservation, the unconstrained closed-loop response provides a closed-loop behavior for the anti-windup augmented closed-loop system to emulate. When it is possible to reproduce this response exactly, the augmented closed-loop system should do so. However, anti-windup augmentation is needed exactly because it isn't always possible to reproduce the unconstrained closed-loop response. In this case, the additional objectives of anti-windup augmentation, like internal and external stability, become relevant. However, good anti-windup augmentation algorithms will have at least one additional qualitative objective: to recover the unconstrained closed-loop response asymptotically whenever this is possible. This objective has not received much attention in the anti-windup literature. One reason is that, for plants that are not exponentially stable, it is difficult to characterize when an unconstrained closed-loop response can be recovered. Another reason is that unconstrained response recovery is more difficult to guarantee than internal and external stability. These issues are discussed next.

For loops that contain exponentially stable plants, it is easy to characterize unconstrained closed-loop responses that can be recovered. In this case an unconstrained closed-loop response is recoverable if the unconstrained input response tends asymptotically to the linear region of the saturation nonlinearity,[1] i.e.,

$$u_\ell(t) - \text{sat}(u_\ell(t)) \to 0 \quad \text{as} \quad t \to \infty .$$

When the plant is not exponentially stable, the characterization of responses that are recoverable is more subtle; indeed, the simple condition given above for exponentially stable plants is no longer sufficient. To illustrate this point, consider the

---

[1]This condition is sufficient but not necessary for unconstrained response recovery. It is possible for the unconstrained input response to converge to the linear region in a sense ($\mathcal{L}_p$ convergence) weaker than asymptotic convergence and still have the possibility of unconstrained response recovery. For more details, see Section 2.4.

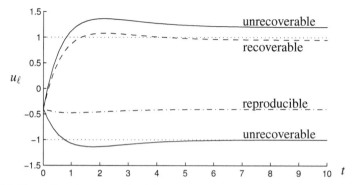

Figure 2.8 Examples of unconstrained input responses corresponding to reproducible, re-coverable, and unrecoverable unconstrained closed-loop trajectories for an inte-grator plant.

unconstrained closed-loop response resulting from the negative feedback intercon-nection of a single integrator plant

$$\dot{x}_{p\ell} = u_\ell + w, \qquad x_{p\ell}(0) = 0.2$$
$$y_\ell = x_{p\ell},$$

and a PI controller with gains $k_P = 2$ and $k_I = 1$. Due to the integral action in the controller, for any constant disturbance input the controller drives to zero the plant's state $x_{p\ell}$, and consequently also drives to zero its output $y_\ell$. When $w$ is a unit step disturbance input, the curves in Figure 2.8 show the unconstrained input $u_\ell$ corresponding to the following selections for $w$: $-1.2, -0.95, 0.4, 1$ (from top to bottom). When considering a unitary saturation at the plant input (the dotted lines in the figure represent the saturation limits), the upper solid curve corresponds to an unconstrained closed-loop input response that is unrecoverable by the anti-windup augmented closed loop because its steady state corresponds to input values that exceed the saturation limits. The dashed curve corresponds to a response that is recoverable but not reproducible because it exceeds the saturation limits during the transient response. The dashed-dotted curve is permanently within the saturation limits; hence it is reproducible by the saturated system. Finally, the lower solid curve makes the point: it also corresponds to an unconstrained closed-loop response that is unrecoverable, even though it converges asymptotically to the linear region of the saturation function. To see that it is unrecoverable, consider the evolution equation for the saturated plant with the input disturbance $w = 1$, i.e.,

$$\dot{x}_{p\ell} = \text{sat}(u) + 1 .$$

Because of the unity saturation limit, the derivative of the state can never be neg-ative, no matter what the input signal $u$ is. Then, since $x_{p\ell}(0) = 0.2$, necessarily $x_p(t) \geq 0.2$ for all positive times. On the other hand, the plant state converges to zero in the absence of constraints. This means that the unconstrained closed-loop response cannot be recovered asymptotically in the presence of saturation, regard-less of the controller used.

This example shows, for plants that are not exponentially stable, that it is not enough for the unconstrained control input to converge asymptotically to the linear region of the saturation nonlinearity. Nevertheless, for marginally stable plants (like a single integrator) or marginally unstable plants (like a double integrator), the unconstrained closed-loop response is recoverable when the unconstrained input response converges asymptotically[2] to a closed set in the interior of the linear region of the saturation nonlinearity. For example, introducing the family of functions

$$\text{sat}_\delta(u) := \begin{cases} u_M - \delta, & \text{if } u \geq u_M - \delta \\ u, & \text{if } u_m + \delta \leq u \leq u_M - \delta \\ u_m + \delta, & \text{if } u \leq u_m + \delta \end{cases} \qquad (2.3)$$

parameterized by $\delta > 0$, it is enough to have

$$u_\ell(t) - \text{sat}_\delta(u_\ell(t)) \to 0 \quad \text{as} \quad t \to \infty$$

for some $\delta > 0$. It is still more complicated to characterize recoverable unconstrained closed-loop responses for plants that are exponentially unstable. This characterization will not be given here.

Unconstrained response recovery is guaranteed for constrained linear systems that match (in an appropriate sense) the unconstrained linear system when saturation is not present and satisfy $u(t) - \text{sat}(u(t)) \to 0$ as $t \to \infty$, where $u(t)$ is the plant input in the constrained closed-loop system. This condition can be checked a posteriori to ascertain whether unconstrained response recovery has occurred. Thus, unconstrained response recovery will be guaranteed a priori for any anti-windup augmentation algorithm that can guarantee the implication

$$\lim_{t \to \infty} [u_\ell(t) - \text{sat}(u_\ell(t))] \to 0 \qquad \Longrightarrow \qquad \lim_{t \to \infty} [u(t) - \text{sat}(u(t))] \to 0 \,.$$
$$(\text{or } \lim_{t \to \infty} [u_\ell(t) - \text{sat}_\delta(u_\ell(t))] \to 0)$$

Many anti-windup augmentation algorithms are able to guarantee this implication either directly or indirectly. In fact, many of the algorithms described here, especially those discussed in the last part of the book, are derived specifically with the objective of minimizing the difference between the unconstrained closed-loop response and the anti-windup augmented closed-loop response. Since the unconstrained closed loop is assumed to be internally and externally stable, those properties (regional or global) accrue for the anti-windup augmented closed-loop system as a product of attempting to recover the unconstrained closed-loop response. In particular, if the unconstrained closed loop has certain $\mathcal{L}_2$ input-output stability properties,[3] and if the anti-windup augmented response converges, in an $\mathcal{L}_2$ sense, to the unconstrained closed-loop response, then the anti-windup augmented closed-loop system will also enjoy certain $\mathcal{L}_2$ input-output stability properties. Some of the algorithms described later will aim to optimize directly the $\mathcal{L}_2$ input-output stability properties for the anti-windup augmented closed-loop system.

---

[2] Again, weaker ($\mathcal{L}_p$) convergence is allowed.
[3] This stability notion will be discussed in Section 2.4.

## 2.3 ANTI-WINDUP AUGMENTATION

### 2.3.1 The augmented closed-loop system

The qualitative objectives associated with the anti-windup problem suggest some structure for anti-windup augmentation. For example, the small signal preservation property suggests that the output of the anti-windup augmentation should be identically equal to zero in the case where the output of the unconstrained controller is always within the limits of the actuator and the state of the anti-windup augmentation, if it exists, is initialized to zero. The most common way to achieve this feature is to make the input to the anti-windup augmentation equal to the difference between the input commanded to the actuator and the input supplied by the actuator to the plant. This scenario is illustrated in Figure 2.9, and it is the architecture that will be used typically in this book. The corresponding closed-loop system is called

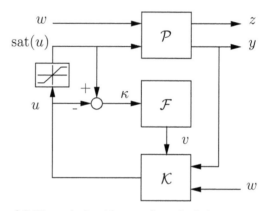

Figure 2.9 The typical architecture for anti-windup augmentation.

the *anti-windup augmented closed-loop system*. The following terminology is introduced for this system. The internal state trajectory of the closed loop is the *anti-windup augmented response*. The signals $u$, $z$, and $y$ are called, respectively, the *anti-windup augmented control input response*, the *anti-windup augmented performance output response*, and the *anti-windup augmented measured output response*.

The block $\mathcal{F}$, which is called the *anti-windup augmentation* or *filter*, represents a system that is possibly dynamic and possibly nonlinear. In its state-space representation, anti-windup augmentation has the general form

$$\begin{aligned} \dot{x}_{aw} &= f_{aw}(x_{aw}, \kappa) \\ v &= h_{aw}(x_{aw}, \kappa). \end{aligned} \quad (2.4)$$

The state $x_{aw}$ will not exist for the case where $\mathcal{F}$ has no dynamics. Very often, the functions $f_{aw}$ and $h_{aw}$ will be linear functions. If asymptotic stability is achieved and the anti-windup augmentation has dynamics, then, when $\kappa = 0$, the filter must necessarily have a constant steady-state value $x_{aw}^*$ that is asymptotically stable. Without loss of generality, the condition $x_{aw}^* = 0$ is assumed. In particular, $f_{aw}(0,0) = 0$. Moreover, if small signal preservation is achieved, then typically this dictates that $h_{aw}(0,0) = 0$.

**Example 2.3.1**    For the example reported in Section 1.2.1, the anti-windup closed-loop system of Figure 2.9 has the specific realization shown in Figure 2.10. As depicted there, Algorithm 11, which is used to synthesize the anti-windup filter used for this example, generates a stable, first-order linear anti-windup filter whose output is added to the unconstrained controller's input and output. The filter has the state-space realization

$$\dot{x}_{aw} = -x_{aw} + \kappa$$
$$v = \begin{bmatrix} 1 \\ 1 \end{bmatrix} x_{aw} .$$

□

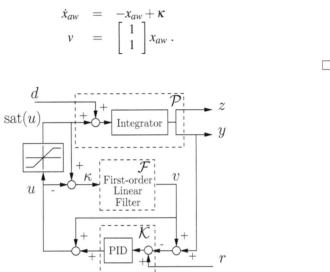

Figure 2.10  The anti-windup closed-loop system for the example described in Section 1.2.1.

### 2.3.2  Measurement requirements

It may appear, due to Figure 2.9, that a measurement of the output of the actuator is needed to implement anti-windup augmentation. In fact, this is not the case and is not even desirable, especially if the actuator has any fast, unmodeled dynamics. Indeed, it is important for meeting the objectives of anti-windup augmentation that the input to the anti-windup augmentation filter be identically zero when the commanded inputs are small. Instead of measuring the output of the actuator, it is best to simulate the saturation of the actuator. In other words, the composite controller comprises the anti-windup filter $\mathcal{F}$, the unconstrained controller $\mathcal{K}$, *and* the saturation nonlinearity depicted in Figure 2.9. As long as the simulated saturation level does not exceed the true saturation level of the actuator, the true saturation level will not affect the constrained closed loop.

One strong appeal of the anti-windup architecture in Figure 2.9 is that it only requires the measurements that were already being used by the unconstrained controller. Typically, this architecture is sufficient for achieving high performance with anti-windup augmentation. However, occasionally, especially for the case of exponentially unstable plants, it can be helpful to have additional plant state measurements available when synthesizing anti-windup augmentation.

### 2.3.3 Linear vs. nonlinear

In the case where $\mathcal{F}$ is a linear dynamical system, the anti-windup augmentation is said to be *linear*. Otherwise, it is said to be *nonlinear*. Computationally, it is easiest to automate the synthesis of linear anti-windup augmentation. However, because the anti-windup control problem is inherently nonlinear, there are situations where nonlinear anti-windup augmentation can provide superior anti-windup performance. This will be illustrated by examples later.

### 2.3.4 Continuous-time vs. sampled data

The focus of this book is anti-windup augmentation synthesis for continuous-time closed-loop systems. Most of the algorithms presented generate anti-windup filters that are continuous-time systems, whether static or dynamic, as contrasted in the next subsection. However, one algorithm in the last part of the book generates an anti-windup filter part of which is updated continuously and part of which is updated through samples. The reason for this is that the idea behind the algorithm's development is most tractable in discrete time. It is important to recognize that anti-windup augmentation does not necessarily need to be updated continuously in order to be effective at solving the anti-windup problem for continuous-time systems.

### 2.3.5 Static vs. dynamic

In the case where $\mathcal{F}$ is a block with no memory, the anti-windup augmentation is said to be *static*. Otherwise it is said to be *dynamic*. Static, linear anti-windup augmentation can be quite appealing because of its simplicity, and it can be very effective on some systems. At times it is inadequate. There is seldom a need to use anti-windup augmentation that has a dynamic order larger than that of the plant. When the dynamic order of the anti-windup augmentation is equal to that of the plant, it is called *plant-order anti-windup augmentation*. When the dynamic order is greater than zero but less than that of the plant, it is called *reduced-order anti-windup augmentation*.

### 2.3.6 External vs. full authority

Figure 2.9 shows the anti-windup augmentation block $\mathcal{F}$ affecting the control system by injecting its output signals into the unconstrained controller block. In all of the algorithms considered herein for linear unconstrained controllers, the output signals enter the unconstrained controller block additively. In some instances, the state equations for the unconstrained controller can be affected directly. In other situations, the unconstrained controller can be affected only at its input and output. These two different scenarios are illustrated in Figure 2.11. The diagram on the left side of Figure 2.11 shows the case where the anti-windup augmentation injects signals only at the input and output of the unconstrained controller. This is referred to as *external anti-windup augmentation*. The diagram on the right side of Figure 2.11 shows the case where the anti-windup augmentation affects the state equations of the unconstrained controller. This is referred to as *full-authority anti-windup aug-mentation*.

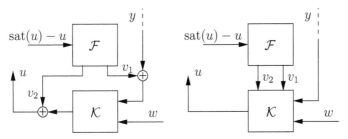

Figure 2.11 External anti-windup augmentation (left) and full-authority anti-windup aug-mentation (right).

To describe the effect of the signal $s$ on the unconstrained controller, let the unconstrained controller have state-space realization $(A_c, B_c, C_c, D_c)$. Then the external anti-windup augmentation in the left-hand diagram of Figure 2.11 affects the controller by injecting its output $v = (v_1, v_2)$ as follows:

$$\mathcal{K} \begin{cases} \dot{x}_c &= A_c x_c + B_c \begin{bmatrix} y + v_1 \\ w \end{bmatrix} \\ u &= C_c x_c + D_c \begin{bmatrix} y + v_1 \\ w \end{bmatrix} + v_2 . \end{cases}$$

Alternatively, the full-authority anti-windup augmentation in the right-hand diagram of Figure 2.11 affects the controller by injecting its outputs $v_1$ and $v_2$ as follows:

$$\mathcal{K} \begin{cases} \dot{x}_c &= A_c x_c + B_c \begin{bmatrix} y \\ w \end{bmatrix} + v_1 \\ u &= C_c x_c + D_c \begin{bmatrix} y \\ w \end{bmatrix} + v_2 . \end{cases}$$

These equations are also represented graphically in the block diagram of Figure 2.12.

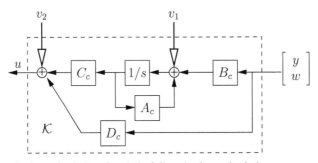

Figure 2.12 Anti-windup signals in full-authority anti-windup augmentation.

Any full-authority anti-windup compensator can be implemented as an external anti-windup compensator when the matrix multiplying $y$ in the $\dot{x}_c$ equation has full row rank. This situation holds for PID controllers, for example. Indeed, the full row rank condition implies the existence of a matrix $M$, with the number of columns equal to the state dimension of $x_c$ and the number of rows equal to the number of components of $u_c$, such that the square matrix

$$B_c \begin{bmatrix} M \\ 0 \end{bmatrix}$$

is the identity matrix. So, if $\tilde{v} = (\tilde{v}_1, \tilde{v}_2)$ is the output of a full-authority anti-windup compensator, then

$$v = (v_1, v_2) := (M\tilde{v}_1, \tilde{v}_2)$$

is the output of an external anti-windup compensator such that the trajectories of the two anti-windup augmented closed-loop systems coincide.

Conversely, any external anti-windup compensator can be implemented as a full-authority anti-windup compensator. Indeed, if $\tilde{v} = (\tilde{v}_1, \tilde{v}_2)$ is the output of an external anti-windup compensator, then

$$v = (v_1, v_2) := \left( B_c \begin{bmatrix} \tilde{v}_1 \\ 0 \end{bmatrix}, \tilde{v}_2 \right)$$

is the output of a full-authority anti-windup compensator such that the trajectories of the two anti-windup augmented closed-loop systems coincide. It follows that the achievable results when using full-authority anti-windup are no worse, and often better, than results that are obtained when using external anti-windup. On the other hand, for control systems where it is difficult to act directly on the unconstrained controller states, as in some analog controllers, for example, results for external anti-windup augmentation are useful. Also, as the dynamic order of the anti-windup augmentation increases, the advantage of full-authority anti-windup over external anti-windup decreases. The last part of this book is dedicated to anti-windup augmentation that is of plant order and has the external anti-windup structure.

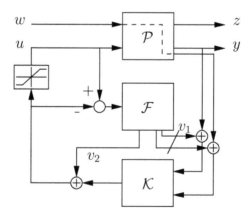

Figure 2.13 Extension of the external anti-windup scheme to allow anti-windup authority through the reference input.

The way the left diagram in Figure 2.11 has been drawn, the external anti-windup compensator is allowed to modify only the input coming from the plant measurement and not the input due to a reference command. Recall that the reference signal has been combined into the signal $w$. If anti-windup filter signals can be injected into the unconstrained controller at locations where the reference command enters, this can be addressed within the diagram of Figure 2.11. This is done by redefining the output $y$ of the plant to include an extra channel corresponding to the reference value fed through the plant $\mathcal{P}$, as shown in the block diagram of Figure 2.13. This allows the anti-windup augmentation to enforce modifications on both controller inputs by way of an augmented version of its first output $v_1$. In the common case of unity feedback, where the reference and the plant output enter the controller equations at the same location, this modification provides no enhancements.

### 2.3.7 Algebraic loops introduced by proper anti-windup filters

The general form of the anti-windup filter $\mathcal{F}$ is given in equation (2.4). When the function $h$ in (2.4) can be taken to be independent of $\kappa$, the anti-windup filter is said to be *strictly proper*. Otherwise, the anti-windup filter is said to be *proper*. Notice that the specific anti-windup filter used in Figure 2.10 is strictly proper.

Unless the anti-windup filter is strictly proper, it is possible that the anti-windup augmentation in Figure 2.9 introduces an algebraic loop into the anti-windup augmented control system. An *algebraic loop* is a closed path that does not involving passing through any dynamic element while traversing the path. When it exists, the algebraic loop introduced into the anti-windup augmented controller in Figure 2.9 occurs in the loop from the input of the saturation element through the anti-windup filter to the unconstrained controller and back to the input of the saturation element.

An algebraic loop generates an implicit equation that may or may not have a unique solution. An algebraic loop is said to be *well-posed* if the implicit equation it creates has a unique solution. Otherwise, the algebraic loop is said to be *ill-posed*. All of the anti-windup augmentation algorithms given later will guarantee that the algebraic loops in the anti-windup augmented controller are well-posed. However, such guarantees do not give insight into how to solve the implicit equation in order to generate an explicit commanded input to send to the plant's actuators. At least three different approaches exist for solving this problem:

1. In the case of simulations, the solvers used by the simulation software to generate the explicit control signals may be relied upon. Usually, these are based on a Newton method and are quite efficient.
2. A hand-crafted solver, which typically would be an iterative algorithm that approximates the explicit solution after a reasonable number of steps, can be used.
3. Especially when the number of input variables is small, one may consider relying on a lookup table to generate the explicit control values. Based on the material in the preceding sections, when the feedthrough term of the anti-windup augmentation is linear, the algebraic loops in the anti-windup aug-

mented controller have the form

$$u = \zeta + \Lambda(u - \text{sat}(u)) \tag{2.5}$$

where $\zeta$ is some function of the plant output and the unconstrained controller states. The saturation nonlinearity is assumed to be a decentralized saturation function here. When well-posed, the solution to the equation (2.5) has a piecewise affine form with respect to the variable $\zeta$ with $3^n$ possibly distinct regions, where $n$ is the number of components of $u$. This is because equation (2.5) can be rewritten as $\Xi(u) = \zeta$, where the function $\Xi$ is piecewise affine with $3^n$ regions, corresponding to all of the different combinations of the three conditions i) $u_i < (u_m)_i$, ii) $(u_m)_i \le u_i \le (u_M)_i$, and iii) $(u_M)_i < u_i$ for $i$ ranging from 1 to $n$. Let the matrix $M_i$ and the vector $b_i$ be such that $\Xi(u) = M_i u + b_i$ for $u$ values in the $i$th region. Since the algebraic loop is well-posed, the matrix $M_i$ is invertible for each $i$. The solution to the equation $\Xi(u) = \zeta$ is then given by $M_j^{-1}(\zeta - b_j)$, where $j$ is such that $M_j^{-1}(\zeta - b_j)$ is a vector belonging to region $j$. Due to well-posedness, there will always exist such an index $j$ and, when there is more than one index, each index returns the same answer for the solution $u$.

The last two approaches are especially relevant when attempting to implement the overall anti-windup augmented controller on a computer in a real-time environment.

There is one warning that applies in all cases: "almost ill-posed" algebraic loops can cause problems. Simulations typically run more slowly when the system contains an almost ill-posed algebraic loop. This behavior occurs because the closer to ill-posed the equations are, the smaller the simulation steps must be to make sure that no important features in the closed-loop trajectories are missed. Explicit solutions to almost ill-posed algebraic loops are more sensitive to small perturbations to the vector $\zeta$. For these reasons, most of the anti-windup augmentation algorithms to appear later attempt to produce solutions with algebraic loops that are far from ill-posed, or *strongly well-posed*. In terms of the discussion above, this is related to ensuring that the matrices $M_i$ are not very close to being singular.

It is also worth noting that the unconstrained closed loop may exhibit an algebraic loop if neither the plant nor the unconstrained controller is strictly proper. It will always be assumed that any algebraic loops in the unconstrained closed loop are well-posed. This section concludes with an example demonstrating that the following scenario is possible: the unconstrained closed-loop system has a well-posed algebraic loop, the insertion of the saturation nonlinearity destroys well-posedness, and the anti-windup augmentation restores well-posedness. The example also provides more exposure to the idea of algebraic loops.

Consider the unconstrained plant given by the equations

$$\begin{aligned} \dot{x}_{p\ell} &= x_{p\ell} + u_\ell \\ y_\ell &= x_{p\ell} + u_\ell \end{aligned}$$

and the unconstrained, static controller $y_{c\ell} = k u_{c\ell}$ where $k$ is a feedback gain.[4] Con-

---

[4]Admittedly, the closed-loop exponential stability enabled by this feedback architecture will not be robust to fast, stable unmodeled plant dynamics. This simple architecture is used here to simplify an illustration about algebraic loops.

necting these systems via the equations $u_\ell = y_{c\ell}$ and $u_{c\ell} = y_\ell$ produces an algebraic loop and the implicit equation $u_\ell = kx_{p\ell} + ku_\ell$. This equation has a unique solution for each value of $x_{p\ell}$ as long as $k \neq 1$. In this case, the explicit solution is given by

$$u_\ell = \frac{k}{1-k}x_{p\ell},$$

which makes the unconstrained closed loop exponentially stable when $k > 1$.

Now suppose the unit saturation element is introduced at the input to the plant. In this case, the implicit equation generated by the algebraic loop becomes $u = kx_p + k\mathrm{sat}(u)$, which can also be written as

$$u - k\mathrm{sat}(u) = kx_p. \tag{2.6}$$

Taking $k > 1$ so that the unconstrained closed loop is exponentially stable, for small values of $x_p$ there will be three values of $u$ satisfying the equation (2.6), and thus the feedback system is not well-posed. For example, when $x_p = 0$, the following values for $u$ satisfy the equation: $u = -k$, $y_p = 0$, and $y_p = k$. The existence of these three solutions can be understood by plotting the function $u \mapsto u - k\mathrm{sat}(u)$ for $k > 1$, as done in Figure 2.14, and noting that there are three intersections with each horizontal line passing near the origin. In other words, for each small number $c$ there are three different values of $u$ that satisfy $u - k\mathrm{sat}(u) = c$.

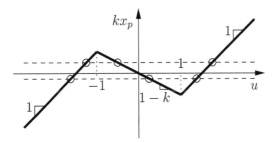

Figure 2.14 The algebraic loop equation for $k > 1$, when the feedback system is not well-posed.

Finally, introduce the static anti-windup filter

$$v = \Lambda(u - \mathrm{sat}(u)),$$

where $\Lambda$ is the anti-windup gain. The implicit equation in this case is now $u = kx_p + k\mathrm{sat}(u) + \Lambda u - \Lambda\mathrm{sat}(u)$, which can be rewritten as

$$(1 - \Lambda)u - (k - \Lambda)\mathrm{sat}(u) = kx_p.$$

As long as $\Lambda > 1$ and $k > 1$, the function $u \mapsto (1 - \Lambda)u - (k - \Lambda)\mathrm{sat}(u)$ will be monotonically decreasing, as shown in Figure 2.15. Indeed, for small $u$ the slope will be $1 - k$, whereas for large $u$ the slope will be $1 - \Lambda$. Thus, the implicit equation will have a unique solution, which is given by the piecewise affine function

$$u = \begin{cases} \dfrac{k}{1-k}x_p & \left|\dfrac{k}{1-k}x_p\right| \leq 1 \\[2mm] \dfrac{1}{1-\Lambda}(kx_p + k - \Lambda) & \dfrac{k}{1-k}x_p > 1 \\[2mm] \dfrac{1}{1-\Lambda}(kx_p - k + \Lambda) & \dfrac{k}{1-k}x_p < -1 . \end{cases} \tag{2.7}$$

In the special case where $\Lambda = k$, this expression becomes $u = \frac{k}{1-k} x_p$ for all $x_p$. As long as $k > 1$ and $\Lambda > 1$, the feedback (2.7) passed through the saturation nonlinearity results in local exponential stability. For this particular problem, the parameters $k > 1$ and $\Lambda > 1$ do not affect the set of initial conditions from which convergence to the origin occurs; this set is always equal to the interval $(-1, 1)$.

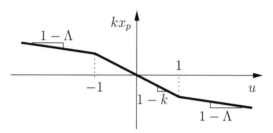

Figure 2.15 The algebraic loop equation with anti-windup compensation guarantees well-posedness.

## 2.4 QUANTITATIVE PERFORMANCE OBJECTIVES

### 2.4.1 Motivations

Recall that the qualitative anti-windup augmentation objectives are 1) small signal preservation, 2) internal stability, 3) external stability, and 4) unconstrained performance recovery. The small signal preservation property is enforced by the structure of anti-windup augmentation. In particular it comes from the fact that the anti-windup augmentation is driven by the difference between the commanded input and the saturated input and, with zero initial condition, it maps the zero input to the zero output. The remaining three objectives can be enforced by pursuing related quantitative objectives.

The idea of "convergence" comes up multiple times in the description of the three remaining qualitative objectives. First, there is the asymptotic stability objective for the case where the exogenous inputs are zero. This objective corresponds to convergence of the state to zero. Second, there is the unconstrained response recovery objective, which corresponds to convergence of the anti-windup augmented closed-loop response to the unconstrained closed-loop response. Instead of pursuing asymptotic convergence directly for these objectives, a different measure of signals, one that implies asymptotic convergence in many situations, is considered. That measure corresponds to the energy in a signal as measured by integrating the square of its size. The key fact to know in this case is the following:

> For a scalar, nonnegative, uniformly continuous signal $t \mapsto f(t)$, if the infinite integral of the square of the function is bounded, then necessarily the function values converge to zero as $t$ becomes arbitrarily large.[5]

---

[5] A slight generalization of this fact is known as Barbalat's lemma in the control literature.

The opposite assertion is not true: convergence of $f(t)$ to zero as $t$ becomes arbitrarily large does not imply that the infinite integral of the square of $f$ is bounded. A simple example that illustrates this is the function $f(t) = 1/\sqrt{(1+t)}$. A bounded integral automatically rules out signals that converge too slowly to zero. A bounded integral is a desirable property since slowly converging signals are usually associated with poor performance. Some examples of scalar nonnegative signals and their squared integrals are reported in Figure 2.16. Note that the third signal is not square integrable despite its convergence to zero.

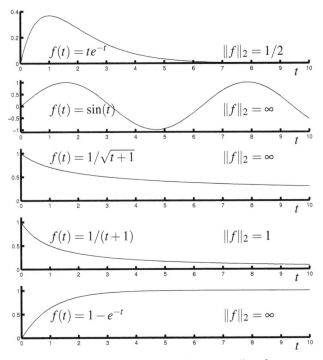

Figure 2.16 Examples of signals and corresponding $\mathcal{L}_2$ norms.

How might scalar, nonnegative signals arise in anti-windup augmented closed-loop systems? One example is the time domain function $t \mapsto |x(t)|$, where $x(t)$ represents the state of the closed-loop system at time $t$ and $|x(t)|$ represents the Euclidean norm of the vector $x(t)$, i.e., $|x(t)| = \sqrt{\sum_{i=1}^{n} x_i(t)^2}$, where $x_i(t)$ represents the $i$th component of the vector $x(t)$. Another example is $t \mapsto |w(t)|$, where $w$ represents an exogenous signal. A third example is

$$t \mapsto |u_\ell(t) - \mathrm{sat}(u_\ell(t))|$$

(or

$$t \mapsto |u_\ell(t) - \mathrm{sat}_\delta(u_\ell(t))|\,),$$

where $u_\ell$ is the unconstrained closed-loop input response. A final example is the function

$$t \mapsto |x(t) - x_\ell(t)|,$$

where $x$ is the response of the plant for the anti-windup augmented closed-loop system and $x_\ell$ is the response of the plant for the unconstrained closed-loop system. Whenever the square of the norm of a signal can be integrated locally,[6] the square root of the corresponding infinite integral is called the $\mathcal{L}_2$ norm of the signal. For a signal $t \mapsto x(t)$ the symbol $||x||_2$ will be used to denote the $\mathcal{L}_2$ norm. In particular,

$$||x||_2 := \lim_{t \to \infty} \int_0^t |x(\tau)|^2 d\tau.$$

When the $\mathcal{L}_2$ norm of a signal $t \mapsto x(t)$ is finite, the signal is said to belong to $\mathcal{L}_2$, written $x(\cdot) \in \mathcal{L}_2$, and its size is said to be square integrable. By slight abuse of standard terminology, the $\mathcal{L}_2$ norm of a signal will sometimes be referred to as the energy of the signal.

When might one of these scalar, nonnegative signals be uniformly continuous so that it is possible to conclude that it converges asymptotically to zero? A fact that is relevant for all of the anti-windup augmented closed-loop systems in this book is the following:

*If*

1. *the size of the signal $t \mapsto x(t)$ is square integrable,*
2. *the signal $t \mapsto x(t)$ is the solution to a differential equation*

$$\dot{x} = f(x, w_1, w_2)$$

   *where, for some $L > 0$ and all $(x, w_1, w_2)$,*

$$|f(x, w_1, w_2)| \leq L(1 + |x| + |w_1| + |w_2|),$$

3. *the signal $t \mapsto w_1(t)$ is bounded and the size of the signal $t \mapsto w_2(t)$ is square integrable, and*
4. *the map $x \mapsto g(x)$ is continuous,*

*then the signal $t \mapsto |g(x(t))|^2$ is uniformly continuous.*

This fact and the qualitative objectives of anti-windup augmentation motivate pursuing a solution that guarantees the size of the state of the anti-windup augmented closed-loop system is square integrable, and also that the size of the difference between this state and the state of the unconstrained closed-loop system is square integrable. Indeed, this property will guarantee internal stability, external stability, and unconstrained response recovery, among other things.

It is acknowledged here that square integrability is pursued also because it is a mathematically tractable problem that admits convenient synthesis algorithms. In the end, the suitability of an anti-windup design will be judged not by its $\mathcal{L}_2$ performance but rather by the subjective evaluation of the time response by those who have commissioned the control system.

---

[6]In general, the notion of integration used can be that due to Lebesgue, which generalizes the notion of integration by Riemann used in standard calculus. Lebesgue's notion of integration is defined for a more general class of time domain signals.

### 2.4.2 Internal and external stability via the standard $\mathcal{L}_2$ gain

It is reasonable to assume that the size of the performance plant output $z$ is related to the size of the state of the plant and, based on the discussion above, that if $z$ belongs to $\mathcal{L}_2$, then this state will converge to zero. This condition is a type of detectability assumption. Therefore, one way to pursue internal and $\mathcal{L}_2$ external stability is the following: guarantee, by means of anti-windup augmentation, that there exist $\gamma > 0$ and $\beta > 0$ satisfying

$$||z||_2 \le \beta|x_{cl}(0)| + \gamma||w||_2 \tag{2.8}$$

for initial conditions and disturbances that are expected during operation. This is a standard performance characterization for closed-loop control systems with exogenous inputs. The smallest possible value of $\gamma$ that can be used is typically called the $\mathcal{L}_2$ gain from $w$ to $z$. During synthesis, it is desirable to make this $\mathcal{L}_2$ gain as small as possible since this will decrease the energy in the performance plant output. It will not be possible, even with anti-windup augmentation, to induce a global, finite $\mathcal{L}_2$ gain globally unless the plant is exponentially stable. On the other hand, it will always be possible to induce a regional finite $\mathcal{L}_2$ gain over some region. The aim would be to make this region as large as possible while making the $\mathcal{L}_2$ gain as small as possible. It will turn out that the region can be made to be the entire state space when the plant is exponentially stable. However, in general, the larger the guaranteed region of successful operation, the larger is the guaranteed $\mathcal{L}_2$ gain. In Section 2.4.4 we will elaborate further on this tradeoff.

### 2.4.3 Internal and external stability and URR via the URR gain

Like in the previous subsection, it is reasonable to suppose that the size of the difference between the performance plant output $z$ and the performance plant output in the unconstrained closed-loop system $z_\ell$ is related to the size of the plant state mismatch $x_p - x_{p\ell}$ and that if $z - z_\ell$ belongs to $\mathcal{L}_2$, then this state mismatch will converge to zero. Therefore, one way to pursue unconstrained response recovery (and, from the properties of the unconstrained closed-loop system, also internal and external stability for exogenous inputs that cause $u_\ell(t)$ to converge in an $\mathcal{L}_2$ sense to the region where "sat" is linear) is the following: guarantee, by means of anti-windup augmentation, that there exist $\gamma > 0$ and $\beta > 0$ satisfying[7]

$$||z - z_\ell||_2 \le \beta|x_{aw}(0)| + \gamma||u_\ell - \text{sat}(u_\ell)||_2 \tag{2.9}$$

for initial conditions and unconstrained controller signals that are expected during operation. The smallest possible $\gamma$ that can be used in this inequality is called the unconstrained response recovery (URR) gain. During synthesis, it is desirable to make the number $\gamma$ as small as possible since this will decrease the energy in the mismatch between performance plant output in the anti-windup augmented closed-loop system and the performance plant output in the unconstrained closed-loop system. Clearly, this objective bears a strong resemblance to the objective of minimizing the standard $\mathcal{L}_2$ gain. However, it is aimed much more directly at the URR

---

[7]In this inequality, it may be desirable to replace "sat" by "sat$_\delta$" in certain situations.

objective of anti-windup synthesis. It will not be possible, even with anti-windup augmentation, to induce a finite URR gain globally unless the plant is exponentially stable. On the other hand, it will always be possible to induce a regional finite URR gain. Moreover, for plants that are not exponentially unstable, it will be possible to induce a *nonlinear* URR gain *globally* when "sat" is replaced by "sat$_\delta$" in the inequality above. Many of the anti-windup augmentation algorithms presented later are derived based on the principle of minimizing the URR gain and maximizing the region over which this gain applies. As is the case for standard $\mathcal{L}_2$ gains, there will be a tradeoff between the guaranteed region of successful operation and the URR gain over this region. This tradeoff and the nonlinear gains are the topic of the next section.

### 2.4.4 Nonlinear anti-windup performance measures

Using a single factor to characterize the energy attenuation or amplification in a feedback loop with saturation can be misleading. This is because the limited authority the feedback controller has to attenuate low-energy exogenous inputs is not adequate to attenuate high-energy exogenous inputs by the same factor. Instead, it makes more sense to characterize energy attenuation or amplification through a nonlinear gain function.

In general, for linear feedback loops with input saturation, there are three qualitatively different shapes for the nonlinear $\mathcal{L}_2$ gain function, which quantifies the maximum possible energy value for the system's output for a given level of energy in the system's exogenous input. In the following descriptions, the function $r \mapsto \Gamma(r)$ will denote this gain function. The function $r \mapsto \bar{\Gamma}(r)$, defined by $\bar{\Gamma}(r) := \Gamma(r)/r$ for $r > 0$, is called the gain at the energy level $r$. Typically, $\bar{\Gamma}(r)$ is nondecreasing in $r$ and thus the function $\Gamma$ is convex.

Among the three shapes, the first shape is associated with a feedback loop that is externally $\mathcal{L}_2$ stable for low-energy exogenous inputs but not for high-energy exogenous inputs. This situation necessarily holds when the plant is exponentially unstable and has been stabilized by feedback. In general, it applies to the situation where the feedback controller achieves local but not global asymptotic stability. The shape of the nonlinear gain curve $\Gamma$ in this case is depicted in Figure 2.17. Particular to this case is the fact that the output energy may approach an infinite value for an exogenous input with a finite energy value. The slope of the linear asymptote for $\Gamma$ at low input energies corresponds to the $\mathcal{L}_2$ gain for the linear system without input saturation.

The second shape is associated with a feedback loop that is externally $\mathcal{L}_2$ stable for exogenous inputs of all energy levels, but not with a global, finite $\mathcal{L}_2$ gain. This situation may hold when the plant has marginally stable or marginally unstable modes and has been globally asymptotically stabilized by feedback. The shape of the nonlinear gain curve $\Gamma$ in this case is depicted in Figure 2.18. Particular to this case is the fact that the nonlinear curve cannot be upper bounded globally by a linear function. Thus, the ratio between input energy and output energy may grow without bound as the input energy increases. This is depicted in the plot of the function $\bar{\Gamma}$ in Figure 2.19, which is unbounded.

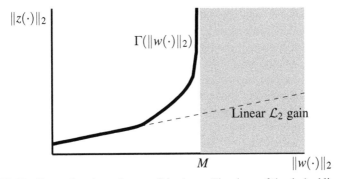

Figure 2.17 Nonlinear $\mathcal{L}_2$ gain: a first possible shape. The slope of the dashed line is the $\mathcal{L}_2$ gain of the linear loop.

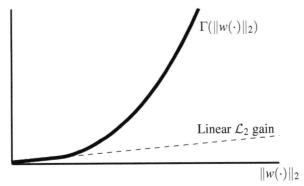

Figure 2.18 Nonlinear $\mathcal{L}_2$ gain: a second possible shape. The slope of the dashed line is the $\mathcal{L}_2$ gain of the linear loop.

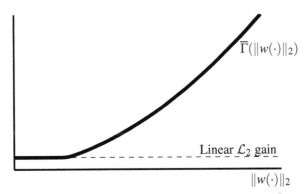

Figure 2.19 Nonlinear $\mathcal{L}_2$ gain of Figure 2.18 represented by the function $\bar{\Gamma}$. The asymptotic value of $\bar{\Gamma}$ for low energy is the $\mathcal{L}_2$ gain of the linear loop.

The third shape is associated with a feedback loop that is externally $\mathcal{L}_2$ stable for exogenous inputs of all energy levels, with a global, finite $\mathcal{L}_2$ gain. This situation can hold only if the plant is (globally) exponentially stable and the feedback loop is also globally exponentially stable. The shape of the nonlinear gain curve $\Gamma$ in this case is depicted in Figure 2.20. In this case, the nonlinear curve can be upper bounded globally by a linear function. Thus, the ratio between input energy and output energy cannot grow without bound as the input energy increases. Nevertheless, this ratio is typically much smaller for inputs with low energy than it is for inputs with high energy. These facts are indicated in the plot of the function $\bar{\Gamma}$ in Figure 2.21. The slope of the linear asymptote to $\Gamma$ for low input energy (equivalently, the asymptotic value of $\bar{\Gamma}$ for low energy) corresponds to the $\mathcal{L}_2$ gain for the linear system without input saturation. The slope of the linear asymptote to $\Gamma$ for high input energy (equivalently, the asymptotic value of $\bar{\Gamma}$ for high energy) cannot be any better than the $\mathcal{L}_2$ gain for the system without feedback. It's possible that the asymptotic value of $\bar{\Gamma}$ for high energy is larger than the $\mathcal{L}_2$ gain for the system without feedback.

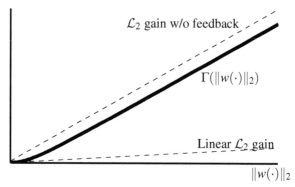

Figure 2.20  Nonlinear $\mathcal{L}_2$ gain: a third possible shape. The slope of the two dashed lines are the $\mathcal{L}_2$ gain of the linear loop and the $\mathcal{L}_2$ gain without feedback.

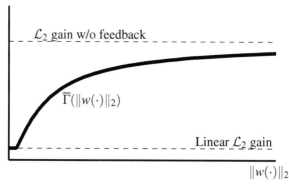

Figure 2.21  Nonlinear $\mathcal{L}_2$ gain of Figure 2.20 represented by the function $\bar{\Gamma}$. The asymptotic value of $\bar{\Gamma}$ for low energy is the $\mathcal{L}_2$ gain of the linear loop while its asymptotic value for high energy is the $\mathcal{L}_2$ gain without feedback.

All of these plots suggest that in order to guarantee $\mathcal{L}_2$ stability for a larger class of finite energy inputs a higher $\mathcal{L}_2$ gain for worst case performance must be accepted. In principle, it should be possible to design nonlinear feedback to achieve a specified nonlinear gain curve rather than just to achieve a certain finite gain over a specified operating range. Unfortunately, it seems that control design for linear systems with input saturation has not progressed to this point yet. At least with anti-windup synthesis, the gain curve for low energy inputs is specified independent of what one is able to achieve for the gain curve for high energy inputs.

In all of the plots provided in this section, the nonlinear gain curves plotted are actually upper bounds on the true gain curves. This is because they were certified with certain mathematically tractable but somewhat conservative tools, like quadratic energy storage functions and sector characterizations of the saturation nonlinearity. However, there is no conservatism in the gain curve for low energy inputs, where the saturation nonlinearity has very little effect.

In the next chapter, the mathematical tools that were used to generate the curves of this section will be described. These same tools serve as the theory behind many of the anti-windup synthesis algorithms that appear in the later parts of the book.

## 2.5 NOTES AND REFERENCES

Historically, anti-windup schemes arose from practical industrial needs and later on inspired the theoretical research community with precise problem statements and formal stability and performance guarantees. Chapter 11 contains an annotated bibliography commenting on the core anti-windup literature based on what could be seen as different maturity phases of the anti-windup research field. Within those references, much effort has been devoted at different times to surveying the existing schemes and establishing a unifying framework for anti-windup design. These references are listed in Section 11.4 on page 271.

The concept of unconstrained response recovery in anti-windup synthesis, as described in Section 2.2.4, appears in [MR3] and [OS4]. It has been also considered more recently in [OS10] and [MI23].

The important role of algebraic loops when using static anti-windup was first brought out by Mulder, Kothare, and Morari in [MI11]. Syaichu-Rohman and Middleton investigated the robustness of algebraic loops that appear in anti-windup synthesis in [SAT12].

The role of nonlinear performance, through nonlinear $\mathcal{L}_2$ gains, in anti-windup analysis and synthesis was brought into focus in [SUR8]. It has its roots in the nonlinear $\mathcal{L}_2$ analysis of Megretski [SAT7], which also inspired the nonlinear $\mathcal{L}_2$ gain results in [SAT8]. Similar quantification of performance appears in [SAT14], [SAT15], and [MI33]

A proof of the italicized fact on page 40 was published by Barbalat in 1959. A more recent reference with the proof is [G20, Lemma 8.2]. A proof of the italicized fact on page 42 can be found in [G12, G14], for example.

# Chapter Three

## Analysis and Synthesis of Feedback Systems:
## Quadratic Functions and LMIs

### 3.1 INTRODUCTION

#### 3.1.1 Description of feedback loop

This chapter starts by addressing the analysis of feedback loops with saturation. In particular, it develops tools for verifying internal stability and quantifying $\mathcal{L}_2$ external stability for the well-posed feedback interconnection of a linear system with a saturation nonlinearity. Such a system is depicted in Figure 3.1 and has state-space representation

$$
\tilde{\mathcal{H}} \quad \begin{cases}
\dot{x} &= \tilde{A}x + \tilde{B}\sigma + \tilde{E}w \\
z &= \tilde{C}x + \tilde{D}\sigma + \tilde{F}w \\
u &= \tilde{K}x + \tilde{L}\sigma + \tilde{G}w \\
\sigma &= \operatorname{sat}(u).
\end{cases}
\tag{3.1}
$$

It is more convenient for analysis and synthesis purposes to express this system in terms of the *deadzone* nonlinearity $q = u - \operatorname{sat}(u)$ as shown in Figure 3.2. The state-space representation of (3.1) using the deadzone nonlinearity is obtained by first solving for $u$ in the equation

$$
u = \tilde{K}x + \tilde{L}u - \tilde{L}q + \tilde{G}w.
\tag{3.2}
$$

The solution can be obtained when $I - \tilde{L}$ is invertible, which is a necessary condition for well-posedness of the feedback system (3.1). For more information on well-posedness, see Section 3.4.2. The solution is used to write the system (3.1) as

$$
\mathcal{H} \quad \begin{cases}
\dot{x} &= Ax + Bq + Ew \\
z &= Cx + Dq + Fw \\
u &= Kx + Lq + Gw \\
q &= u - \operatorname{sat}(u),
\end{cases}
\tag{3.3}
$$

where

$$
\begin{bmatrix} A & B & E \\ C & D & F \\ K & L & G \end{bmatrix} = \begin{bmatrix} \tilde{A} & -\tilde{B} & \tilde{E} \\ \tilde{C} & -\tilde{D} & \tilde{F} \\ 0 & 0 & 0 \end{bmatrix} + \begin{bmatrix} \tilde{B} \\ \tilde{D} \\ I \end{bmatrix} (I - \tilde{L})^{-1} \begin{bmatrix} \tilde{K} & -\tilde{L} & \tilde{G} \end{bmatrix}.
\tag{3.4}
$$

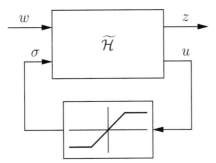

Figure 3.1  A closed-loop system involving a saturation nonlinearity.

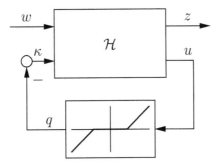

Figure 3.2  A closed-loop system involving a saturation nonlinearity written in compact form in negative feedback with a deadzone nonlinearity.

### 3.1.2  Quadratic functions and semidefinite matrices

The tools used here rely on nonnegative quadratic functions for analysis. Such functions will lead to numerical algorithms that involve solving a set of linear matrix inequalities (LMIs) in order to certify internal stability or quantify external performance. Efficient commercial LMI solvers are widely available.

A nonnegative quadratic function is a mapping $x \mapsto x^T P x$ where $P$ is symmetric, in other words, $P$ is equivalent to its transpose, and $x^T P x \geq 0$ for all $x$. In general, a symmetric matrix $P$ that satisfies $x^T P x \geq 0$ for all $x$ will be called a *positive semidefinite matrix*, written mathematically as $P \geq 0$. If $x^T P x > 0$ for all $x \neq 0$, then $P$ will be called a *positive definite matrix*, written mathematically as $P > 0$. A symmetric matrix $Q$ is *negative semidefinite*, written mathematically as $Q \leq 0$, if $-Q$ is positive semidefinite. A similar definition applies for a *negative definite matrix*. All of these terms are reserved for symmetric matrices. The reason for this is that a general square matrix $Z$ can be written as $Z = S + N$ where $S$ is symmetric, and $N$ is anti-symmetric, i.e., $N = -N^T$, and then it follows that $x^T Z x = x^T S x$. In other words, the anti-symmetric part plays no role in determining the sign of $x^T Z x$. The notation $P_1 > P_2$, respectively $P_1 \geq P_2$, indicates that the matrix $P_1 - P_2$ is positive definite, respectively positive semidefinite. Note that if $P > 0$, then there exists $\varepsilon > 0$ sufficiently small so that $P > \varepsilon I$.

## 3.2 UNCONSTRAINED FEEDBACK SYSTEMS

To set the stage for results regarding the system (3.3), consider using quadratic functions to analyze unconstrained feedback systems, where the saturation nonlinearity in (3.3) is replaced by the identity function, in other words, $q \equiv 0$, so that the system (3.3) becomes

$$\begin{aligned} \dot{x} &= Ax + Ew \\ z &= Cx + Fw. \end{aligned} \qquad (3.5)$$

### 3.2.1 Internal stability

When checking internal stability, $w$ is set to zero, $z$ plays no role, and (3.5) becomes simply $\dot{x} = Ax$. To certify exponential stability for the origin of this system, one method that generalizes to feedback loops with input saturation involves finding a nonnegative quadratic function that strictly decreases along solutions, except at the origin. Quadratic functions are convenient because they lead to stability tests that involve only linear algebra.

To determine whether a function is decreasing along solutions, it is enough to check whether, when evaluated along solutions, the function's time derivative is negative. For a continuously differentiable function, the time derivative can be obtained by computing the directional derivative of the function in the direction $Ax$ and then evaluating this function along the solution. This corresponds to the mathematical equation

$$\overbrace{\dot{V(x(t))}} = \langle \nabla V(x(t)), Ax(t) \rangle,$$

where $V$ represents the function, $\overbrace{\dot{V(x(t))}}$ represents its time derivative along solutions at time $t$, $\nabla V(x)$ is the gradient of the function, and $\langle \nabla V(x), Ax \rangle$ is the directional derivative of the function in the direction $Ax$. For a quadratic function $V(x) = x^T P x$, where $P$ is a symmetric matrix, the classic chain rule gives that this directional derivative equals

$$x^T (PA + A^T P) x.$$

Thus, in order for the time derivative to be negative along solutions, except at the origin, it should be the case that the directional derivative satisfies

$$x^T (A^T P + PA) x < 0 \qquad \forall x \neq 0,$$

in other words, $A^T P + PA < 0$. In summary, to certify internal stability for the system (3.5) with a given matrix $A$, one looks for a symmetric matrix $P$ satisfying

$$\begin{aligned} P &\geq 0 \\ 0 &> A^T P + PA. \end{aligned} \qquad (3.6)$$

This is a particular example of a set of LMIs, which will be discussed in more detail in Section 3.3. Software for checking the feasibility of LMIs is widely available. It turns out that the LMIs in (3.6) are feasible if and only if the system (3.5) is internally stable.

### 3.2.2 External stability

Now the external disturbance $w$ is no longer constrained to be zero. Thus, the relevant equation is (3.5). The goal is to determine the $\mathcal{L}_2$ gain from disturbance $w$ to the performance output variable $z$ and simultaneously establish internal stability. It is possible to give an arbitrarily tight upper bound on this gain by again exploiting nonnegative quadratic functions. However, the quadratic function will not always be decreasing along solutions. Instead, an upper bound on the time derivative will be integrated to derive a relationship between the energy in the disturbance $w$ and the energy in the performance output variable $z$. Again, the directional derivative of the function $x^T P x$ in the direction $Ax + Ew$ generates the time derivative of the function along solutions. This time the directional derivative is given by

$$x^T(A^T P + PA)x + 2x^T PEw.$$

In order to guarantee an $\mathcal{L}_2$ gain less than a number $\gamma > 0$ and to establish internal stability at the same time, it is sufficient to have

$$x^T(A^T P + PA)x + 2x^T PEw < -\gamma\left(\frac{1}{\gamma^2}z^T z - w^T w\right) \qquad \forall (x, w) \neq 0. \qquad (3.7)$$

Using the definition of $z$ in (3.5), this condition is the same as having, for all $(x, w) \neq 0$,

$$\begin{bmatrix} x \\ w \end{bmatrix}^T \left( \begin{bmatrix} A^T P + PA & PE \\ E^T P & -\gamma I \end{bmatrix} + \frac{1}{\gamma}\begin{bmatrix} C^T \\ F^T \end{bmatrix}\begin{bmatrix} C & F \end{bmatrix} \right) \begin{bmatrix} x \\ w \end{bmatrix} < 0.$$

In other words, the large matrix in the middle of this expression is negative definite. In summary, to certify internal stability and $\mathcal{L}_2$ external stability with gain less than $\gamma > 0$ for the system (3.5) with a given set of matrices $(A, B, E, F)$, it suffices to find a symmetric matrix $P$ satisfying

$$\begin{aligned} P &\geq 0 \\ 0 &> \begin{bmatrix} A^T P + PA & PE \\ E^T P & -\gamma I \end{bmatrix} + \frac{1}{\gamma}\begin{bmatrix} C^T \\ F^T \end{bmatrix}\begin{bmatrix} C & F \end{bmatrix}. \end{aligned} \qquad (3.8)$$

This is another set of LMIs, the feasibility of which can be tested with standard commercial software. Moreover, such software can be used to approximate the smallest possible number $\gamma$ that makes the LMIs feasible. The feasibility of the LMIs in (3.8) is also necessary for the $\mathcal{L}_2$ gain to be less than $\gamma$ with internal stability.

## 3.3 LINEAR MATRIX INEQUALITIES

As the preceding sections show, the analysis of dynamical systems benefits greatly from the availability of software to solve LMIs. The sections that follow show that LMIs also arise when using quadratic functions to analyze feedback loops with saturation. LMIs also appear in most of the anti-windup synthesis algorithms given in this book. The goal of this section is to provide some familiarity with LMIs and to highlight some aspects to be aware of when using LMI solvers.

Linear matrix inequalities are generalizations of linear scalar inequalities. A simple example of a linear scalar inequality is

$$2za + q < 0,$$ (3.9)

where $a$ and $q$ are known parameters and $z$ is a free variable. In contrast to the linear scalar equality $2za + q = 0$, which either admits no solution (if $a = 0$ and $q \neq 0$), an infinite number of solutions (if $a = 0$ and $q = 0$), or one solution (if $a \neq 0$), the linear scalar inequality (3.9) either admits no solutions (if $a = 0$ and $q \geq 0$) or admits a convex set of solutions given by $\{z \in \mathbb{R} : z < -q/(2a)\}$. In the former case the inequality is said to be *infeasible*. In the latter case it is said to be *feasible*.

Linear matrix inequalities generalize linear scalar inequalities by allowing the free variables to be matrices, allowing the expressions in which the free variables appear to be symmetric matrices, and generalizing negativity or positivity conditions to negative or positive definite matrix conditions.

Replacing the quantities in (3.9) with their matrix counterparts and insisting that the resulting matrix be symmetric gives the linear matrix inequality

$$A^T Z^T + ZA + Q < 0.$$ (3.10)

In this LMI, $A$ and $Q$ are known matrices and $Q$ is symmetric. The matrix $Z$ comprises $m$ times $n$ free variables where $m$ denotes the number of columns of $Z$ and $n$ denotes the number of rows of $Z$. Characterizing the solution set of (3.10) is not as easy as before, because the solution space will be delimited by several hyperplanes that depend on the entries of the matrices $A$ and $Q$. However, one important property of this solution set is that it is convex. In particular, if the matrices $\{Z_1, \ldots, Z_k\}$ all satisfy the LMI (3.10) then for any set of numbers $\{\lambda_1, \ldots, \lambda_k\}$, where each of these numbers is between zero and one, inclusive, and the sum of the numbers is one, the matrix

$$Z := \sum_{i=1}^{k} \lambda_i Z_i$$

also satisfies the LMI (3.10). The convexity property arises from the fact that (3.10) is linear in the free variable $Z$.

Now consider the case where $Q$ is taken to be zero and the free variable $Z$ is required to be symmetric and positive semidefinite. The variable $Z$ will be replaced by the variable $P$ for this case. With the free variable being symmetric, the parameter $A$ is now required to be a square matrix. Now, recall that the matrix condition $A^T P + PA < 0$ is equivalent to the existence of $\varepsilon > 0$ such that $A^T P + PA + \varepsilon I \leq 0$. Therefore, an extra free variable $\varepsilon$ can be introduced to write the overall conditions as the single LMI

$$\begin{bmatrix} P & 0 \\ 0 & -(A^T P + PA + \varepsilon I) \end{bmatrix} \geq 0.$$ (3.11)

The implicit constraint that $P$ is symmetric reduces the number of free variables in the matrix $P$ to the quantity $n(n+1)/2$, where $n$ is the size of the square matrix $P$.

The feasibility of the LMI (3.11) is equivalent to the simultaneous feasibility of the two LMIs

$$\begin{aligned} P &\geq 0 \\ 0 &> A^T P + PA \end{aligned} \qquad (3.12)$$

which match the LMIs (3.6) that appeared in the analysis of internal stability for linear systems. It is worth noting that most LMI solvers have difficulty with inequalities that are not strict. This is because sometimes, in this case, the feasibility is not robust to small changes in the parameters of the LMI. For example, with the choice

$$A = \begin{bmatrix} 0 & 1 \\ -1 & 0 \end{bmatrix},$$

the LMIs $P > 0$ and $0 \geq A^T P + PA$ are feasible (note that the strictly inequality and the nonstrict inequality have been exchanged relative to (3.12)) as can be seen by taking $P = I$. However, the LMIs are not feasible if one adds to $A$ the matrix $\varepsilon I$ for any $\varepsilon > 0$. On the other hand, strict LMIs, if feasible, are always robustly feasible. Fortunately, the feasibility of the LMIs (3.12) is equivalent to the feasibility of the LMIs

$$\begin{aligned} P &> 0 \\ 0 &> A^T P + PA . \end{aligned} \qquad (3.13)$$

This can be verified by letting $\widehat{P}$ denote a feasible solution to (3.12) and observing that $\widehat{P} + \varepsilon I$ must be a feasible solution to (3.13) for $\varepsilon > 0$ sufficiently small. From the discussion in Section 3.2.1, it follows that the LMIs (3.13) are feasible if and only if the system $\dot{x} = Ax$ is internally stable.

Next consider the matrix conditions that appeared in Section 3.2.2, the feasibility of which was equivalent to having $\mathcal{L}_2$ external stability with gain less than $\gamma > 0$ and internal stability for (3.5). Using the same idea as above to pass to a strict inequality for the matrix $P$, the feasibility of the matrix conditions in (3.8) is equivalent to the feasibility of the matrix conditions

$$\begin{aligned} P &> 0 \\ 0 &> \begin{bmatrix} A^T P + PA & PE \\ E^T P & -\gamma I \end{bmatrix} + \frac{1}{\gamma} \begin{bmatrix} C^T \\ F^T \end{bmatrix} \begin{bmatrix} C & F \end{bmatrix} . \end{aligned} \qquad (3.14)$$

The matrices $(A, E, C, F)$ are parameters that define the problem. If the value $\gamma$ is specified, then the feasibility of the resulting LMIs in terms of the free variable $P$ can be checked with an LMI solver. If the interest is in finding values for $\gamma > 0$ to make the matrix conditions feasible, then $\gamma$ can be taken to be a free variable. However, the matrix conditions do not constitute LMIs because of the nonlinear dependence on the free parameter $\gamma$ through the factor $1/\gamma$ that appears. Fortunately, there is a way to convert the matrix conditions above into LMIs in the free variables $\gamma$ and $P$ using the following fact:

> (Schur complements) *Let $Q$ and $R$ be symmetric matrices and let $S$ have the same number of rows as $Q$ and the same number of columns as $R$. Then the matrix condition*
> $$\begin{bmatrix} Q & S \\ S^T & R \end{bmatrix} > 0$$

*is equivalent to the matrix conditions*

$$R > 0$$
$$Q - SR^{-1}S^T > 0.$$

This fact can be applied to the matrix conditions (3.14) to obtain the matrix conditions

$$\gamma > 0$$
$$P > 0$$
$$0 > \begin{bmatrix} A^T P + PA & PE & C^T \\ E^T P & -\gamma I & F^T \\ C & F & -\gamma I \end{bmatrix} \quad (3.15)$$

which are LMIs in the free variables $P$ and $\gamma$. If the system $\dot{x} = Ax$ is internally stable, then these LMIs will be feasible. This follows from the fact that there will exist $P > 0$ satisfying $A^T P + PA < 0$, which is the matrix that appears in the upper left-hand corner of the large matrix in (3.15), and a consequence of Finsler's lemma, which is the following:

> (Finsler's lemma) *Let $Q$ be symmetric and let $H$ have the same number of columns as $Q$. If $\zeta^T Q \zeta < 0$ for all $\zeta \neq 0$ satisfying $H\zeta = 0$, then, for all $\gamma > 0$ sufficiently large, $Q - \gamma H^T H < 0$.*

If one applies this fact with the matrix

$$Q := \begin{bmatrix} A^T P + PA & PE & C^T \\ E^T P & 0 & F^T \\ C & F & 0 \end{bmatrix}$$

and

$$H = \begin{bmatrix} 0 & 0 & 0 \\ 0 & I & 0 \\ 0 & 0 & I \end{bmatrix},$$

one sees that the LMIs in (3.15) will be feasible for an appropriate $P$ matrix and large enough $\gamma > 0$.

To determine a tight upper bound on the $\mathcal{L}_2$ gain, one is interested in making $\gamma$ as small as possible. The task of minimizing $\gamma$ subject to satisfying the LMIs (3.15) is an example of an *LMI eigenvalue problem* and can be written as:

$$\min_{P,\gamma} \gamma, \quad \text{subject to:}$$

$$P > 0, \quad (3.16a)$$

$$0 > \begin{bmatrix} A^T P + PA & PE & C^T \\ E^T P & -\gamma I & F^T \\ C & F & -\gamma I \end{bmatrix}. \quad (3.16b)$$

Since large block matrices that appear in LMIs must always be symmetric, the entries below the diagonal must be equal to the transposes of the entries above the diagonal. For this reason, such matrices can be replaced with the "$\star$" symbol without any loss of information. For example, (3.16) can be written as

$$\min_{P,\gamma} \gamma, \quad \text{subject to:}$$

$$P > 0, \tag{3.17a}$$

$$0 > \begin{bmatrix} A^T P + PA & PE & C^T \\ \star & -\gamma I & F^T \\ \star & \star & -\gamma I \end{bmatrix}, \tag{3.17b}$$

with no loss of information.

An alternative notation that may simplify (3.16) relies on the use of the function "He" which, given any square matrix $X$, is defined as $\text{He}X := X + X^T$, so that (3.16) can be written as

$$\min_{P,\gamma} \gamma, \quad \text{subject to:}$$

$$P > 0, \tag{3.18a}$$

$$0 > \text{He} \begin{bmatrix} PA & PE & 0 \\ 0 & -\gamma I/2 & 0 \\ C & F & -\gamma I/2 \end{bmatrix}. \tag{3.18b}$$

The LMI feasibility and eigenvalue problems can be solved efficiently using modern numerical software packages. As an example, the code needed in MAT-LAB's LMI control toolbox to implement the search for the optimal solution to (3.16) or, equivalently, of (3.17) and (3.18), is given next.

**Example 3.3.1**    Implementing LMIs using MATLAB's LMI control toolbox requires first defining the LMI constraints structure and then running the solver on those constraints. The LMI constraints are specified by a start line (setlmis([]);) and an end line (mylmisys=getlmis;) which also gives a name to the LMI constraints. Then the LMI constraints consist of a first block where the LMI variables are listed and of a second block where the LMI constraints are described in terms of those variables. The following code gives a rough idea of how this should be implemented. Comments within the code provide indications of where the different blocks are located. For more details, the reader should refer to MATLAB's LMI control toolbox user's guide.

```
% Initialize the LMI system
setlmis([]);

% Specify the variables of the LMI
P     = lmivar(1,[length(A) 1]);
gamma = lmivar(1,[1 0]);

% Describe the LMI constraints

% 1st LMI (P>0)
Ppos = newlmi;
```

```
lmiterm([-Ppos 1 1 P],1,1);

% 2nd LMI (L2 gain)
L2lmi = newlmi;
lmiterm([L2lmi 1 1 P],1,A,'s');
lmiterm([L2lmi 1 2 P],1,E);
lmiterm([L2lmi 2 2 gamma],-1/2,eye(size(E,2)),'s');
lmiterm([L2lmi 3 1 0],C);
lmiterm([L2lmi 3 2 0],F);
lmiterm([L2lmi 3 3 gamma],-1/2,eye(size(C,1)),'s');

% Assign a name to the LMI system
mylmisys=getlmis;

% Solve the LMI

% Choose the function to be minimized (gamma)
n = decnbr(mylmisys);
cost = zeros(n,1);
for j=1:n,
  cost(j)=defcx(mylmisys,j,gamma);
end

% Run the LMI solver
[copt,xopt]=mincx(mylmisys,cost);

% Decode the solution (if any)
if not(isempty(copt)),
  Psol     = dec2mat(mylmisys,xopt,P);
  gammasol = dec2mat(mylmisys,xopt,gamma);

  % compare to alternative Hinf norm computation
  sys = pck(A,E,C,F);
  out = hinfnorm(sys);
  disp([out(2) gammasol])
end
```

$\square$

**Example 3.3.2** The MATLAB code illustrated in the previous example is quite streamlined and using the LMI control toolbox in such a direct way can many times become quite complicated in terms of actual MATLAB code. An alternative to this is to indirectly specify the LMI constraints and use the LMI control toolbox solver by way of the YALMIP (=Yet Another LMI Parser) front-end. The advantages of using YALMIP mainly reside in the increased simplicity of the code (thereby significantly reducing the probability of typos) and in the code portability to alternative solvers to the classic LMI control toolbox (SeDuMi is a much used alternative). The same calculation reported in the previous example is computed using the YALMIP front-end in the following code:

```
% compute the size of the matrices
n       = length(A);
[ny,nu] = size(F);

% decision variables
P=sdpvar(n);     % symmetric n-x-n
gamma=sdpvar(1); % scalar

% define the inequality constraints
M = [    P*A             P*E            zeros(n,ny);
      zeros(nu,n) -gamma/2*eye(nu) zeros(nu,ny);
           C              F          -gamma/2*eye(ny)];

constr = set(M+M'<0) + set(P>0);

% set the solver and its options
% here we use the LMI control toolbox
opts=sdpsettings;
opts.solver='lmilab';

% solve the LMI minimizing gamma
yalmipdiagnostics=solvesdp(constr,gamma,opts)

% evaluate solution variables (if any)
Psol=double(P);
gammasol=double(gamma);

% compare to alternative Hinf norm computation
sys = pck(A,E,C,F);
out = hinfnorm(sys);
disp([out(2) gammasol])
```

□

**Example 3.3.3**   CVX is another useful program for solving structured convex optimization problems, including LMIs. The previous calculations in CVX are as follows:

```
cvx_begin sdp

% compute the size of the matrices
n       = length(A);
[ny,nu] = size(F);
```

```
% decision variables
variable P(n,n) symmetric;
variable gamma;

% define the inequality constraints
M = [      P*A             P*E              zeros(n,ny);
        zeros(nu,n)  -gamma/2*eye(nu)  zeros(nu,ny);
            C              F             -gamma/2*eye(ny)];

minimize gamma
subject to
      gamma>0;
      P>0;
      M+M'<0;

cvx_end
```

□

There are a few points to make about the variability in solutions to LMI eigen-value problems returned by using different commercial solvers. First, notice that the LMI eigenvalue problem in (3.17) involves an optimization over an open set of matrices. So, technically, it is not possible to achieve the minimum. It would be more appropriate to say that the optimization problem is looking for the infimum. Indeed, if a minimum $\gamma^*$ existed and satisfied the LMIs, then it would also be the case that $\gamma - \varepsilon$ satisfied the LMIs for $\varepsilon > 0$ sufficiently small, contradicting the fact that $\gamma^*$ is a minimum. A consequence of this fact is that, since it is not possible to reach an infimum, each solver will need to make its own decision about the path to take toward the infimum and at what point to stop. Different paths to the infimum and different stopping conditions will cause different solvers to return different solutions. Second, unless the optimization is strictly convex, the solution to the optimization problem may be nonunique. This fact may also contribute to variability in the solutions returned by different solvers. In each of these cases, the different minima returned should be quite close to one another, whereas the matrices returned that satisfy the LMIs may be quite different. In anti-windup synthesis, these differences may lead to different anti-windup compensator matrices. Therefore, when trying to reproduce the results in the later examples, one may find that the compensators determined with a particular software package are very different from the ones found in the book. Nevertheless, the responses induced by the compensators should be very similar to the ones reported herein. In particular, the performance metric for which the compensator was designed should be very nearly the same.

Before moving on to the analysis of constrained feedback systems, one additional matrix manipulation, which will be used to analyze systems with saturation, will be introduced. It is closely related to Finsler's lemma.

(S-procedure) *Let $M_0$ and $M_1$ be symmetric matrices and suppose there exists $\zeta^*$ such that $\zeta^* M_1 \zeta^* > 0$. Then the following statements are equivalent:*

    *i. There exists $\tau > 0$ such that $M_0 - \tau M_1 > 0$.*
    *ii. $\zeta^T M_0 \zeta > 0$ for all $\zeta \neq 0$ such that $\zeta^T M_1 \zeta \geq 0$.*

The implication from (i) to (ii) is simple to see, and does not require the existence of $\zeta^*$ such that $\zeta^* M_1 \zeta^* > 0$. The opposite implication is nontrivial.

## 3.4 CONSTRAINED FEEDBACK SYSTEMS: GLOBAL ANALYSIS

### 3.4.1 Sector characterizations of nonlinearities

In order to arrive at LMIs when checking the internal stability and $\mathcal{L}_2$ external stability for feedback loops with saturations or deadzones, one typically inscribes the saturation or deadzone into a conic region and applies the S-procedure. To understand the idea behind inscribing a nonlinearity into a conic region, consider a scalar saturation function. Figure 3.3 contains, on the left, the block diagram of the saturation function and, on the right, the graph of the saturation function. The figure emphasizes that the graph of the saturation function is contained in a conic sector delimited by the line passing through the origin with slope zero and the line passing through the origin with slope one. The deadzone nonlinearity, as shown in Figure 3.4, is also contained in this sector. In fact, the sector contains any scalar nonlinearity with the property that its output $y$ always has the same sign as its input $u$ and $y$ has a magnitude that is never bigger than that of $u$. This condition can be expressed mathematically using the quadratic inequality $y(u - y) \geq 0$, which, when focusing on the deadzone nonlinearity where $y = q$, becomes

$$q(u - q) \geq 0 .$$

This condition says that $qu \geq q^2$, which captures the sign and magnitude information described above.

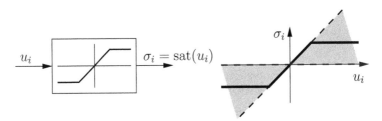

Figure 3.3  The scalar saturation function and its sector properties.

For a decentralized nonlinearity where each component of the nonlinearity is inscribed in the sector described above, the quadratic condition

$$\sum_{i=1}^{n_u} w_i q_i (u_i - q_i) \geq 0$$

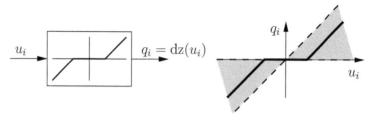

Figure 3.4 The scalar deadzone function and its sector properties.

holds, where $q_i$ are the components of the output vector $y$, $u_i$ are the components of the input vector $u$, and $w_i$ are arbitrary positive weightings. This can also be written as

$$q^T W(u-q) \geq 0, \tag{3.19}$$

where $W$ is a diagonal matrix consisting of the values $w_i$.

A sector characterization of nonlinearities introduces some conservativeness since the analysis using sectors will apply to any nonlinearity inscribed in the sector. The payoff in using sector characterizations is that the mathematical description, in terms of quadratic inequalities, is compatible with the analysis of feedback systems using quadratic functions. Indeed, the S-procedure described earlier permits combining the quadratic inequality describing the sector with the quadratic inequalities involved in the directional derivative to arrive at LMIs for the analysis of feedback loops with sector nonlinearities. This will be done subsequently.

### 3.4.2 Guaranteeing well-posedness

Before analyzing the feedback loop given by (3.3), shown in Figure 3.2, it must be established that the feedback loop is well-posed. In particular, it should be verified that the equation

$$u - L(u - \mathrm{sat}(u)) = v$$

admits a solution $u$ for each $v$ and that the solution $u(v)$ is a sufficiently regular function of $v$. When $\mathrm{sat}(u)$ is decentralized, the equation admits a solution that is a Lipschitz function of $v$ when each of the matrices

$$I - L\Delta \qquad \Delta = \mathrm{diag}(\delta_1, \ldots, \delta_{n_u}), \qquad \delta_i \in \{0,1\} \quad \forall i \in \{1, \ldots, n_u\}$$

is invertible. It turns out that each of these matrices is invertible when there exists a diagonal, positive definite matrix $W$ such that

$$L^T W + WL - 2W < 0 . \tag{3.20}$$

This LMI in $W$ will appear naturally in the analysis LMIs in the following sections. The LMI can be strengthened by replacing the "2" with a smaller positive number in order to guarantee that the feedback loop is not too close to being ill-posed. In particular, in many of the algorithms proposed later in this book, the following LMI will be employed:

$$L^T W + WL - 2(1-v)W < 0 , \tag{3.21}$$

or equivalent versions of it. This LMI is referred to as a *strong well-posedness* constraint and enforces a bound on the speed of variation of the closed-loop state (by bounding the Lipschitz constant of the right-hand side of the dynamic equation).

### 3.4.3 Internal stability

To analyze internal stability for (3.3), set $w = 0$ and consider the resulting system, which is given by

$$\begin{aligned}
\dot{x} &= Ax + Bq \\
z &= Cx + Dq \\
u &= Kx + Lq \\
q &= u - \text{sat}(u) .
\end{aligned} \tag{3.22}$$

Again, a nonnegative quadratic function $V(x) = x^T P x$ will be used. Like in the case of linear systems, it is desirable for the time derivative of $V(x(t))$ to be negative except at the origin. The time derivative is obtained from the directional derivative of $V(x)$ in the direction $Ax + Bq$, which is given by

$$\langle \nabla V(x), Ax + Bq \rangle = x^T (A^T P + PA)x + 2x^T PBq. \tag{3.23}$$

Since $q$ is the output of the deadzone nonlinearity, it follows from the discussion in Section 3.4.1 that the property

$$q^T W(u - q) \geq 0$$

holds for any diagonal positive semidefinite matrix $W$. Then, using the definition of $u$ in (3.22), the requirement on the time derivative translates to the condition

$$q^T W(Kx + Lq - q) \geq 0, \; (x, q) \neq 0 \quad \Longrightarrow \quad x^T (A^T P + PA)x + 2x^T PBq < 0 . \tag{3.24}$$

In order to guarantee this implication, it is enough to check that

$$x^T (A^T P + PA)x + 2x^T PBq + 2q^T W(Kx + Lq - q) < 0 \quad \forall (x, q) \neq 0 . \tag{3.25}$$

This is the easy implication of the S-procedure, where the $\tau$ has been absorbed into the free variable $W$. In fact, the S-procedure can be used to understand that there is no loss of generality in replacing (3.24) by (3.25). The condition (3.25) can be written equivalently as

$$\begin{bmatrix} x \\ q \end{bmatrix}^T \begin{bmatrix} A^T P + PA & PB + K^T W \\ B^T P + WK & WL + L^T W - 2W \end{bmatrix} \begin{bmatrix} x \\ q \end{bmatrix} < 0 \quad \forall (x, q) \neq 0 . \tag{3.26}$$

In other words,

$$\begin{aligned}
0 &> \begin{bmatrix} A^T P + PA & PB + K^T W \\ \star & WL + L^T W - 2W \end{bmatrix} \\
&= \text{He} \begin{bmatrix} PA & PB \\ WK & WL - W \end{bmatrix} .
\end{aligned} \tag{3.27}$$

This condition is an LMI in the free variables $P$, which should be positive semidefinite, and $W$, which should be diagonal and positive definite. The matrices $(A, B, K, L)$ are parameters that define the system (3.22). Notice that, because of the lower right-hand entry in the matrix, the LMI condition for well-posedness given in Section 3.4.2 is automatically guaranteed when (3.27) is satisfied. In addition, because of the upper left-hand entry in the matrix, the matrix $A$ must be such that the linear system $\dot{x} = Ax$ is internally stable. Finally, note that the LMI condition is not necessary for internal stability. For example, consider the system

$$
\begin{aligned}
\dot{x} &= -x+q \\
u &= x \\
q &= u - \text{sat}(u) .
\end{aligned}
\tag{3.28}
$$

The LMI for internal stability becomes

$$
0 > \begin{bmatrix} -2p & -p-w \\ \star & -2w \end{bmatrix}
\tag{3.29}
$$

or, equivalently,

$$
He \begin{bmatrix} p & w \\ p & w \end{bmatrix} > 0 .
\tag{3.30}
$$

In order for this matrix to be positive definite, the determinant of the matrix, given by $4pw - (p+w)^2$, must be positive. However, $4pw - (p+w)^2 = -(p-w)^2$ which is never positive. Therefore, the internal stability LMI is not feasible. Nevertheless, the system is internally stable since the system is equivalent to the system

$$
\dot{x} = -\text{sat}(x)
\tag{3.31}
$$

for which the quadratic function $V(x) = x^2$ decreases along solutions but not at a quadratic rate.

In general, the LMI for internal stability for the system (3.3) will not be feasible if the linear system $\dot{x} = \left[A + B(I-L)^{-1}K\right]x$ is not internally stable. This is the system that results from (3.3) by setting $w = 0$ and $\text{sat}(u) \equiv 0$. To put it another way, when the system (3.3) is expressed as a feedback interconnection of a linear system with a saturation nonlinearity rather than a deadzone, the resulting system is

$$
\begin{aligned}
\dot{x} &= \left[A + B(I-L)^{-1}K\right]x - B\left[I + (I-L)^{-1}L\right]\sigma + \left[E + B(I-L)^{-1}G\right]w \\
z &= \left[C + D(I-L)^{-1}K\right]x - D\left[I + (I-L)^{-1}L\right]\sigma + \left[F + D(I-L)^{-1}G\right]w \\
u &= (I-L)^{-1}\left[Kx - L\sigma + Gw\right] \\
\sigma &= \text{sat}(u) .
\end{aligned}
\tag{3.32}
$$

Then, for the LMI-based internal stability test in (3.27) to be feasible, it is necessary that (3.32) with $\sigma = 0$ and $w = 0$ be internally stable.

### 3.4.4 External stability

Now consider establishing $\mathcal{L}_2$ external stability for the system (3.3). Again relying on a nonnegative quadratic function, and combining the ideas in Sections 3.2.2 and 3.4.3, it is sufficient to have that

$$
q^T W(Gw + Kx + Lq - q) \geq 0 , \ (x,q,w) \neq 0
$$

implies

$$x^T (A^T P + PA)x + 2x^T P(Bq + Ew) < -\gamma \left( \frac{1}{\gamma^2} z^T z - w^T w \right). \tag{3.33}$$

Then, using the definition of $z$, and applying the S-procedure and Schur complements, produces the condition

$$\begin{bmatrix} A^T P + PA & PB + K^T W & PE & C^T \\ \star & -2W + WL + L^T W & WG & D^T \\ \star & \star & -\gamma I & F^T \\ \star & \star & \star & -\gamma I \end{bmatrix} < 0, \tag{3.34}$$

which is an LMI in the unknowns $P = P^T > 0$, $W > 0$ diagonal and $\gamma > 0$. The four blocks in the upper left corner correspond to the LMI for internal stability given in (3.27). It follows that if the system (3.3) is internally stable, then the LMI (3.34) is feasible using the solutions $P$ and $W$ from the internal stability LMI and then picking $\gamma > 0$ sufficiently large. Of course, the goal is to see how small $\gamma$ can be chosen. This objective corresponds to solving the eigenvalue problem

$$\min_{P,W,\gamma} \gamma, \quad \text{subject to:}$$

$$P = P^T > 0, W > 0 \text{ diagonal} \tag{3.35a}$$

$$\text{He} \begin{bmatrix} PA & PB & PE & 0 \\ WK & WL - W & WG & 0 \\ 0 & 0 & -\gamma I/2 & 0 \\ C & D & F & -\gamma I/2 \end{bmatrix} < 0. \tag{3.35b}$$

Since a necessary condition for global internal and external stability for the system (3.3) is that the matrix $A + B(I - L)^{-1}K$ be exponentially stable, there will be many situations where the LMIs given in this section will not be feasible. For this reason, it is helpful to have a generalization of these results for the case where only regional internal and external stability can be established. This is the topic of the next section.

## 3.5 CONSTRAINED FEEDBACK SYSTEMS: REGIONAL ANALYSIS

### 3.5.1 Regional sectors

To produce LMI results that are helpful for a regional analysis of systems with saturation, it is necessary to come up with a tighter sector characterization of the saturation nonlinearity while still using quadratic inequalities. This appears to be impossible to do globally. However, for analysis over a bounded region, there is a way to make progress. For the scalar saturation function, one fruitful idea is to take $H$ to be an arbitrary row vector and note that the quadratic inequality

$$(\sigma + Hx)(u - \sigma) \geq 0 \quad \forall x \text{ satisfying } \text{sat}(Hx) = Hx \tag{3.36}$$

will hold for any input-output pairs $u$ and $\sigma$ generated by the saturation nonlinearity. This can be verified by checking two cases:

1. If $\sigma := \text{sat}(u) = u$, then $\sigma = u$, and so the quadratic form on the left-hand side is zero.
2. If $\text{sat}(u) \neq u$, then the sign of $(u - \sigma)$ is equal to the sign of $u$ and also the sign of $\sigma + Hx$ is equal to the sign of $u$ so that the product is not negative. The condition $\text{sat}(Hx) = Hx$ is used in this step.

The corresponding condition for the deadzone nonlinearity having input $u$ and output $q$ can be derived from (3.36) by using the definition $q := \text{dz}(u) = u - \text{sat}(u) = u - \sigma$. The resulting quadratic condition is

$$(u - q + Hx)q \geq 0 \qquad \forall x \text{ satisfying } \text{sat}(Hx) = Hx. \qquad (3.37)$$

The decentralized vector version of this inequality is

$$(u - q + Hx)^T Wq \geq 0 \qquad \forall x \text{ satisfying } \text{sat}(Hx) = Hx, \qquad (3.38)$$

where $H$ is now a matrix of appropriate dimensions and $W$ is a diagonal, positive definite matrix. In order to have $\text{sat}(Hx) = Hx$ for all possible values of $x$, it must be the case that $H = 0$. In this case, the sector condition in (3.38) reduces to the global sector condition used previously.

### 3.5.2 Internal stability

In this section, the sector characterization of the previous section is exploited. The saturation nonlinearity is taken to be decentralized, the saturation limits are taken to be symmetric, and the $i$th function is limited in range to $\pm \bar{u}_i$. In order to guarantee the condition $\text{sat}(Hx(t)) = Hx(t)$ for solutions to be considered, which is needed to exploit the sector condition of the previous section, the condition

$$x^T H_i^T H_i x / \bar{u}_i^2 < x^T Px \qquad \forall x \neq 0 \qquad (3.39)$$

is imposed, where $H_i$ denotes the $i$th row of $H$. Using Schur complements, the condition (3.39) can be written as the matrix condition

$$\begin{bmatrix} P & H_i^T \\ H_i & \bar{u}_i^2 \end{bmatrix} > 0, \qquad i = 1, \ldots, n_u. \qquad (3.40)$$

According to (3.39), $x^T Px \leq 1$ implies $\text{sat}(Hx) = Hx$. So, the analysis of internal stability can now proceed like before but restricting attention to values of $x$ for which $x^T Px \leq 1$ and using the sector condition from the previous section. Indeed, if in the set $\mathcal{E}(P) := \{x : x^T Px \leq 1\}$ the quadratic function $x^T Px$ is decreasing along solutions, then the set $\mathcal{E}$ will be forward invariant and convergence to the origin will ensue.

Picking up the analysis from Section 3.4.3 immediately after the description of the directional derivative of the quadratic function $x^T Px$ in the direction $Ax + Bq$ in equation (3.23), the condition (3.24) is now replaced by the condition

$$q^T W(Hx + Kx + Lq - q) \geq 0, \ (x, q) \neq 0,$$
$$\implies \quad x^T(A^T P + PA)x + 2x^T PBq < 0. \qquad (3.41)$$

In order to guarantee this implication, it is enough to check that, for all $(x, q) \neq 0$,

$$x^T(A^T P + PA)x + 2x^T PBq + 2q^T W(Hx + Kx + Lq - q) < 0. \qquad (3.42)$$

The condition (3.42) can be written equivalently as

$$\begin{bmatrix} x \\ q \end{bmatrix}^T \begin{bmatrix} A^TP+PA & PB+K^TW+H^TW \\ B^TP+WK+WH & WL+L^TW-2W \end{bmatrix} \begin{bmatrix} x \\ q \end{bmatrix} < 0. \qquad (3.43)$$

In other words,

$$0 > \text{He} \begin{bmatrix} PA & PB \\ WH+WK & WL-W \end{bmatrix}. \qquad (3.44)$$

The matrix conditions that can then be used to establish internal stability over the region $\mathcal{E}(P)$ are given by (3.40) and (3.44) together with $P > 0$, $W > 0$ and the condition that $W$ is diagonal. These conditions are not linear in the free variables $P$, $W$, and $H$. In particular, notice the term $H^TW$ that appears in the upper right-hand corner of the matrix in (3.44). Nevertheless, there is a nonlinear transformation of the free variables that results in an LMI condition. The transformation exploits the fact that the condition $S < 0$ is equivalent to the condition $R^T S R < 0$ for any invertible matrix $R$. Make the definitions $Q := P^{-1}$, $U := W^{-1}$, and $Y := HQ$. These definitions make sense since $W$ and $P$ must be positive definite. Moreover, $P$, $W$, and $H$ can be recovered from $Q$, $U$, and $Y$. Now observe that

$$\begin{bmatrix} P^{-1} & 0 \\ 0 & I \end{bmatrix} \begin{bmatrix} P & H_i^T \\ H_i & \bar{u}_i^2 \end{bmatrix} \begin{bmatrix} P^{-1} & 0 \\ 0 & I \end{bmatrix} = \begin{bmatrix} Q & QH_i^T \\ H_iQ & \bar{u}_i^2 \end{bmatrix}$$
$$= \begin{bmatrix} Q & Y_i^T \\ Y_i & \bar{u}_i^2 \end{bmatrix} \qquad (3.45)$$

and

$$\begin{bmatrix} P^{-1} & 0 \\ 0 & W^{-1} \end{bmatrix} \begin{bmatrix} A^TP+PA & PB+K^TW+H^TW \\ \star & WL+L^TW-2W \end{bmatrix} \begin{bmatrix} P^{-1} & 0 \\ 0 & W^{-1} \end{bmatrix}$$
$$= \begin{bmatrix} QA^T+AQ & BU+QK^T+QH^T \\ \star & LU+UL^T-2U \end{bmatrix} \qquad (3.46)$$
$$= \begin{bmatrix} QA^T+AQ & BU+QK^T+Y^T \\ \star & LU+UL^T-2U \end{bmatrix}.$$

Thus, internal stability over the region $\mathcal{E}(Q^{-1})$ results if the following LMIs are feasible in the free parameters $Q$, $U$, and $Y$:

$$\begin{aligned} Q &> 0 \\ U &> 0 \quad \text{diagonal} \\ 0 &> \text{He} \begin{bmatrix} AQ & BU \\ Y+KQ & LU-U \end{bmatrix} \qquad (3.47) \\ \begin{bmatrix} Q & Y_i^T \\ Y_i & \bar{u}_i^2 \end{bmatrix} &> 0, \quad i=1,\ldots,n_u. \end{aligned}$$

Moreover, it is possible to use these LMIs as constraints for optimizing the set $\mathcal{E}(Q^{-1})$ in some way. In the next section, LMIs will be given for establishing internal stability and minimizing the $\mathcal{L}_2$ external gain over a region. Numerical examples will be given there.

### 3.5.3 External stability

Consider establishing $\mathcal{L}_2$ external stability for the system (3.3) over a region. The initial condition will be taken to be zero and the size of the solution $x(t)$ will be limited by limiting the energy in the disturbance input $w$. In particular, if it is true that

$$\dot{V}(x(t)) \leq w^T(t)w(t) \tag{3.48}$$

whenever $x(t)^T P x(t) \leq s^2$, and if $||w||_2 \leq s$, then it follows by integrating this inequality that

$$x(t)^T P x(t) = V(x(t)) \leq ||w||_2^2 \leq s^2 \qquad \forall t \geq 0. \tag{3.49}$$

Thus, changing the bound in (3.39) to

$$s^2 x^T H_i^T H_i x / \bar{u}_i^2 < x^T P x \qquad \forall x \neq 0 \tag{3.50}$$

will guarantee that $\text{sat}(Hx(t)) = Hx(t)$ for all disturbances $w$ with $||w||_2 \leq s$. The condition (3.50) corresponds to the matrix condition

$$\begin{bmatrix} P & H_i^T \\ H_i & \bar{u}_i^2/s^2 \end{bmatrix} > 0, \qquad i = 1, \dots, n_u. \tag{3.51}$$

The appropriate condition on the derivative of the function $x^T P x$ is guaranteed by the fact that condition

$$q^T W (Hx + Gw + Kx + Lq - q) \geq 0, \quad (x, q, w) \neq 0$$

implies

$$x^T (A^T P + PA)x + 2x^T P(Bq + Ew) < -\frac{1}{\gamma^2} z^T z + w^T w, \tag{3.52}$$

which generalizes the inequalities (3.7) and (3.33) addressing the same problem, respectively for the case without saturation and for the case with a global sector bound on the saturation. As compared to (3.7) and (3.33), the right-hand side of the inequality above is divided by $\gamma$. This is a key fact which allows writing inequality (3.49) independently of $\gamma$ and therefore to derive the LMIs (3.54) below for the regional $\mathcal{L}_2$ gain computation.

Using the definition of $z$, and applying the S-procedure and Schur complements, produces the matrix conditions

$$0 > \begin{bmatrix} A^T P + PA & PB + K^T W + H^T W & PE & C^T \\ \star & WL + L^T W - 2W & WG & D^T \\ \star & \star & -I & F^T \\ \star & \star & \star & -\gamma^2 I \end{bmatrix} \tag{3.53}$$

$$= \text{He} \begin{bmatrix} PA & PB & PE & 0 \\ WH + WK & WL - W & WG & 0 \\ 0 & 0 & -I/2 & 0 \\ C & D & F & -\gamma^2 I/2 \end{bmatrix}.$$

The four blocks in the upper left corner correspond to the LMI for internal stability given in (3.44). As before, the matrix conditions (3.51) and (3.53) are not linear in the free variables $P$, $W$, and $H$. Again, notice the $H^T W$ term that appears but that the matrix conditions can be reformulated as LMIs through a nonlinear transformation on the free variables. Using the same transformation as in the case for internal stability, the following LMIs are obtained in the free variables $Q$, $U$, and $Y$:

$$
\begin{aligned}
Q &> 0 \\
U &> 0
\end{aligned}
$$

$$
0 \;>\; \text{He} \begin{bmatrix}
AQ & BU & E & 0 \\
Y + KQ & LU - U & G & 0 \\
0 & 0 & -I/2 & 0 \\
CQ & DU & F & -\gamma^2 I/2
\end{bmatrix} \tag{3.54}
$$

$$
\begin{bmatrix} Q & Y_i^T \\ Y_i & \bar{u}_i^2/s^2 \end{bmatrix} > 0, \qquad i = 1,\dots,n_u \,.
$$

The minimum $\gamma^2 > 0$ for which the LMIs are feasible is a nondecreasing function of $s$. An upper bound on the nonlinear gain from $\|w\|_2$ to $\|z\|_2$ can be established by solving the LMI eigenvalue problem of minimizing $\gamma^2$ subject to the LMIs (3.54) for a wide range of values for $s$ and plotting $\gamma$ as a function of $s$. This procedure generates the function $\bar{\Gamma}$ mentioned in Section 2.4.4. The function $\Gamma$ described in Section 2.4.4 is given by $s\bar{\Gamma}(s)$.

## 3.6 ANALYSIS EXAMPLES

In this section, a few examples of nonlinear gain computation are provided to illustrate the use of the LMIs (3.54). The second example is taken from subsequent chapters, where anti-windup problems are solved.

**Example 3.6.1**   Consider a one-dimensional plant stabilized by negative feedback through a unit saturation and subject to a scalar disturbance $w$. The closed-loop dynamic equation is given by

$$
\begin{aligned}
\dot{x}_p &= a x_p + \text{sat}(u) + w \\
u &= -(a + 10) x_p,
\end{aligned}
$$

which can be written in terms of the deadzone function as

$$
\begin{aligned}
\dot{x}_p &= -10 x_p + q + w \\
q &= \text{dz}((a + 10) x_p).
\end{aligned} \tag{3.55}
$$

System (3.55) will be analyzed in three main cases:

1. $a = 1$, which resembles an exponentially unstable plant stabilized through a saturated loop; this case has been considered on page 28 when discussing regional external stability;
2. $a = 0$, which represents an integrator, namely a marginally stable plant stabilized through a saturated loop; this case has been considered on page 30 when characterizing unrecoverable responses;
3. $a = -1$, which represents an exponentially stable plant whose speed of convergence is increased through a saturated loop; this case has been considered on page 62 when illustrating LMI-based internal stability tests.

The three cases listed above have been used to generate, respectively, the nonlinear gains in Figures 2.17, 2.18, and 2.20 on page 45 when first introducing nonlinear performance measures.

When adopting the representation (3.3), system (3.55) is described by the following selection, with $z = x_p$:

$$
\left[ \begin{array}{c|c|c} A & B & E \\ \hline C & D & F \\ \hline K & L & G \end{array} \right] = \left[ \begin{array}{c|c|c} -10 & -1 & 1 \\ \hline 1 & 0 & 0 \\ \hline -a-10 & 0 & 0 \end{array} \right], \tag{3.56}
$$

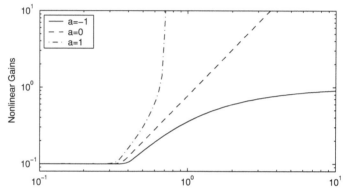

Figure 3.5  Nonlinear $\mathcal{L}_2$ gains for the three cases considered in Example 3.6.1.

The matrices (3.56) can be used in the LMI conditions (3.54) for different values of $s$, to compute the nonlinear gains for the three cases under consideration. The resulting curves are shown in Figure 3.5, where it is possible to appreciate the three different behaviors noted in Section 2.4.4 on page 44 and characterizing exponentially unstable plants, nonexponentially unstable plants with poles on the imaginary axis, and exponentially stable plants.                                                    □

**Example 3.6.2**   (SISO academic example no. 2) This example will receive attention in the following chapter. Different solutions to the corresponding windup problem are given in later chapters. Here, the nonlinear gains characterizing the different solutions will be computed by employing the LMI conditions (3.54).

The nonlinear $\mathcal{L}_2$ gains corresponding to the following closed loops are comparatively shown in Figure 3.6:

1. Saturated closed loop without anti-windup compensation. Based on the problem data given in Example 4.3.4 on page 89, the closed loop corresponds to:

$$
\left[ \begin{array}{c|c|c} A & B & E \\ \hline C & D & F \\ \hline K & L & G \end{array} \right] = \left[ \begin{array}{cc|c|cc} -0.6 & -1.1 & 1 & -0.5 & -1 \\ 1 & 0 & 0 & 0 & 0 \\ \hline -0.2 & -0.45 & -0.5 & 0.25 & -0.5 \\ \hline -0.2 & -0.45 & -0.5 & 0.25 & -0.5 \\ \hline -0.4 & -0.9 & 1 & 0.5 & -1 \end{array} \right],
$$

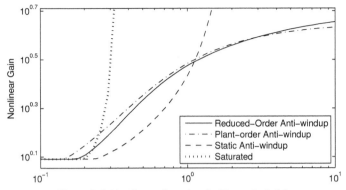

Figure 3.6 Nonlinear $\mathcal{L}_2$ gains for Example 3.6.2.

and the resulting nonlinear gain is represented by the dotted curve in Figure 3.6.

2. Closed loop with static regional anti-windup compensation. Based on the problem solution given in Example 4.4.2 on page 104, the closed loop corresponds to:

$$
\left[\begin{array}{c|c|c}
A & B & E \\
\hline
C & D & F \\
\hline
K & L & G
\end{array}\right] =
\left[\begin{array}{ccc|c|c}
-0.6 & -1.1 & 1 & -0.2 & -1 \\
1 & 0 & 0 & 0 & 0 \\
-0.2 & -0.45 & -0.5 & -0.9732 & -0.5 \\
\hline
-0.2 & -0.45 & -0.5 & 0.1 & -0.5 \\
\hline
-0.4 & -0.9 & 1 & 0.8 & -1
\end{array}\right],
$$

and the resulting nonlinear gain is represented by the dashed curve in Figure 3.6.

3. Closed loop with full-order dynamic global anti-windup compensation. Based on the problem solution given in Example 5.4.3 on page 119, the closed loop corresponds to:

$$
\left[\begin{array}{c|c|c}
A & B & E \\
\hline
C & D & F \\
\hline
K & L & G
\end{array}\right] =
$$

$$
\left[\begin{array}{ccc|ccc|c|c}
-0.6 & -1.1 & 1 & 0.699 & 0.0049 & -0.508 & -1 \\
1 & 0 & 0 & 0 & 0 & 0 & 0 \\
-0.2 & -0.45 & -0.5 & 0.286 & -0.0043 & -0.674 & -0.5 \\
0 & 0 & 0 & -0.196 & -0.0093 & -2.104 & 0 \\
0 & 0 & 0 & 0 & -2.838 & 0.0073 & 0 \\
\hline
-0.2 & -0.45 & -0.5 & -0.349 & -0.0024 & 0.2538 & -0.5 \\
\hline
-0.4 & -0.9 & 1 & 0.699 & 0.0049 & 0.4923 & -1
\end{array}\right],
$$

and the resulting nonlinear gain is represented by the dashed-dotted curve in Figure 3.6.

4. Closed loop with reduced-order dynamic global anti-windup compensation. Based on the problem solution given in Example 5.4.5 on page 128, the closed

loop corresponds to:

$$\left[ \begin{array}{c|c|c} A & B & E \\ \hline C & D & F \\ \hline K & L & G \end{array} \right] = \left[ \begin{array}{cccc|c|c} -0.6 & -1.1 & 1 & 1.093 & -0.202 & -1 \\ 1 & 0 & 0 & 0 & 0 & 0 \\ -0.2 & -0.45 & -0.5 & 1.050 & -0.0585 & -0.5 \\ 0 & 0 & 0 & -0.359 & -0.3405 & 0 \\ \hline -0.2 & -0.45 & -0.5 & -0.546 & 0.1008 & -0.5 \\ -0.4 & -0.9 & 1 & 1.093 & 0.7984 & -1 \end{array} \right],$$

and the resulting nonlinear gain is represented by the solid curve in Figure 3.6.

$\square$

## 3.7 REGIONAL SYNTHESIS FOR EXTERNAL STABILITY

This section addresses synthesis in feedback loops with saturation based on the LMIs that were derived earlier in this chapter. Only the case of regional $\mathcal{L}_2$ external stability is described. The purpose of this section is to give an indication of the types of calculations that arise in the synthesis of anti-windup algorithms.

### 3.7.1 LMI formulations of anti-windup synthesis

In typical anti-windup synthesis, the designer can inject the deadzone nonlinearity, driven by the control input, at various places in the feedback loop and can also inject the state of a filter driven by the deadzone nonlinearity. Letting $x$ denote the composite state of the plant having $n_p$ components, unconstrained controller having $n_c$ components, and anti-windup filter having $n_{aw}$ components, the synthesis problem can be written as

$$
\begin{aligned}
\dot{x} &= \overline{A}x + \overline{B}_0(u - q) + \overline{B}_1 \left( K_1 \overline{C}_1 x + \Phi_1 q \right) + \overline{E}w \\
z &= \overline{C}_2 x + \overline{D}q + \overline{F}w \\
u &= \overline{K}_0 x + \overline{L}q + K_2 \overline{C}_1 x + \Phi_2 q + \overline{G}w \\
q &= u - \mathrm{sat}(u)
\end{aligned}
\tag{3.57}
$$

where the design parameters are $K_1$, $K_2$, $\Phi_1$, and $\Phi_2$. All of the other matrices are fixed by the problem description, coming either from the plant model or the model of the unconstrained controller and are overlined for notational convenience. Typically, $\overline{C}_1 x$ represents the states of the anti-windup augmentation filter. The design parameters $K_1$ and $K_2$ determine how those states are used to determine the characteristics of the filter and the injected terms in the controller. When using static anti-windup augmentation, which corresponds to the case where there are no states in the anti-windup augmentation filter, the matrices $K_1$ and $K_2$ are set to zero. The design parameters $\Phi_1$ and $\Phi_2$ determine how the deadzone nonlinearity is injected into the state equations and controller. The matrix $\overline{B}_1$ limits where the states of the anti-windup filter and the deadzone nonlinearity can be injected into the dynamical equations.

With the definitions

$$\left[\begin{array}{cc|c} A & B & E \\ \hline C & D & F \\ \hline K & L & G \end{array}\right] = \tag{3.58}$$

$$\left[\begin{array}{c|c|c} \bar{A}+\bar{B}_0\bar{K}_0 + (\bar{B}_1 K_1 + \bar{B}_0 K_2)\bar{C}_1 & \bar{B}_0(\bar{L}+\Phi_2 - I) + \bar{B}_1\Phi_1 & \bar{E}+\bar{B}_0\bar{G} \\ \hline \bar{C}_2 & \bar{D} & \bar{F} \\ \hline \bar{K}_0 + K_2\bar{C}_1 & \bar{L}+\Phi_2 & \bar{G} \end{array}\right],$$

the system (3.57) agrees with the system (3.3) and the corresponding regional performance analysis LMI from Section 3.5.3 is

$$\begin{aligned} Q &> 0 \\ U &> 0 \text{ diagonal} \\ 0 &> \left[\begin{array}{cccc} QA^T + AQ & BU + QK^T + Y^T & E & QC^T \\ \star & LU + UL^T - 2U & G & UD^T \\ \star & \star & -I & F^T \\ \star & \star & \star & -\gamma^2 I \end{array}\right] \end{aligned} \tag{3.59}$$

$$\left[\begin{array}{cc} Q & Y_i^T \\ Y_i & \bar{u}_i^2/s^2 \end{array}\right] > 0 , i = 1,\dots, n_u .$$

Since some of the components of $A$, $B$, $K$, and $L$ are free for design, these matrices are replaced in (3.59) by their definitions from (3.58). The additional definitions

$$\begin{aligned} \Theta_1 &:= \Phi_1 U \\ \Theta_2 &:= \Phi_2 U \end{aligned} \tag{3.60}$$

give relationships between $\Phi_i$ and $\Theta_i$ for $i \in \{1,2\}$ that are invertible because $U$ is positive definite by assumption. Using these definitions, the matrix conditions (3.59) become

$$Q > 0$$
$$U > 0 \text{ diagonal}$$

$$0 > \mathrm{He}\left[\begin{array}{c} (\bar{A}+\bar{B}_0\bar{K}_0)Q + (\bar{B}_1 K_1 + \bar{B}_0 K_2)\bar{C}_1 Q \\ \bar{K}_0 Q + K_2\bar{C}_1 Q + Y \\ 0 \\ 0 \end{array}\right. \tag{3.61}$$

$$\left.\begin{array}{ccc} \bar{B}_1\Theta_1 + \bar{B}_0((\bar{L}-I)U + \Theta_2) & \bar{B}_0\bar{G}+\bar{E} & Q\bar{C}_2^T \\ \Theta_2 + (\bar{L}-I)U & \bar{G} & U\bar{D}^T \\ 0 & -\frac{1}{2}I & \bar{F}^T \\ 0 & 0 & -\frac{1}{2}\gamma^2 I \end{array}\right]$$

$$0 > -\left[\begin{array}{cc} Q & Y_i^T \\ Y_i & \bar{u}_i^2/s^2 \end{array}\right] , i = 1,\dots, n_u .$$

In the case of static anti-windup augmentation, where $K_1 = 0$ and $K_2 = 0$, the conditions (3.61) constitute LMIs in the free variables $U$, $Q$, $\Theta_1$, $\Theta_2$, and $\gamma$. Note that in (3.61) all the fixed parameters are overlined and all the free variables are not.

For some problems related to anti-windup synthesis, $\overline{C}_1$ is invertible and then, with the definitions

$$\begin{aligned} X_1 &:= K_1 \overline{C}_1 Q \\ X_2 &:= K_2 \overline{C}_1 Q, \end{aligned} \qquad (3.62)$$

which give an invertible relationship between $K_i$ and $X_i$, $i \in \{1,2\}$, the conditions (3.61) constitute LMIs in the free variables $U$, $Q$, $\Theta_1$, $\Theta_2$, $X_1$, $X_2$, and $\gamma$.

In the more typical situation where $\overline{C}_1$ is not invertible, a different approach can be taken to eliminate the nonlinear terms that involve products of $Q$ and $K_i$, $i \in \{1,2\}$, at least when the size of the anti-windup augmentation filter has the same number of states as the plant model. In this situation, define

$$\Lambda := \begin{bmatrix} K_1 & \Theta_1 \\ K_2 & \Theta_2 \end{bmatrix} \qquad (3.63)$$

and construct matrices $\Psi$, $H$, and $G$ such that the conditions (3.61) become

$$\begin{aligned} Q &> 0 \\ U &> 0 \\ 0 &> \mathrm{He}\,(\Psi + H\Lambda G) \\ 0 &> \begin{bmatrix} Q & Y_i^T \\ Y_i & \bar{u}_i^2/s^2 \end{bmatrix}, i = 1, \ldots, n_u. \end{aligned} \qquad (3.64)$$

According to the "elimination lemma" from linear algebra, there exists a value $\Lambda$ satisfying $0 > \mathrm{He}\,(\Psi + H\Lambda G)$ if and only if

$$\begin{aligned} 0 &> W_H^T \Psi W_H, \\ 0 &> W_G^T \Psi W_G \end{aligned} \qquad (3.65)$$

where $W_H$ and $W_G$ are any full-rank matrices satisfying $W_H^T H = 0$ and $G W_G = 0$.

Then, exploiting the special structure of anti-windup problems, the matrices $Q$ and $Y$ can be partitioned as

$$Q = \begin{bmatrix} \begin{bmatrix} R_{11} & R_{12} \\ R_{12}^T & R_{22} \end{bmatrix} & N \\ N^T & M \end{bmatrix}, P = Q^{-1} = \begin{bmatrix} S^{-1} & P_{12} \\ P_{12}^T & P_{22} \end{bmatrix}, Y = \begin{bmatrix} Z & Y_b & Y_c \end{bmatrix},$$

where $R_{11}$ is an $n_p \times n_p$ matrix, $R_{22}$ is an $n_c \times n_c$ matrix, and $M$ is an $n_{aw} \times n_{aw}$ matrix, and it can be verified that $W_H^T \Psi W_H < 0$ is an LMI in $R_{11}$ and $Z$, while $W_G^T \Psi W_G < 0$ is an LMI in $S$. Then, as long as $R_{11}$ and $S$ satisfy the condition $R_{11} - S_{11} > 0$, which is another LMI in $R_{11}$ and $S$, it is always possible to pick $R_{12}$, $R_{22}$, $N$, and $M$ so that $PQ = I$. Finally, as long as

$$0 < \begin{bmatrix} R_{11} & Z_i^T \\ Z_i & \bar{u}_i^2/s^2 \end{bmatrix}, i = 1, \ldots, n_u,$$

which is yet another LMI in $R_{11}$ and $Z$, it is always possible to pick $Y_b$ and $Y_c$ so that

$$0 < \begin{bmatrix} Q & Y_i^T \\ Y_i & \bar{u}_i^2/s^2 \end{bmatrix}, i = 1, \ldots, n_u.$$

Finally, with $Y$ and $Q$ generated, it is possible to find $\Lambda$ and $U$ such that $0 > \mathrm{He}\,(\Psi + H\Lambda G)$.

### 3.7.2 Restricting the size of matrices

It is sometimes convenient to restrict the size of the matrices determined by the LMI solver when seeking for optimal solutions minimizing the gain $\gamma^2$ in (3.61). Indeed, it may sometimes happen that minimizing the gain leads the LMI solver in a direction where certain parameters become extremely large providing very little performance increase. To avoid this undesirable behavior, it is often convenient to restrict the size of the free variables in (3.61) by incorporating extra constraints in the LMI optimization.

Given any free matrix variable $M$, one way to restrict its entries is to impose a bound $\kappa$ on its maximum singular value, namely, impose

$$M^T M < \kappa^2 I.$$

This can be done by dividing the equation above by $\kappa$ and applying a Schur complement to $\kappa I - M^T \frac{1}{\kappa} M > 0$, which gives:

$$\begin{bmatrix} \kappa I & M^T \\ M & \kappa I \end{bmatrix} > 0.$$

This type of solution can be used, for example, to restrict the size of the anti-windup matrices $K_1$, $K_2$, $\Phi_1$ and $\Phi_2$ in (3.61). In particular, this is achieved by imposing

$$\begin{bmatrix} \kappa I & \Lambda^T \\ \Lambda & \kappa I \end{bmatrix} > 0, \quad U > I, \tag{3.66}$$

where $\Lambda$ is defined in equation (3.63). Then, since by (3.60)

$$\begin{bmatrix} K_1 & \Phi_1 \\ K_2 & \Phi_2 \end{bmatrix} = \begin{bmatrix} K_1 & \Theta_1 \\ K_2 & \Theta_2 \end{bmatrix} \begin{bmatrix} I & 0 \\ 0 & U^{-1} \end{bmatrix},$$

the anti-windup parameters will satisfy $\left| \begin{bmatrix} K_1 & \Phi_1 \\ K_2 & \Phi_2 \end{bmatrix} \right| \leq |\Lambda| \, |U^{-1}| \leq |\Lambda| \leq \kappa$, where $|\cdot|$ denotes the maximum singular value of its argument.

Notice that imposing the extra constraints (3.66) reduces the feasibility set of the synthesis LMIs and conservatively enforces the bound on the anti-windup matrices (because it also restricts the set of allowable free variables $U$). However, it works well in many practical cases, as illustrated in some of the examples discussed in the following chapters.

## 3.8 NOTES AND REFERENCES

LMIs have played a foundational role in analysis and control of dynamical systems for several decades. A comprehensive book on this topic is the classic [G17], where an extensive list of references can be found. That book also contains the facts quoted herein concerning Schur complements, the Finsler lemma, the S-procedure, and the elimination lemma. The analysis of Section 3.4 corresponds to the MIMO version of the classical circle criterion in state-space form. The regional analysis of Section 3.5 is primarily due to the generalized sector condition introduced in [SAT11] and [MI27].

The global and regional analysis discussed in Sections 3.4 and 3.5 corresponds to the quadratic results of [SAT15], where nonquadratic stability and $\mathcal{L}_2$ performance estimates are also given. The synthesis method of Section 3.7 corresponds to the regional techniques of [MI33] for the most general case, but previous papers followed that approach for anti-windup synthesis: the static global design of [MI11], the dynamic global design of [MI18, MI26] and its external extension in [MI22], the discrete-time results of [MI24] and [MI30]. The elimination lemma used in Section 3.7 can be found in [G17]. It was first used as shown here in the context of LMI-based $\mathcal{H}_\infty$ controller synthesis. The corresponding techniques appeared simultaneously and independently in [G5] and [G6] (see also [G8] for an explicit solution to the second design step when applying the elimination lemma).

Well-posedness for feedback loops involving saturation nonlinearities has been first addressed in [MR15], where results from [G18] were used to establish sufficient conditions for well-posedness. Similar tools were also used in [MI18], where the well-posedness of the earlier schemes of [MI11] was also proved. The strong well-posedness constraint discussed in Section 3.4.2 arises from the results of [MI20]. More recently, in [SAT15] a further characterization of sufficient only and necessary and sufficient conditions has been given.

Regarding the LMI solvers mentioned in Examples 3.3.1–3.3.3, the LMI control toolbox [G19] of MATLAB can be purchased together with MATLAB. YALMIP [G15] is a modeling language for defining and solving advanced optimization problems. CVX [G1] is a package for specifying and solving convex programs. Both [G15] and CVX [G1] are extensions of MATLAB which can be downloaded from the web for free and easily installed as toolboxes on a MATLAB installation.

PART 2

Direct Linear Anti-windup Augmentation

# Chapter Four

Static Linear Anti-windup Augmentation

## 4.1 OVERVIEW

This chapter introduces the first constructive tools for anti-windup augmentation. According to the characterization of Section 2.3, the chapter addresses the simplest possible augmentation scheme that may induce on the closed loop the qualitative objectives discussed in Section 2.2, possibly in addition to some of the quantitative performance objectives of Section 2.4.

This chapter focuses on the "static linear anti-windup" augmentation architecture, wherein the difference between the input and the output of the saturation block drives a static linear system that injects modification signals into the unconstrained controller dynamics. The corresponding anti-windup filter structure is the simplest possible because it only consists of linear matrix gains of suitable dimensions.

Although the linear static anti-windup architecture is extremely simple, such is not the case for the algorithms that can be employed for the selection of the corresponding matrix gains. The class of algorithms considered in this chapter is based on linear matrix inequalities (which were introduced in Section 3.3) that, when appropriately solved, provide gain selections that correspond to optimal values of certain performance measures. The reason why only LMI-based algorithms are reported (despite a large literature on alternative rigorous or qualitative approaches) is that the resulting constructions have been proven to perform extremely well in many application cases and that once the reader gains confidence with the LMI mathematical tool and with a preferred LMI solver, the construction of optimal anti-windup gains becomes quite straightforward as the computational burden is carried out automatically by the software.

The performance measure optimized by all the algorithms discussed in this chapter is the input-output gain from the input signal $w(\cdot)$ to the performance output signal $z(\cdot)$ (recall that $w(\cdot)$ can contain both disturbances and references acting on the closed-loop system). Therefore, the algorithms reported next will aim at minimizing $\gamma$ in the following inequality

$$\|z\|_2 \leq \gamma \|w\|_2 \,,$$

either globally, namely, for arbitrarily large selections of $w(\cdot)$, or regionally, namely, for selections of $w(\cdot)$ below a certain bound.

Throughout the chapter, both full-authority and external anti-windup architectures will be addressed (according to the characterization in Section 2.3.6), although full-authority algorithms will be receiving more attention.

All the algorithms given in this chapter are based on the use of LMIs; thus, the corresponding stability and performance guarantees will rely upon the techniques outlined in Chapter 3.

## 4.2 KEY STATE-SPACE REPRESENTATIONS

Before introducing the anti-windup design algorithms, it is mandatory to introduce suitable state-space representations of the control systems under consideration. These representations will be both necessary to compute certain parameters needed in the anti-windup design algorithm and to allow use of the nonlinear gain calculation tools of Section 3.5.3 on page 66 on the compensated closed-loop systems determined from the algorithms.

The baseline saturated control system before linear anti-windup augmentation consists in the following linear systems: a plant with input saturation:

$$
\mathcal{P} \quad
\begin{cases}
\dot{x}_p &=& A_p x_p + B_{p,u}\operatorname{sat}(u) + B_{p,w}\, w \\
z &=& C_{p,z} x_p + D_{p,zu}\operatorname{sat}(u) + D_{p,zw}\, w \\
y &=& C_{p,y} x_p + D_{p,yu}\operatorname{sat}(u) + D_{p,yw}\, w \,,
\end{cases}
\tag{4.1}
$$

and a linear unconstrained controller:

$$
\mathcal{K} \quad
\begin{cases}
\dot{x}_c &=& A_c x_c + B_{c,y}\, y + B_{c,w}\, w \\
u &=& C_c x_c + D_{c,y}\, y + D_{c,w}\, w \,.
\end{cases}
\tag{4.2}
$$

For these two systems it will be useful to construct the matrices characterizing their interconnection in the absence of saturation. This interconnection can be represented in a compact way by stacking the two (plant + controller) states in a single state vector as $x_{cl} := \begin{bmatrix} x_p \\ x_c \end{bmatrix}$. The corresponding state-space equations are given by the following *compact closed-loop representation*:

$$
\begin{aligned}
\dot{x}_{cl} &=& A_{cl} x_{cl} + B_{cl,q}\, q + B_{cl,w}\, w \\
z &=& C_{cl,z} x_{cl} + D_{cl,zq}\, q + D_{cl,zw}\, w \\
u &=& C_{cl,u} x_{cl} + D_{cl,uq}\, q + D_{cl,uw}\, w \,,
\end{aligned}
\tag{4.3}
$$

where $q = \operatorname{dz}(u)$ encompasses the general effect of the nonlinearity on the otherwise linear control system.

Under the baseline assumption that the unconstrained interconnection between the plant (4.1) and the controller (4.2) is well-posed, the matrices appearing in (4.3) are uniquely determined by the values of the matrices in (4.1) and (4.2) as follows:

$$
\left[
\begin{array}{c}
A_{cl} \\
\hline
C_{cl,z} \\
\hline
C_{cl,u}
\end{array}
\right]
=
\left[
\begin{array}{c|c}
A_p + B_{p,u}\Delta_u D_{c,y} C_{p,y} & B_{p,u}\Delta_u C_c \\
B_{c,y}\Delta_y C_{p,y} & A_c + B_{c,y}\Delta_y D_{p,yu} C_c \\
\hline
D_{p,zu}\Delta_u D_{c,y} C_{p,y} + C_{p,z} & D_{p,zu}\Delta_u C_c \\
\hline
\Delta_u D_{c,y} C_{p,y} & \Delta_u C_c
\end{array}
\right]
\tag{4.4a}
$$

$$
\left[
\begin{array}{c|c}
B_{cl,q} & B_{cl,w} \\
\hline
D_{cl,zq} & D_{cl,zw} \\
\hline
D_{cl,uq} & D_{cl,uw}
\end{array}
\right]
=
\left[
\begin{array}{c|c}
-B_{p,u}\Delta_u & B_{p,w} + B_{p,u}\Delta_u(D_{c,y}D_{p,yw} + D_{c,w}) \\
-B_{c,y}\Delta_y D_{p,yu} & B_{c,w} + B_{c,y}\Delta_y(D_{p,yu}D_{c,w} + D_{p,yw}) \\
\hline
-D_{p,zu}\Delta_u & D_{p,zw} + D_{p,zu}\Delta_u(D_{c,y}D_{p,yw} + D_{c,w}) \\
\hline
I - \Delta_u & \Delta_u(D_{c,w} + D_{c,y}D_{p,yw})
\end{array}
\right],
\tag{4.4b}
$$

where the matrices $\Delta_u := (I - D_{c,y}D_{p,yu})^{-1}$ and $\Delta_y := (I - D_{p,yu}D_{c,y})^{-1}$ are well-defined as long as the unconstrained closed loop is well-posed.

The goal of static linear anti-windup augmentation is to design a suitable matrix gain $D_{aw}$ that is driven by the function $q = dz(u) = u - \text{sat}(u)$ and that injects modifications on the controller equations (4.2), either based on a full-authority architecture or based on an external architecture. The resulting compensated closed loop is a simple extension of system (4.3) which incorporates the additional anti-windup signals, as follows:

$$
\begin{aligned}
\dot{x}_{cl} &= A_{cl}x_{cl} + (B_{cl,q} + B_{cl,v}D_{aw})q + B_{cl,w}w \\
z &= C_{cl,z}x_{cl} + (D_{cl,zq} + D_{cl,zv}D_{aw})q + D_{cl,zw}w \\
u &= C_{cl,u}x_{cl} + (D_{cl,uq} + D_{cl,uv}D_{aw})q + D_{cl,uw}w \,,
\end{aligned}
\tag{4.5}
$$

where the matrices $B_{cl,v}$, $D_{cl,uv}$, and $D_{cl,zv}$ depend on the architecture, full authority or external, of the compensation scheme under consideration. The values of these matrices are detailed separately for the two architectures in the following two sections.

### 4.2.1 Closed-loop representation with full-authority anti-windup

Full-authority static linear anti-windup augmentation corresponds to modifying the controller equations (4.2) as follows:

$$
\mathcal{K} \left\{
\begin{aligned}
\dot{x}_c &= A_c x_c + B_{c,y} y + B_{c,w} w + D_{aw,1}q \\
u &= C_c x_c + D_{c,y} y + D_{c,w} w + D_{aw,2}q \,,
\end{aligned}
\right.
\tag{4.6}
$$

where the two signals $v_1 = D_{aw,1}q$ and $v_2 = D_{aw,2}q$ are injected in the state and output equation, respectively, with the goal of recovering closed-loop stability and performance. The matrix $D_{aw} = \begin{bmatrix} D_{aw,1} \\ D_{aw,2} \end{bmatrix} \in \mathbb{R}^{n_c+n_u \times n_u}$ (namely, having $n_c + n_u$ rows and $n_u$ columns) is the design parameter.

When adopting the full-authority architecture, the matrices $B_{cl,v}$, $D_{cl,uv}$, and $D_{cl,zv}$ in (4.5) correspond to the following values:

$$
\begin{bmatrix} B_{cl,v} \\ \hline D_{cl,uv} \\ \hline D_{cl,zv} \end{bmatrix} =
\begin{bmatrix} 0 & B_{p,u}\Delta_u \\ I_{n_c} & B_{c,y}\Delta_y D_{p,yu} \\ 0 & \Delta_u \\ 0 & D_{p,zu}\Delta_u \end{bmatrix}.
\tag{4.7}
$$

After the anti-windup gains are designed by way of any of the full-authority algorithms reported in this chapter, the nonlinear performance of the resulting anti-windup compensation scheme can be evaluated relying on the LMI-based tools introduced in Section 3.5.3. To this end, it will be useful to use the following equations which show in explicit form the selection of the matrices in (3.54) that should be carried out to represent the static full-authority linear anti-windup architecture with that notation:

$$
\left[\begin{array}{cc|c} A & B & E \\ C & D & F \\ \hline K & L & G \end{array}\right] =
\left[\begin{array}{c|c|c} A_{cl} & B_{cl,q} + B_{cl,v}D_{aw} & B_{cl,w} \\ C_{cl,z} & D_{cl,zq} + B_{cl,zv}D_{aw} & D_{cl,zw} \\ C_{cl,u} & D_{cl,uq} + B_{cl,uv}D_{aw} & D_{cl,uw} \end{array}\right],
\tag{4.8}
$$

where $D_{aw} = \begin{bmatrix} D_{aw,1} \\ D_{aw,2} \end{bmatrix}$. Most of the examples provided in this chapter will incorporate a nonlinear $\mathcal{L}_2$ gain analysis curve determined using transformation (4.8)
and the tools of Section 3.5.3.

### 4.2.2 Closed-loop representation with external anti-windup

Different from the full-authority case, in external static linear anti-windup augmentation, the modification signals denoted by $v$ and enforced on the controller by the
anti-windup gain are not allowed to access all the controller states but can only
affect the external signals, namely, the controller input and output. The controller
equations (4.2) then become:

$$\mathcal{K} \quad \begin{cases} \dot{x}_c & = & A_c x_c + B_{c,y} (y + D_{aw,1} q) + B_{c,w} w \\ u & = & C_c x_c + D_{c,y} (y + D_{aw,1} q) + D_{c,w} w + D_{aw,2} q , \end{cases} \tag{4.9}$$

where the two signals $v_1 = D_{aw,1} q$ and $v_2 = D_{aw,2} q$ are injected, respectively, at
the controller input and at the controller output. The matrix $D_{aw} = \begin{bmatrix} D_{aw,1} \\ D_{aw,2} \end{bmatrix} \in$
$\mathbb{R}^{n_y + n_u \times n_u}$ (namely, having $n_y + n_u$ rows and $n_u$ columns) is the design parameter.

When adopting the external architecture, the matrices $B_{cl,v}$, $D_{cl,uv}$, and $D_{cl,zv}$ in
(4.5) correspond to the following values:

$$\begin{bmatrix} B_{cl,v} \\ \hline D_{cl,uv} \\ \hline D_{cl,zv} \end{bmatrix} = \begin{bmatrix} -B_{p,u}\Delta_u D_{c,y} & B_{p,u}\Delta_u \\ B_{c,y}\Delta_y & B_{c,y}\Delta_y D_{p,yu} \\ \hline \Delta_u D_{c,y} & \Delta_u \\ \hline D_{p,zu}\Delta_u D_{c,y} & D_{p,zu}\Delta_u \end{bmatrix} . \tag{4.10}$$

Similar to the static full-authority case, static linear external anti-windup schemes
can also be analyzed by way of the nonlinear performance estimation tools of Section 3.5.3 on page 66. As in the full-authority case, the matrices in (3.54) should
be selected using equation (4.8). However, for the external case, the matrices $B_{cl,v}$,
$D_{cl,uv}$, and $D_{cl,zv}$ must be selected as in (4.10).

In the external anti-windup construction reported later in this chapter, certain
LMIs will depend on the matrices corresponding to the state-space representation
of the plant and of the unconstrained controller when they are disconnected. The
corresponding dynamical system is referred to as *compact open-loop representation* and is described by the equations

$$\begin{aligned} \dot{x}_{ol} & = & A_{ol} x_{ol} + B_{ol,w} w + B_{ol,v} v_1 \\ z_{ol} & = & C_{ol,z} x_{ol} + D_{ol,zw} w, \end{aligned} \tag{4.11}$$

where the input $v_1$ resembles the first signal injected by the external anti-windup
action and where the matrices correspond to the following selections:

$$\begin{bmatrix} A_{ol} & B_{ol,w} & B_{ol,v} \\ \hline C_{ol,z} & D_{ol,zw} \end{bmatrix} = \begin{bmatrix} A_p & 0 & B_{p,w} & 0 \\ 0 & A_c & B_{c,w} & B_{c,y} \\ \hline C_{p,z} & 0 & D_{p,zw} \end{bmatrix} . \tag{4.12}$$

## 4.3 ALGORITHMS PROVIDING GLOBAL GUARANTEES

The first family of algorithms considered in this chapter provides global stability and performance guarantees for the anti-windup closed loop. In particular, an optimized input-output gain will be guaranteed for the augmented closed loop. The desirable global guarantees given by these algorithms are obtained at the cost of a restricted applicability. Indeed, static direct linear global anti-windup will only apply to a strict subclass of the control systems with exponentially stable plants. More generic plants can be handled either by trading the global properties for regional ones, as shown later in this chapter, or by adopting dynamic compensators, as shown in the next chapter.

### 4.3.1 Global full-authority augmentation

Full-authority linear static anti-windup compensation synthesis amounts to selecting the matrices $D_{aw,1}$ and $D_{aw,2}$ of the static anti-windup compensator in Figure 4.1. An algorithm to construct such compensators is now offered.

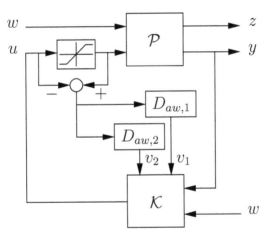

Figure 4.1 Full-authority static anti-windup compensation scheme.

Note that the simplicity in the anti-windup structure (a simple linear gain) is a tradeoff for the lack of general applicability of the algorithm. In particular, the feasibility conditions at step 2 of the algorithm should be verified beforehand and if these don't admit a solution the reader should use alternative constructions (such as the regional static ones of Section 4.4 or the dynamic algorithms given in the next chapter).

**Algorithm 1** (Static full-authority global DLAW)

| Applicability | | | | Architecture | | | Guarantee |
|---|---|---|---|---|---|---|---|
| Exp Stab | Marg Stab | Marg Unst | Exp Unst | Lin/ NonL | Dyn/ Static | Ext/ FullAu | Global/ Regional |
| $\sqrt{}$* | | | | L | S | FA | G |

> *Comments*: Simple architecture, very commonly used but not neces-
> sarily feasible for any exponentially stable plant. Global input-output
> gain is optimized.
> * Only those plants and unconstrained controllers satisfying step 2.

**Step 1.** Given a plant and a controller of the form (4.1), (4.2), construct the matrices of the compact closed-loop representation using (4.4) and (4.7). Select a constant $v \in [0, 1)$ used to enforce a strong well-posedness constraint (see Section 3.4.2 on page 60).

**Step 2.** Verify the applicability of this algorithm by checking the following feasibility conditions in the variable $R$:

$$R = R^T = \begin{bmatrix} R_{11} & R_{12} \\ R_{12}^T & R_{22} \end{bmatrix} > 0 ,$$

$$R_{11}A_p^T + A_p R_{11} < 0 ,$$

$$RA_{cl}^T + A_{cl}R < 0 .$$

If these conditions are not feasible, this algorithm is not applicable.

**Step 3.** Based on the matrices determined in step 1, find the optimal solution $(R, \gamma) \in \mathbb{R}^{n_{cl} \times n_{cl}} \times \mathbb{R}$ to the following LMI eigenvalue problem:

$\min\limits_{R,\gamma} \gamma$ *subject to*:

*Positivity*:

$$R = R^T = \begin{bmatrix} R_{11} & R_{12} \\ R_{12}^T & R_{22} \end{bmatrix} > 0 ,$$

*Open-loop condition*:

$$\begin{bmatrix} R_{11}A_p^T + A_p R_{11} & B_{p,w} & R_{11}C_{p,z}^T \\ B_{p,w}^T & -\gamma I_{n_w} & D_{p,zw}^T \\ C_{p,z}R_{11} & D_{p,zw} & -\gamma I_{n_z} \end{bmatrix} < 0 ,$$

*Closed-loop condition*:

$$\begin{bmatrix} RA_{cl}^T + A_{cl}R & B_{cl,w} & RC_{cl,z}^T \\ B_{cl,w}^T & -\gamma I_{n_w} & D_{cl,zw}^T \\ C_{cl,z}R & D_{cl,zw} & -\gamma I_{n_z} \end{bmatrix} < 0 .$$

**Step 4.** Based on the matrices determined at step 1, construct the matrices $\Psi \in \mathbb{R}^{m \times m}$, $G_U \in \mathbb{R}^{n_u \times m}$, and $H \in \mathbb{R}^{m \times (n_u + n_c)}$, with $m := n_{cl} + n_u + n_w + n_z$, as follows:

$$\Psi(U, \gamma) = \mathrm{He} \begin{bmatrix} A_{cl}R & B_{cl,q}U + QC_{cl,u}^T & B_{cl,w} & QC_{cl,z}^T \\ 0 & D_{cl,uq}U - U & D_{cl,uw} & UD_{cl,zq}^T \\ 0 & 0 & -\frac{\gamma}{2}I & D_{cl,zw}^T \\ 0 & 0 & 0 & -\frac{\gamma}{2}I \end{bmatrix} ,$$

$$H = \begin{bmatrix} B_{cl,v}^T & 0 \mid D_{cl,uv}^T \mid 0 \mid D_{cl,zv}^T \end{bmatrix} ,$$

$$G_U = \begin{bmatrix} 0 & 0 \mid I \mid 0 \mid 0 \end{bmatrix} ,$$

where for any square matrix $X$, $\mathrm{He}(X) = X + X^T$.

**Step 5.** Find the optimal solution $(\Lambda_U, U, \gamma)$ of the LMI eigenvalue problem:

$$\min_{\Lambda_U, U, \gamma} \gamma \ subject \ to:$$

*Strong well-posedness* (if $\nu = 0$, then redundant):

$$-2(1-\nu)U + \mathrm{He}\left(D_{cl,uq}U + \begin{bmatrix} 0_{n_u \times n_c} & I_{n_u} \end{bmatrix} \Lambda_U\right) < 0,$$

*Anti-windup gains LMI*:

$$\Psi(U,\gamma) + G_U^T \Lambda_U^T H^T + H\Lambda_U G_U < 0.$$

**Step 6.** Select the static full-authority anti-windup compensator as: $\nu = D_{aw}q$ with the selection $D_{aw} = \Lambda_U U^{-1}$.

$\star$

### 4.3.1.1 Interpretation of feasibility conditions*

The feasibility conditions at step 2 of Algorithm 1 characterize the class of control systems for which the static global anti-windup augmentation is applicable. A first fact that appears from the feasibility conditions is that both the plant and the unconstrained closed-loop system need to satisfy a Lyapunov equation; therefore, exponential stability of the plant is a necessary, and quite restrictive, assumption for Algorithm 1 to be applicable. Moreover, since the matrix $R$ is the same in the LMI involving the plant and in the LMI involving the unconstrained closed loop, extra constraints are actually imposed by the feasibility conditions about a special type of compatibility between open-loop and closed-loop systems, as clarified next.

In the case where $n_c = 1$, so that $n_{cl} = n_p$, namely, the order of the unconstrained closed loop is the same as that of the plant, these conditions can be understood as requiring a single quadratic Lyapunov function, characterized by the matrix $R^{-1}$, to decrease both along the trajectories of the closed-loop system and along the trajectories of the open-loop plant. In the more general case when $n_c > 0$, the condition generalizes to what could be denoted as a "quasi-common" quadratic Lyapunov function between the unconstrained closed loop and the open-loop plant. To give further insight in the meaning of this feasibility condition, an example with $n_c = 0$ and $n_p = 2$ is reported next so that the closed-loop trajectories can be visualized on the open-loop and closed-loop state space.

**Example 4.3.1**   Consider an exponentially stable linear plant $\mathcal{P}$ of the form (4.1) with matrices

$$
\begin{aligned}
A_p &= \begin{bmatrix} -0.05 & 1 \\ -10 & -0.5 \end{bmatrix}, & B_{p,u} &= \begin{bmatrix} 0 \\ 1 \end{bmatrix}, \\
C_{p,y} &= \begin{bmatrix} 9.9 & 0.495 \end{bmatrix}, & D_{p,yu} &= \begin{bmatrix} 0 \\ 0 \end{bmatrix},
\end{aligned}
$$

and the following static unconstrained controller $\mathcal{K}$:

$$\mathcal{K} \ \left\{ \ y_c \ = \ D_{c,y}u_c + D_{c,w}w \right. ,$$

with $D_{c,y} = 1$. The remaining plant and controller matrices are irrelevant to the feasibility discussion carried out in this example.

The unconstrained closed-loop matrix $A_{cl}$ constructed according to equation (4.4) is exponentially stable and corresponds to:

$$A_{cl} = A_p + B_{p,u} D_{c,y} C_{p,y} = \begin{bmatrix} -0.05 & 1 \\ -0.1 & -0.005 \end{bmatrix}.$$

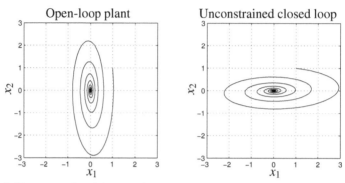

Figure 4.2  Phase portraits of the open-loop plant and unconstrained closed-loop systems in Example 4.3.1.

For this simple example, the feasibility conditions at step 2 of Algorithm 1 are not satisfied. Inspecting the phase portraits of the open loop and unconstrained closed loop, reported in Figure 4.2, reveals that it is not possible to define ellipses centered at the origin for which both the left and right trajectories are pointing inward. This is a graphical interpretation of the fact that there is no common quadratic Lyapunov function for the two systems, which is a necessary condition for Algorithm 1 to be applicable.                                                                         □

### 4.3.1.2  Case studies

Even though Algorithm 1 has a restricted applicability, as illustrated in Example 4.3.1, whenever the plant is exponentially stable and the feasibility conditions at step 2 are satisfied, it constitutes a very simple and effective construction. Next, four examples are reported where, whenever feasible, this static DLAW design leads to extremely improved closed-loop responses (as compared to the saturated responses without anti-windup compensation).

The first example is an academic MIMO system. This example is useful to show how the direct linear anti-windup augmentation technique directly applies to systems with multiple inputs. For these systems, qualitative solutions may face severe problems related to the fact that the saturation nonlinearity changes the so-called "input directionality" thus causing undesired effects at the plant output. Since the approach of Algorithm 1 optimizes a performance measure involving the plant output, the resulting anti-windup compensation implicitly accounts for the input allocation and no extra effort is needed when dealing with MIMO plants. Indeed, as shown next, applying the algorithm in a straightforward manner leads to very desirable results.

**Example 4.3.2** (MIMO academic example no. 2) Consider an exponentially stable plant that can be written in the form of (4.1) with matrices

$$
\left[\begin{array}{c|c} A_p & B_{p,u} \\ \hline C_{p,y} & D_{p,yu} \end{array}\right] = \left[\begin{array}{cc|cc} -0.01 & 0 & 1 & 0 \\ 0 & -0.01 & 0 & 1 \\ -0.4 & 0.5 & 0 & 0 \\ 0.3 & -0.4 & 0 & 0 \end{array}\right]
$$

and $B_{p,w} = D_{p,yw} = 0_{2\times2}$ and suppose a controller has been designed of the form (4.2) with matrices

$$
\left[\begin{array}{c|c|c} A_c & B_{c,y} & B_{c,w} \\ \hline C_c & D_{c,y} & D_{c,w} \end{array}\right] = \left[\begin{array}{cc|cc|cc} 0 & 0 & 1 & 0 & -1 & 0 \\ 0 & 0 & 0 & 1 & 0 & -1 \\ 0.02 & 0.025 & 2 & 2.5 & -2 & -2.5 \\ 0.015 & 0.02 & 1.5 & 2 & -1.5 & -2 \end{array}\right]
$$

such that the unconstrained closed-loop system has a desirable decoupled unconstrained response (here the role of $w$ is to be a vector reference). However, if a decentralized unit saturation limits the plant input then the plant has a very poor response. See the dotted line in Figure 4.3 for the plant output response to a reference changing from $[0, 0]^T$ to $[0.63, 0.79]^T$ at time $t = 0$. The saturated closed-loop system peak magnitude output for $y_1$ is $-4.72$ and for $y_2$ is $4.75$.

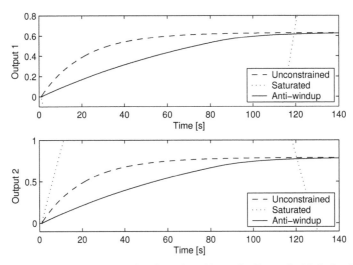

Figure 4.3 Plant output responses of various closed loops for Example 4.3.2. See legend in Table 4.1.

By choosing the performance output $z = y - w$, the remaining matrices of the plant are

$$
C_{p,z} = C_{p,y}, \qquad D_{p,zu} = D_{p,yu}, \qquad D_{p,zw} = D_{p,yw} - I_2 .
$$

For this example, the feasibility conditions at step 2 of Algorithm 1 hold, and with the selection $\nu = 0.001$, the algorithm yields the static full-authority anti-windup compensator gains

$$
D_{aw,1} = \left[\begin{array}{cc} -0.5987 & -0.7364 \\ -0.7484 & -0.9819 \end{array}\right], \quad D_{aw,2} = \left[\begin{array}{cc} -2.068 & -3.9271 \\ -2.3945 & -2.0681 \end{array}\right], \quad (4.13)
$$

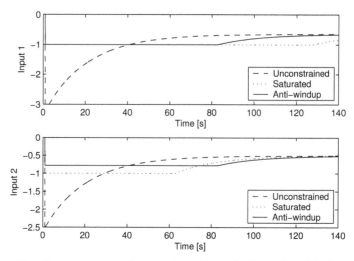

Figure 4.4 Plant input responses of various closed loops for Example 4.3.2. See legend in Table 4.1.

| System | Line |
|---|---|
| unconstrained | dashed |
| saturated | dotted |
| full-authority static anti-windup | solid |

Table 4.1 Legend for the time histories in Figures 4.3 and 4.4.

which guarantee the performance level $\gamma = 1.56$. The step response of the resulting anti-windup closed-loop system is very much improved, as shown by the solid curves in Figures 4.3 and 4.4.

It is useful to point out that the algebraic loop resulting from the presence of a nonzero selection of $D_{aw,2}$ in (4.13) becomes a key element to enforce a desirable output response. Indeed, this matrix is responsible for the input behavior of Figure 4.4, where the second input of the anti-windup response (solid curve in the lower plot) is forced to lie in the interior of the saturated region. This type of behavior would not be achievable by strictly proper compensators (namely, compensators with $D_{aw,2} = 0$) because the controller output exceeds the saturation limits on both inputs (this is evident from the saturated response [dotted curves in Figure 4.4], and only an implicit equation arising from an algebraic loop can instantaneously force the actual plant input within the interior of the linear region of the saturation.

The closed-loop performance of the static anti-windup compensator can be appreciated by inspecting the nonlinear gain of the augmented closed loop with anti-windup, which is represented in Figure 4.5 and compared to the gain before anti-windup compensation. These gains have been computed using the transformation (4.8) and the tools of Section 3.5.3. From this figure it is evident that the com-

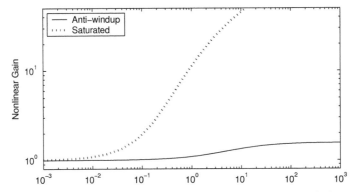

Figure 4.5 Nonlinear gains for Example 4.3.2 characterizing the system before and after anti-windup augmentation.

pensation leads to a dramatic gain improvement (note that the scale in the figure is logarithmic). It is also instructive to observe that the anti-windup compensation was designed to optimize the global $\mathcal{L}_2$ gain, namely, the gain for very large values of the input size, which led to the optimal value $\gamma = 1.561$, corresponding to the far right of the solid curve in Figure 4.5.  □

The next example represents a model of an electrical network where static anti-windup is feasible and performs well. For this example, it will be shown in the next chapter that dynamic anti-windup can lead to improved performance.

**Example 4.3.3** (Electrical network)  Consider the passive electrical circuit in

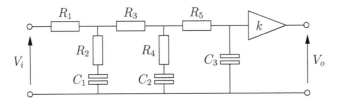

Figure 4.6 The passive electrical circuit of Example 4.3.3.

Figure 4.6, where the control input is the voltage $V_i$ and the measurement output coincides and corresponds to the voltage $V_o$. The resistors and capacitor values are selected as $R_1 = 313\Omega$, $R_2 = 20\ \Omega$, $R_3 = 315\ \Omega$, $R_4 = 17\ \Omega$, $R_5 = 10\ \Omega$, and $C_1 = C_2 = C_3 = 0.01\ F$. The gain $k$ is selected so that the transfer function from $V_i$ to $V_o$ is monic. This function can be easily computed as

$$\frac{V_o(s)}{V_i(s)} = \frac{(s+5)(s+6)}{(s+0.3)^2(s+10)}.$$

Based on the above transfer function, a state-space representation for this circuit can be easily derived (e.g., using the function ssdata of MATLAB) as

$$\left[\begin{array}{c|c} A_p & B_{p,u} \\ \hline C_{p,y} & D_{p,yu} \end{array}\right] = \left[\begin{array}{ccc|c} -10.6 & -6.09 & -0.9 & 1 \\ 1 & 0 & 0 & 0 \\ 0 & 1 & 0 & 0 \\ \hline -1 & -11 & -30 & 0 \end{array}\right],$$

and since there's no disturbance in the problem setting, $B_{p,w} = 0_{3 \times 1}$ and $D_{p,yw} = 0$. The following approximate PID controller can be used to robustly stabilize the unconstrained closed loop and to induce a desirable step response. The controller induces an infinite gain margin and a phase margin of 89.5 degrees:

$$\left[\begin{array}{c|c|c} A_c & B_{c,y} & B_{c,w} \\ \hline C_c & D_{c,y} & D_{c,w} \end{array}\right] = \left[\begin{array}{cc|c|c} -80 & 0 & 1 & -1 \\ 1 & 0 & 0 & 0 \\ \hline 20.25 & 1600 & 80 & -80 \end{array}\right],$$

where $w = r$ corresponds to the reference input for the desired value of the output voltage $V_o$.

When input saturation is present at the input voltage $V_i$, the saturated closed loop experiences windup and is in need of anti-windup compensation. In particular, the control input is saturated here between the maximum and minimum voltages $\pm 1$ V and a global static full-authority anti-windup compensator is designed following Algorithm 1 selecting $v = 0.1$. For the algorithm to optimize the gain from the reference input $r = w$ and the output tracking error $e = r - y$, the performance output matrices of the plant (4.1) are selected as $C_{p,z} = C_{p,y}$, $D_{p,zu} = D_{p,yu}$, and $D_{p,zw} = D_{p,yw} - 1$. The feasibility conditions at step 2 of Algorithm 1 are satisfied for this example; therefore, the algorithm provides a solution to the windup problem corresponding to the selection

$$D_{aw,1} = \left[\begin{array}{c} 0.283 \\ -0.0036 \end{array}\right], \qquad D_{aw,2} = 0.98,$$

which guarantees the global performance level $\gamma = 85.22$.

Figure 4.7 reports the input and output of the plant when the unconstrained, saturated and anti-windup closed loops are driven by a reference signal of doublet type, switching between $+1.5V$ and $-1.5V$ at increasingly narrow time intervals and finally going back to zero. The simulation results show unacceptable overshoots characterizing the saturated responses and demonstrate the effectiveness of anti-windup augmentation in mitigating this undesired effect. Nevertheless, the anti-windup response is quite slow and it will be shown in the following sections that when using dynamic (see Example 5.4.2 on page 118) or regional static (see Example 4.4.2 on page 104) anti-windup, improvements will be achieved on the speed of convergence of the anti-windup response.

Figure 4.8 reports the nonlinear gains characterizing the closed loop before and after anti-windup. These gains have been computed using the transformation (4.8) and the tools of Section 3.5.3. Inspecting the nonlinear gains it appears that the saturated closed loop before anti-windup admits a regional gain only up to the value $s \approx 0.075$. Above that value, either the saturated closed loop loses external stability or the quadratic tools adopted to establish this gain curve are too conservative.

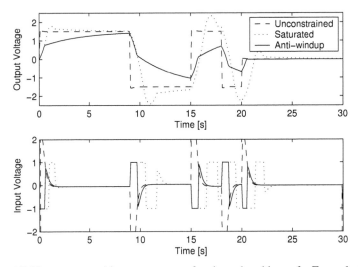

Figure 4.7 Plant output and input responses of various closed loops for Example 4.3.3.

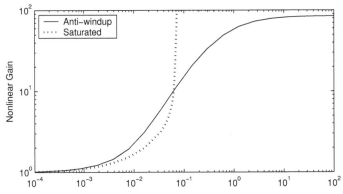

Figure 4.8 Nonlinear gains for Example 4.3.3 characterizing the system before and after anti-windup augmentation.

Conversely, the nonlinear gain computed after anti-windup compensation shows external stability for any reference input size and confirms the global performance level $\gamma = 85.22$ provided by the optimal anti-windup design algorithm and corresponding to the far right value of the solid curve in Figure 4.8. □

In the next example static compensation is not feasible. The simplicity of the example, which involves a SISO plant with two internal states, indicates that there are situations where even elementary anti-windup problems require dynamic compensation for global results based on quadratic Lyapunov functions. The example will be revisited in the following chapter where dynamic DLAW designs will be introduced.

**Example 4.3.4** (SISO academic example no. 2) Consider the following SISO plant without any disturbance input:

$$
\left[\begin{array}{c|c|c} A_p & B_{p,u} & B_{p,w} \\ \hline C_{p,y} & D_{p,yu} & D_{p,yw} \end{array}\right] = \left[\begin{array}{cc|c|c} -0.2 & -0.2 & 1 & 0 \\ 1 & 0 & 0 & 0 \\ \hline -0.4 & -0.9 & -0.5 & 0 \end{array}\right].
$$

Consider the following one-dimensional controller selection for the SISO plant introduced above:

$$
\left[\begin{array}{c|c|c} A_c & B_{c,y} & B_{c,w} \\ \hline C_c & D_{c,y} & D_{c,w} \end{array}\right] = \left[\begin{array}{c|c|c} 0 & 1 & -1 \\ \hline 2 & 2 & -2 \end{array}\right].
$$

When the control input is saturated between $\pm 0.5$, the saturated closed loop exhibits very large oscillations with extremely long settling time. It is therefore of interest to design anti-windup augmentation for this closed loop. Unfortunately, for this example the feasibility conditions at step 2 of Algorithm 1 do not admit a solution. Therefore the algorithm cannot be applied and it is necessary to adopt the alternative tools provided later in the book, namely, either the regional static anti-windup tools illustrated for this problem setting in Example 4.4.2 later in this chapter or the dynamic anti-windup tools illustrated in Examples 5.4.3 and 5.4.5 in the next chapter.                                                                                     □

The example introduced in Section 1.2.3 of the introductory chapter on page 8 is revisited next. For this example, static full-authority anti-windup design is feasible and is effective in reducing the windup effects.

**Example 4.3.5** (F8 longitudinal dynamics) The linearized longitudinal dynamics of an F8 aircraft can be represented by the following selections for the linear plant (4.1):

$$
\left[\begin{array}{c|c} A_p & B_{p,u} \\ \hline C_{p,y} & D_{p,yu} \end{array}\right] = \left[\begin{array}{cccc|cc} -0.8 & -0.0006 & -12 & 0 & -19 & -3 \\ 0 & -0.14 & -16.64 & -32.2 & -0.66 & -0.5 \\ 1 & -0.0001 & -1.5 & 0 & -0.16 & -0.5 \\ 1 & 0 & 0 & 0 & 0 & 0 \\ \hline 0 & 0 & 0 & 1 & 0 & 0 \\ 0 & 0 & -1 & 1 & 0 & 0 \end{array}\right],
$$

where the two inputs correspond to the elevator angle and to the flaperon angle, respectively, and the two outputs correspond to the pitch angle and the flight path angle. No disturbances are taken into account for this example; therefore, the external signal $w$ consists only in two references specifying the desired values at the two plant outputs. As a consequence, the plant matrices related to the input $w$ are zero: $B_{p,w} = 0_{4 \times 2}$, $D_{p,yw} = 0_{2 \times 2}$. The performance output $z$ is selected in such a way that two suitable combinations of the plant state are penalized, thereby leading to desirable responses of the compensated closed loop. In particular, the plant matrices related to $z$ are selected as:

$$
\left[\begin{array}{c|c|c} C_{p,z} & D_{z,yu} & D_{z,yw} \end{array}\right] = \left[\begin{array}{cccc|cc|cc} 0 & 0 & 0 & 3/4 & 0 & 0 & -3/4 & 0 \\ 0.8 & 0 & 12 & 0 & 0 & 0 & 0 & 0 \end{array}\right].
$$

The controller for the F8 dynamics is selected based on an LQG/LTR methodology applied to an augmented plant having two extra integrators to guarantee asymptotic tracking of constant references. The overall controller state-space representation corresponds to:

$$
\left[ \begin{array}{c|c|c} A_c & B_{c,y} & B_{c,w} \\ \hline C_c & D_{c,y} & D_{c,w} \end{array} \right] = \left[ \begin{array}{c|c|c} A_a + B_aG - HC_a & 0_{6 \times 2} & H & -H \\ G & 0_{2 \times 2} & 0_{2 \times 2} & 0_{2 \times 2} \\ \hline 0_{2 \times 6} & I_2 & 0_{2 \times 2} & 0_{2 \times 2} \end{array} \right],
$$

where

$$
\left[ \begin{array}{c|c} A_a & B_a \\ \hline C_a & \end{array} \right] = \left[ \begin{array}{cc|c} 0_{2 \times 2} & 0_{2 \times 4} & I_2 \\ B_{p,u} & A_p & 0_{4 \times 2} \\ \hline 0_{2 \times 2} & C_{p,y} & \end{array} \right],
$$

and the controller parameters $G$ and $H$ are selected as

$$
\left[ \begin{array}{c} G \\ \hline H^T \end{array} \right] = \left[ \begin{array}{cccccc} -52.23 & -3.36 & 73.1 & -0.0006 & -94.3 & 1072 \\ -3.36 & -29.7 & -2.19 & -0.006 & 908.9 & -921 \\ \hline -0.844 & -11.54 & -0.86 & -47.4 & 4.68 & 4.82 \\ 0.819 & 13.47 & 0.25 & 15 & -4.8 & 0.14 \end{array} \right].
$$

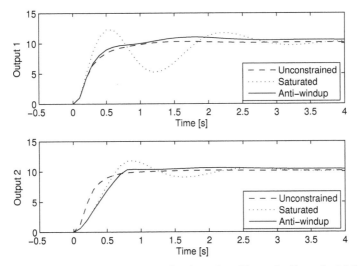

Figure 4.9 Plant output responses of various closed loops for Example 4.3.5.

The unconstrained interconnection of this controller with the F8 longitudinal dynamics model behaves very desirably and guarantees fast and smooth responses to step references for both the regulated outputs. However, if the control inputs are both saturated between $\pm 25$ degrees, namely, $\bar{u} = \dfrac{\pi}{180} \left[ \begin{array}{c} 25 \\ 25 \end{array} \right] \approx \left[ \begin{array}{c} 0.436 \\ 0.436 \end{array} \right]$, the saturated response exhibits unacceptable oscillations. This phenomenon is illustrated in Figures 4.9 and 4.10 representing the plant output and input responses, where the unconstrained behavior corresponds to the dashed curves and the saturated one corresponds to the dotted curves.

Algorithm 1 can be used to design a static full-authority anti-windup compensator for this example because the feasibility conditions in step 2 of the algorithm

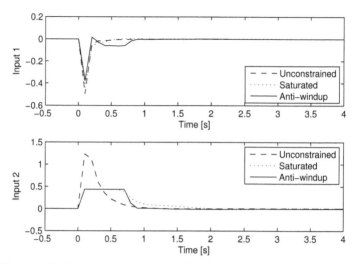

Figure 4.10  Plant input responses of various closed loops for Example 4.3.5.

are satisfied. Selecting $v = 0.005$, the following anti-windup gains are obtained:

$$
\left[ \frac{D_{aw,1}}{D_{aw,2}} \right] =
\begin{bmatrix}
-81.8 & 43.666 \\
313.6 & -309.29 \\
0.90915 & -1.6222 \\
-11613 & -10107 \\
-0.0012891 & -0.37033 \\
-1.8134 & -0.45377 \\
-82.16 & 43.495 \\
310.77 & -310.79 \\
\hline
0.86093 & 0.084357 \\
-0.016786 & 0.86086
\end{bmatrix},
$$

which induce a global performance level $\gamma = 26.44$. The corresponding anti-windup response corresponds to the solid curves in Figures 4.9 and 4.10, that show a significant mitigation of the windup phenomenon, although they present a slowly decaying transient that might be nonacceptable for implementation. This transient will be removed when relying on the dynamic anti-windup schemes of the next chapter (see Example 5.4.4 on page 121 and Example 5.4.7 on page 133).

The nonlinear gains of the closed loop before and after anti-windup augmentation are represented in Figure 4.11. It is worthwhile pointing out that these gains were computed after increasing the default accuracy of the MATLAB LMI control toolbox solver. From the gains it appears that static anti-windup significantly improves the external stability properties of the closed loop.                                    □

### 4.3.2 Global external augmentation

As compared to the full-authority case, external linear static anti-windup compensation corresponds to allowing the linear anti-windup gain to only affect the uncon-

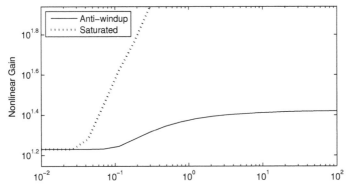

Figure 4.11  Nonlinear gains for Example 4.3.5 characterizing the system before and after anti-windup augmentation.

strained controller external signals, namely, its input and output. The corresponding block diagram is shown in Figure 4.12 where $D_{aw,1}$ and $D_{aw,2}$ are the design parameters.

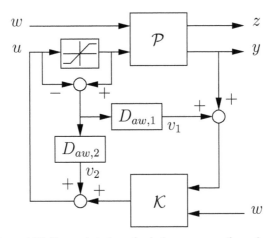

Figure 4.12  External static anti-windup compensation scheme.

Since external anti-windup augmentation is a special case of full-authority augmentation, the degrees of freedom available to external anti-windup for performance and stability purposes are less than those available in the full-authority case. As a consequence, the feasibility conditions of the following algorithm will be more stringent than those of full-authority anti-windup. The payoff for this is that whenever feasible, external anti-windup will lead to compensation gains of smaller dimension, therefore of easier implementation.

**Algorithm 2** (Static external global DLAW)

| Applicability | | | | Architecture | | | Guarantee |
|---|---|---|---|---|---|---|---|
| Exp Stab | Marg Stab | Marg Unst | Exp Unst | Lin/ NonL | Dyn/ Static | Ext/ FullAu | Global/ Regional |
| $\sqrt{}^*$ | | | | L | S | E | G |

| |
|---|
| *Comments*: Useful when some internal states of the controller are inaccessible. Otherwise not more effective than Algorithm 1. Global input-output gain is optimized. |
| * Only those plants and unconstrained controllers satisfying step 2. |

**Step 1.** Given a plant and a controller of the form (4.1), (4.2), construct the matrices of the compact closed-loop representation using (4.4) and (4.10) and the matrices of the compact open-loop representation using (4.12). Select a constant $v \in [0, 1)$ used to enforce a strong well-posedness constraint (see Section 3.4.2, page 60). Moreover, based on the matrix $B_{ol,v}$ in equation (4.12), generate any matrix $B_{ol,v\perp}$ that spans the null space of its transpose, such that $B_{ol,v\perp}^T B_{ol,v} = 0$ and $\text{rank}(B_{ol,v\perp}) + \text{rank}(B_{ol,v}) = nc + np$. (Note that in general there can be infinite selections of $B_{ol,v\perp}$. As an example, using MATLAB, an orthonormal selection of $B_{ol,v\perp}$ is easily computed using `Bolperp = null(Bolv)`.)

**Step 2.** Verify the applicability of this algorithm by checking the following feasibility conditions in the variable $R$:

$$R = R^T > 0 \,,$$
$$B_{ol,v\perp}^T \left( RA_{ol}^T + A_{ol}R \right) B_{ol,v\perp} < 0 \,,$$
$$RA_{cl}^T + A_{cl}R < 0 \,,$$

If these conditions are not feasible, this algorithm is not applicable.

**Step 3.** Define $M := \begin{bmatrix} B_{ol,v\perp} & 0 & 0 \\ 0 & I_{nw} & 0 \\ 0 & 0 & I_{nz} \end{bmatrix}$. Based on the matrices determined at step 1, find the optimal solution $(R, \gamma) \in \mathbb{R}^{ncl \times ncl} \times \mathbb{R}$ of the following eigenvalue problem:

$\min_{R,\gamma} \gamma$ *subject to*:

*Positivity*:
$$R = R^T > 0 \,,$$

*Open-loop condition*:
$$M^T \begin{bmatrix} RA_{ol}^T + A_{ol}R & B_{ol,w} & RC_{ol,z}^T \\ B_{ol,w}^T & -\gamma I_{nw} & D_{ol,zw}^T \\ C_{ol,z}R & D_{ol,zw} & -\gamma I_{nz} \end{bmatrix} M < 0 \,,$$

*Closed-loop condition*:

$$
\begin{bmatrix}
RA_{cl}^T + A_{cl}R & B_{cl,w} & RC_{cl,z}^T \\
B_{cl,w}^T & -\gamma I_{n_w} & D_{cl,zw}^T \\
C_{cl,z}R & D_{cl,zw} & -\gamma I_{n_z}
\end{bmatrix} < 0 .
$$

**Step 4.** Based on the matrices determined at step 1, construct the matrices $\Psi \in \mathbb{R}^{m \times m}$, $G_U \in \mathbb{R}^{n_u \times m}$, and $H \in \mathbb{R}^{m \times n_y + n_u}$, with $m := n_{cl} + n_u + n_w + n_z$, as follows:

$$
\Psi(U,\gamma) = \mathrm{He}
\begin{bmatrix}
A_{cl}R & B_{cl,q}U + QC_{cl,u}^T & B_{cl,w} & QC_{cl,z}^T \\
0 & D_{cl,uq}U - U & D_{cl,uw} & UD_{cl,zq}^T \\
0 & 0 & -\frac{\gamma}{2}I & D_{cl,zw}^T \\
0 & 0 & 0 & -\frac{\gamma}{2}I
\end{bmatrix},
$$

$$
H = \begin{bmatrix} B_{cl,v}^T & 0 \mid D_{cl,uv}^T \mid 0 \mid D_{cl,zv}^T \end{bmatrix}
$$

$$
G_U = \begin{bmatrix} 0 & 0 \mid I \mid 0 \mid 0 \end{bmatrix},
$$

where for any square matrix $X$, $\mathrm{He}(X) = X + X^T$.

**Step 5.** Find the optimal solution $(\Lambda_U, U, \gamma)$ of the LMI eigenvalue problem:

$\min_{\Lambda_U, U, \gamma} \gamma$ *subject to*:

*Strong well-posedness* (if $v = 0$, then redundant):

$$
-2(1 - v)U + \mathrm{He}\left(D_{cl,uq}U + \begin{bmatrix} 0_{n_u \times n_y} & I_{n_u} \end{bmatrix} \Lambda_U\right) < 0 ,
$$

*Anti-windup gains LMI*:

$$
\Psi(U,\gamma) + G_U^T \Lambda_U^T H^T + H \Lambda_U G_U < 0 .
$$

**Step 6.** Select the static external anti-windup compensator as: $v = D_{aw}q$ with the selection $D_{aw} = \Lambda_U U^{-1}$.

&#9733;

### 4.3.2.1 Interpretation of feasibility conditions*

The feasibility conditions for static external anti-windup augmentation with global guarantees correspond to the LMIs listed at step 2 of Algorithm 2. It is instructive to compare these conditions to the corresponding ones for the full-authority case, reported at step 2 of Algorithm 1 on page 82, the latter ones having been commented on on page 83.

First observe that when $B_{c,u}$ is a square invertible matrix, so that all the controller state equations can be independently modified from the controller input, external anti-windup reduces to being full-authority anti-windup. In that case, since the lower block of $B_{ol,v}$ in equation (4.12) is full rank, it is straightforward to verify that $B_{ol,v\perp} = \begin{bmatrix} I_{n_p} \\ 0 \end{bmatrix}$, and that by (4.12), the external feasibility conditions of Algorithm 2 reduce to the full-authority feasibility conditions of Algorithm 1. According to the interpretation already given on page 83, these conditions require that there be a "quasi-common" quadratic Lyapunov function for the open-loop plant and the closed-loop system.

In the more general case where the matrix $B_{c,u}$ is not square, the external feasibility conditions are evidently more stringent than the full-authority feasibility conditions, so that whenever the external static anti-windup construction is feasible, the full-authority one is feasible too. Conversely, whenever the full-authority scheme is not feasible, the external one is not feasible either. This fact becomes quite intuitive when recognizing that external anti-windup has fewer degrees of freedom than full-authority anti-windup because it is unable, in general, to independently modify each one of the internal controller state equations: it can only operate through the input matrix $B_{c,y}$. Once again, exponential stability of the open-loop plant becomes then a necessary assumption for the applicability of Algorithm 2 and even more strict requirements are imposed by the external feasibility conditions, as clarified next.

A deeper understanding of the external feasibility conditions arises from applying the following simple fact that can be proven by applying Finsler's lemma (see page 54):

> Given a symmetric matrix $\Xi$ and a matrix $B$ having the same number of rows, construct any matrix $B_\perp$ orthogonal to $B$ such that $B^T B_\perp = 0$. Then the following statements are equivalent:
>
> 1. $B_\perp \Xi B_\perp < 0$.
> 2. There exists a matrix $K$ of suitable dimensions such that $\Xi + \mathrm{He}(BK) < 0$.

In light of this fact and based on equations (4.12), it is possible to replace the open-loop condition at step 2 of Algorithm 2 by the following equivalent condition in the variables $\tilde{K}$ and $R$:

$$\mathrm{He}(A_{ol}R + B_{ol,v}\tilde{K}) < 0 ,$$

which, based on the structure of $A_{ol}$ and $B_{ol,v}$ in (4.12) and performing the change of variable $\tilde{K} \to K = \tilde{K}R^{-1}$, becomes:

$$\mathrm{He}\left(\left(\begin{bmatrix} A_p & 0 \\ 0 & A_c \end{bmatrix} + \begin{bmatrix} 0 \\ B_{c,y} \end{bmatrix} K\right)R\right) < 0 ,$$

The last condition corresponds to requiring that there exist a static state feedback law $Kx_{ol}$ for the compact open-loop representation which, injected at the controller input, guarantees quadratic stability with the Lyapunov function $V_{ext} := x_{ol}^T R^{-1} x_{ol}$. It is standard notation in systems theory to call the function $V_{ext}$ a "control Lyapunov function" for the compact open-loop representation.

In light of this definition, of the conditions at step 2 of Algorithm 2, and of the interpretation above, the feasibility conditions for static external anti-windup augmentation can be stated as the existence of a single quadratic Lyapunov function $V_{ext}$ which is both a Lyapunov function for the compact unconstrained closed-loop system and a control Lyapunov function for the compact open-loop representation.

### 4.3.2.2 Case studies

The first two examples considered here show that in some cases the static external DLAW scheme performs as well as the full-authority one of Algorithm 1. The advantage in using external designs for those cases is that the anti-windup gains have smaller dimensions; therefore the on-line implementation is less computationally intensive.

**Example 4.3.6** (MIMO academic example no. 2) Consider the problem described in Example 4.3.2 on page 85. For this particular example, the whole controller state is directly reachable from the controller input, so that external anti-windup is in principle equivalent to full-authority anti-windup. This equivalence is confirmed by the numerical optimization of Algorithm 2 which, when applied to this example, gives an external solution completely equivalent to the full-authority one of Example 4.3.2. In particular, picking $v = 0.001$, the gains given by Algorithm 2 are

$$D_{aw,1} = \begin{bmatrix} -0.599 & -0.7375 \\ -0.749 & -0.9835 \end{bmatrix}, \qquad D_{aw,2} = \begin{bmatrix} 1.005 & 0.0067 \\ 0.004 & 1.005 \end{bmatrix},$$

and the guaranteed global performance is $\gamma = 1.56$. The nonlinear gain curve of the compensated system perfectly coincides with that of Figure 4.5 on page 87, characterizing the full-authority solution. Clearly, for this example there's no difference in picking full-authority or external DLAW compensation. □

**Example 4.3.7** (Electrical network) Consider the same problem addressed in Example 4.3.3 on page 87. Similar to the previous example, external anti-windup leads to almost no performance deterioration. However, in this case there is an advantage in using the external architecture as the dimension of the anti-windup gains is in this case smaller. Selecting $v = 0.1$, the optimal gains resulting from Algorithm 2 are the following ones:

$$D_{aw,1} = -0.0279 , \qquad D_{aw,2} = 3.03 ,$$

which guarantee the global performance is $\gamma = 86.32$, slightly deteriorated from the full-authority case of Example 4.3.3. Nevertheless, both the simulations and the nonlinear gain curve using the two schemes are indistinguishable from each other. This makes the external solution more desirable due to its simpler architecture. □

Even though the last two examples suggest that the external anti-windup scheme might be as effective as the full-authority one, this is not the case in general. There are cases, indeed, where static full-authority anti-windup compensation will be feasible and the external one won't be. This is shown in the next example.

**Example 4.3.8** (F8 longitudinal dynamics) Consider the example introduced in Section 1.2.3 and revisited in Example 4.3.5, where static full-authority anti-windup has been designed for it.

For this example, the feasibility conditions at step 2 of Algorithm 2 are not satisfied; therefore global external linear compensators for this problem can only be determined among the family of dynamic compensators whose construction is the subject of the next chapter. Indeed, in the next chapter this example will be revisited and both a plant-order and a reduced-order dynamic linear anti-windup compensator will be designed for it. □

## 4.4 ALGORITHMS PROVIDING REGIONAL GUARANTEES

This section focuses on static linear anti-windup design algorithms that may overcome certain limitations of the algorithms provided in the previous section. These limitations arise from the fact that the previous algorithms only provide solutions enforcing global stability and performance properties. While these properties are very desirable, the price to be paid for them is often too high, sometimes because the resulting performance is not satisfactory enough for the problem under consideration, at other times because the algorithms may in some cases not be applicable at all because the feasibility conditions are often not satisfied.

The algorithms addressed in this section will provide a good alternative when it is known that the system will operate in a subset of the whole space and global guarantees are not necessary for the problem under consideration. It will be shown that giving up on the global guarantees will lead to weaker feasibility conditions and will remove the requirement that the open-loop plant be exponentially stable, which is overkill in many cases of interest. On the other hand, since only regional closed-loop guarantees are provided by these algorithms, large enough signals could drive the anti-windup closed-loop system unstable, or poor performance could be experienced if signals become too large. Nevertheless, in many practical cases, the tools provided here are the only ones suitable for anti-windup augmentation and their use may lead to radical stability and performance improvements in the operating conditions of the system under consideration.

From the nonlinear performance viewpoint (see Section 2.4.4), the algorithms provided next aim at finding anti-windup gains that optimize the value of the nonlinear gain curve at a certain horizontal position (namely, for a certain bound on $\|w\|_2$). Although no stability or performance guarantees for larger values of $\|w\|_2$ arise from the construction algorithms, it is often the case that an analysis carried out after the anti-windup design will reveal that the system is well behaved in very large regions. This type of afterdesign analysis will be carried out for the examples considered in this section.

Only full-authority regional algorithms will be presented here, although external ones can also be formulated slightly extending the results currently available in the literature. External algorithms will possibly be included in a subsequent edition of this book.

### 4.4.1 Regional full-authority augmentation

Regional static full-authority anti-windup augmentation shares the same architecture as the global counterpart addressed in Section 4.3.1 and amounts to selecting two matrices, $D_{aw,1}$ and $D_{aw,2}$, interconnected to the control system as represented in Figure 4.1.

An important difference between the regional and the global approaches is that when only regional properties are guaranteed, it becomes important to characterize the size of the stability region and of the reachable region from bounded inputs. The former region characterizes the set of initial states that lead to exponentially converging responses, while the second region indicates how large the state can grow when the closed loop is excited by an external input $w$ with bounded $\mathcal{L}_2$ norm. In the following algorithm, estimates of both these regions will be automatically given, possibly allowing one to enforce a desired and prescribed guaranteed size in the direction of the plant states.

An interesting peculiarity of the algorithm given next is its dependence on the saturation levels $\bar{u}_i > 0$, $i = 1,\ldots,n_u$, which may all be different from each other. The algorithm leads to solutions with guaranteed stability and performance also for time-varying saturation limits, as long as the vector $\bar{u} = [\bar{u}_1,\ldots,\bar{u}_{n_u}]^T$ is a lower bound for the saturation values at all times. As a consequence, the algorithm is also applicable to nonsymmetric saturations as long as for each input channel the value $u_i, i = \in \{1,\ldots,n_u\}$ is selected as the most stringent saturation limit for that channel. In this case, the stability and performance estimates provided by the algorithm increase in conservativeness.

**Algorithm 3** (Static full-authority regional DLAW)

| Applicability | | | | Architecture | | | Guarantee |
|---|---|---|---|---|---|---|---|
| Exp Stab | Marg Stab | Marg Unst | Exp Unst | Lin/ NonL | Dyn/ Static | Ext/ FullAu | Global/ Regional |
| $\checkmark$ | $\checkmark$ | $\checkmark$ | $\checkmark$ | L | S | FA | R |

| |
|---|
| *Comments*: Extends the applicability of Algorithm 1 to a larger class of systems by requiring only regional properties. However, feasibility conditions still need to hold. Regional input-output gain is optimized. |

**Step 1.** Given a plant and a controller of the form (4.1), (4.2) and input saturation values $\bar{u}_i$, $i = 1,\ldots,n_u$, construct the matrices of the compact closed-loop representation using (4.4) and (4.7). Select a constant $v \in [0,1)$ used to enforce a strong well-posedness constraint (see Section 3.4.2 on page 60) and select the maximum size $s$ of allowable external inputs $\|w\|_2 < s$ over which the gain minimization has to be performed.

Possibly select two matrices $S_p$ and $R_p$ characterizing a desirable guaranteed reachable set contained in $\mathcal{R}_p = \{x_p : x_p^T R_p^{-1} x_p\}$ and a desirable stability region containing $\mathcal{S}_p = \{x_p : x_p^T S_p^{-1} x_p\}$ in the plant state directions, otherwise ignore the conditions involving $S_p$ and $R_p$ below.

**Step 2.** Verify the applicability of this algorithm by checking the following feasibility conditions in the variables $R$ and $Z$:

$$R = R^T = \begin{bmatrix} R_{11} & R_{12} \\ R_{12}^T & R_{22} \end{bmatrix} > 0,$$

$$He(A_p R_{11} + B_{p,u} Z) < 0,$$

$$He(A_{cl} R) < 0,$$

$$\begin{bmatrix} \bar{u}_i^2/s^2 & Z_i \\ Z_i^T & R_{11} \end{bmatrix} \geq 0, \quad i = 1,\ldots,n_u,$$

$$R_p \leq R \leq S_p,$$

where $Z_i$ denotes the $i$th row of the matrix $Z$. If these conditions are not feasible, either this algorithm is not applicable or the parameters at step 1 should be selected differently.

**Step 3.** Based on the matrices determined at step 1, find the optimal solution $(R, Z, \gamma^2) \in \mathbb{R}^{n_{cl} \times n_{cl}} \times \mathbb{R}^{n_u \times n_p} \mathbb{R}$ to the following LMI eigenvalue problem:

$\min\limits_{R,Z,\gamma^2} \gamma^2$ *subject to*:

*Positivity*:

$$R = R^T = \begin{bmatrix} R_{11} & R_{12} \\ R_{12}^T & R_{22} \end{bmatrix} > 0,$$

*Open-loop condition*:

$$\text{He} \begin{bmatrix} A_p R_{11} + B_{p,u} Z & B_{p,w} & 0 \\ 0 & -\frac{I_{n_w}}{2} & 0 \\ C_{p,z} R_{11} + D_{p,zu} Z & D_{p,zw} & -\frac{\gamma^2 I_{n_z}}{2} \end{bmatrix} < 0,$$

*Closed-loop condition*:

$$\begin{bmatrix} RA_{cl}^T + A_{cl}R & B_{cl,w} & RC_{cl,z}^T \\ B_{cl,w}^T & -I_{n_w} & D_{cl,zw}^T \\ C_{cl,z}R & D_{cl,zw} & -\gamma^2 I_{n_z} \end{bmatrix} < 0,$$

*Regional conditions*:

$$\begin{bmatrix} \bar{u}_i^2/s^2 & Z_i \\ Z_i^T & R_{11} \end{bmatrix} \geq 0, \quad i = 1,\dots,n_u,$$

*Guaranteed reachable set and stability region*:

$$R_p \leq R_{11} \leq S_p .$$

**Step 4.** Based on the matrices determined at step 1, construct the matrices $\Psi \in \mathbb{R}^{m \times m}$, $G_U \in \mathbb{R}^{n_u \times m}$, and $H \in \mathbb{R}^{m \times (n_u + n_c)}$, with $m := n_{cl} + n_u + n_w + n_z$, as follows:

$$\Psi(U, \gamma^2) = \text{He} \begin{bmatrix} A_{cl}R & B_{cl,q}U + QC_{cl,u}^T & B_{cl,w} & QC_{cl,z}^T \\ -Y & D_{cl,uq}U - U & D_{cl,uw} & UD_{cl,zq}^T \\ 0 & 0 & -\frac{1}{2}I_{n_w} & D_{cl,zw}^T \\ 0 & 0 & 0 & -\frac{\gamma^2}{2}I_{n_z} \end{bmatrix},$$

$$H = \begin{bmatrix} B_{cl,v}^T & 0 & D_{cl,uv}^T & 0 & D_{cl,zv}^T \end{bmatrix}$$

$$G_U = \begin{bmatrix} 0 & 0 & I & 0 & 0 \end{bmatrix},$$

with $Y = \begin{bmatrix} Z & ZR_{11}^{-1}R_{12} \end{bmatrix}$ and where for any square matrix $X$, $\text{He}(X) = X + X^T$.

**Step 5.** Find the optimal solution $(\Lambda_U, U, \gamma)$ of the LMI eigenvalue problem:

$\min\limits_{\Lambda_U,U,\gamma^2} \gamma^2$ *subject to*:

*Strong well-posedness (if $v = 0$, then redundant)*:

$$-2(1-v)U + \text{He}\left(D_{cl,uq}U + \begin{bmatrix} 0_{n_u \times n_c} & I_{n_u} \end{bmatrix}\Lambda_U\right) < 0 \,,$$

*Anti-windup gains LMI*:

$$\Psi(U,\gamma^2) + G_U^T\Lambda_U^T H^T + H\Lambda_U G_U < 0 \,.$$

**Step 6.** Select the static full-authority anti-windup compensator as: $v = D_{aw}q$, with the selection $D_{aw} = \Lambda_U U^{-1}$.

★

### 4.4.1.1 Interpretation of feasibility conditions*

It is instructive to compare the feasibility conditions at step 2 of Algorithm 3 to the corresponding feasibility conditions for the global static anti-windup design of Algorithm 1 on page 82.

A first important difference lies in the fact that the open-loop condition has now a new degree of freedom consisting in the variable $Z$, which allows one to overcome the severe applicability limit of the global algorithms: exponential stability of the plant. The role played by $Z$ is quite interesting when understood in conjunction with the $n_u$ conditions involving each of the saturation limits $u_i$, $i = 1, \ldots, n_u$. For each $i$, these conditions can be rewritten by using the Schur complement as $Z_i R_{11}^{-1} Z_i^T \leq \frac{\bar{u}_i^2}{s^2}$, so that two facts become evident:

1. the more stringent the saturation limits $\bar{u}_i^2$ are, the smaller the variable $Z$ will be constrained to be,
2. the larger the size $s$ of allowable external signals is, the smaller the variable $Z$ will be constrained to be.

Imposing that $Z$ be small affects its capability to satisfy the open-loop feasibility constraint, so that at the limit when $s$ goes to infinity (global external stability), $Z$ will be zero and the global limitations of Algorithm 1 will be recovered. Nevertheless, choosing a reasonable value for $s$ and assuming that the saturation limits are compatible with its selection, $Z$ will have enough degrees of freedom to allow the open-loop feasibility constraint to be satisfied. [1]

Similar to the external anti-windup conditions commented on on page 4.3.2.1, the open-loop condition in step 2 of Algorithm 3 can be interpreted in terms of control Lyapunov functions; as a matter of fact, with the variable change $Z \to K_p = ZR_{11}^{-1}$, the open-loop feasibility condition becomes:

$$\text{He}((A_p + B_{p,u}K_p)R_{11}) < 0,$$

which, combined with the closed-loop condition, has a nice system theoretic interpretation: if $n_c = 0$ so that $n_{cl} = n_p$, the function $V = x^T R^{-1} x$ is both a quadratic Lyapunov function for the unconstrained closed loop and a quadratic control Lyapunov function for the open-loop plant from the control input. The coupling condition could in some examples impose applicability limitations that are too severe. Those limitations can be overcome by applying the dynamic regional anti-windup compensation algorithms of the next chapter.

---

[1] Requiring that the saturation limits be compatible with the size of the required stability region amounts to requiring that the actuators have a reasonable dimension for the control application under consideration.

### 4.4.1.2 Achieving guaranteed regional properties

An important aspect of the feasibility conditions for Algorithm 3 is that the regional properties of the algorithm can be quantified by imposing a prescribed guaranteed size for the reachable set from bounded inputs and for the region of attraction, namely, the region of initial conditions that lead to exponentially converging responses. This is done by way of the matrices $R_p$ and $S_p$, respectively, as noted in step 1 of the algorithm. It should be emphasized, however, that larger values of $R_p$ and smaller values of $S_p$ make the feasibility constraints less likely to be fulfilled; therefore the selection of $R_p$ and $S_p$ should always be carried out within a certain reasonable range.

### 4.4.1.3 Case studies

The next example is useful to illustrate a situation where the windup effect is not caused by the reference input but by the disturbance input. In many cases problems induced by the reference input can be avoided by adding a suitable prefiltering action. However, when windup is triggered by disturbances, this is hardly possible and anti-windup is even more important to adopt.

**Example 4.4.1** (A disturbance rejection problem) Consider the disturbance rejection problem corresponding to the block diagram representation in Figure 4.13, where the plant characterized by the following transfer function:

$$\frac{Y(s)}{U(s)} = \frac{1}{s(s+1)}$$

is subject to an input disturbance $d_u(t)$ and its measurement $y(t)$ is affected by the output noise $d_y(t)$. Suppose that the plant control input $u(t)$ is subject to a saturation level of $\pm 15$ and that both the input and output disturbance are bandlimited white noises. The input disturbance $d_u(t)$ is a low-frequency noise that needs to be rejected by the active disturbance rejection system: it has noise power equal to 1 and a sample time of 0.2 seconds. The output disturbance $d_y(t)$ instead is a high-frequency noise affecting the measurement signal: it has noise power equal to $5 \cdot 10^{-8}$ and a sample time of $5 \cdot 10^{-3}$ seconds.

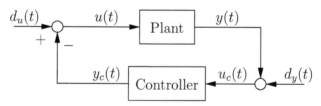

Figure 4.13 The disturbance rejection problem of Example 4.4.1.

For this example, the following controller is designed using a loop shaping technique with the goal of attenuating the low-frequency input disturbance by cranking up the low-frequency gain while preserving a certain level of stability margin:

$$\frac{Y_c(s)}{U_c(s)} = 500\frac{(1+\frac{1}{5}s)^2}{s(1+\frac{1}{100}s)^2}.$$

This controller performs well during normal operation and induces extreme disturbance attenuation at the plant output corresponding to $-60$ dB attenuation for frequencies below $\omega_l = 0.5$ rad/s. Moreover, the loop gain at high frequencies is small enough that the control system is not too sensitive to measurement noise above $\omega_h = 100$ rad/s.

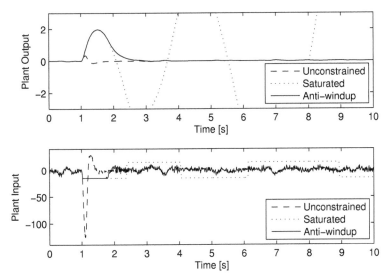

Figure 4.14 Plant output and input responses of various closed loops for Example 4.4.1.

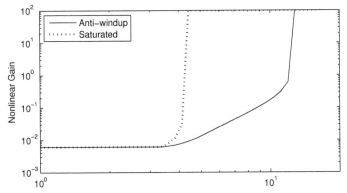

Figure 4.15 Nonlinear gains for Example 4.4.1 characterizing the system before and after anti-windup augmentation.

Assume now that at time $t = 1$ a pulse disturbance of amplitude 100 and length 0.1 seconds is added to the input disturbance $d_u(t)$. Then the closed loop becomes unstable, as shown in Figure 1.28 on page 21 of the introductory chapter. The solution to this problem is easily obtained by transforming the transfer functions above into equivalent state-space representations and running Algorithm 3. Note that since the plant is not exponentially stable (due to the presence of a pole at the origin), all the global anti-windup algorithms provided in this chapter are infeasible. In Algorithm 3, the parameters $v = 0.1$ and $s = 5$ are selected. The algorithm provides the following static anti-windup compensator:

$$D_{aw,1} = \begin{bmatrix} 0.1275 \\ -0.00378 \\ 0.1017 \end{bmatrix}, \qquad D_{aw,2} = 0.8 ,$$

inducing the regional $\mathcal{L}_2$ performance $\gamma = 0.0132$. Figure 4.14 represents the simulation results with anti-windup compensation. It is evident that anti-windup is able to recover stability and fast disturbance rejection recovery. The saturated closed loop instead experiences the onset of persistent oscillations. Figure 4.15 represents the nonlinear $\mathcal{L}_2$ gains before and after anti-windup compensation. Note that unlike in the previous examples, which guaranteed global properties, regional anti-windup here only increases the guaranteed stability region. This is a necessary fact because no anti-windup action can induce global finite $\mathcal{L}_2$ gain for this example where the plant is not exponentially stable.                                                                                           □

**Example 4.4.2** (SISO academic example no. 2) In Example 4.3.4 on page 89 earlier in this chapter, a problem setting has been discussed where global static full-authority anti-windup design is not feasible, and therefore a static solution to the underlying anti-windup problem could not be given. For the same problem statement, namely, a SISO plant with an integrating one-dimensional controller, a regional anti-windup solution is given here. The saturation value is given by $\bar{u} = 0.5$. The plant and controller matrices are reported next for convenience:

$$\left[ \begin{array}{c|c|c} A_p & B_{p,u} & B_{p,w} \\ \hline C_{p,y} & D_{p,yu} & D_{p,yw} \end{array} \right] = \left[ \begin{array}{cc|c|c} -0.2 & -0.2 & 1 & 0 \\ 1 & 0 & 0 & 0 \\ \hline -0.4 & -0.9 & -0.5 & 0 \end{array} \right]$$

$$\left[ \begin{array}{c|c|c} A_c & B_{c,y} & B_{c,w} \\ \hline C_c & D_{c,y} & D_{c,w} \end{array} \right] = \left[ \begin{array}{c|c|c} 0 & 1 & -1 \\ \hline 2 & 2 & -2 \end{array} \right].$$

Algorithm 3 can be used for this example. Selecting $v = 0.1$ and $s = 1$ at step 1 of the algorithm leads to the following solution:

$$D_{aw,1} = -1.07, \qquad D_{aw,2} = 0.6 ,$$

which guarantees the regional performance level $\gamma = 2.712$ for any input having size $\|w\|_2 \leq 1$. Figure 4.16 represents the plant input and plant output responses of the various closed loops when the reference input is selected as a step having amplitude 2. From the plot it is evident that regional anti-windup succeeds in mitigating the windup effects of the saturated closed loop.

Figure 4.17 represents the nonlinear gains characterizing this example before and after anti-windup augmentation. The system after anti-windup compensation is clearly more performing, although both the closed loops with and without anti-windup admit a quadratic linear gain only up to a certain value of the input size (roughly, up to $s = 0.35$ for the saturated closed loop and up to $s = 2.04$ for the anti-windup case). Note that for $s = 1$ the nonlinear gain curve passes through the value $\gamma = 2.712$, which has been established by the optimal design algorithm.                                                                                           □

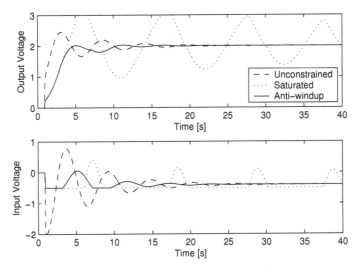

Figure 4.16  Plant output and input responses of various closed loops for Example 4.4.2.

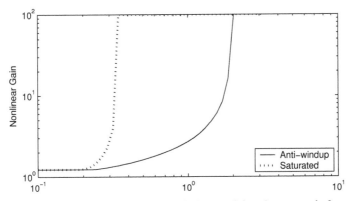

Figure 4.17  Nonlinear gains for Example 4.4.2 characterizing the system before and after
anti-windup augmentation.

**Example 4.4.3**  (Electrical network)  Consider the passive electrical circuit introduced in Example 4.3.3 and the resulting windup-prone control system. For that system both global full-authority and external anti-windup gains could be designed, but the resulting performance was quite sluggish (see the simulations in Figure 4.7 on page 89).

That windup problem can be tackled using a regional approach by way of Algorithm 3. In particular, selecting $v = 0.1$ and $s = 0.01$ in step 1 of the algorithm leads to the following solution:

$$D_{aw,1} = \begin{bmatrix} 0.8824 \\ -0.0111 \end{bmatrix}, \qquad D_{aw,2} = 0.8 ,$$

which guarantees the regional performance level $\gamma = 1.643$ for any input having size $\|w\|_2 \leq 0.01$. Figure 4.18 represents the plant input and plant output responses

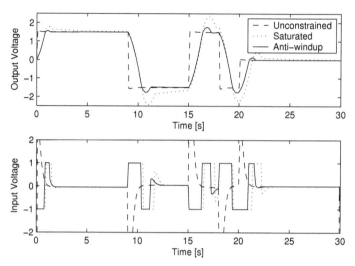

Figure 4.18  Plant output and input responses of various closed loops for Example 4.4.3.

of the various closed loops when using the same reference input that was used in Figure 4.7 on page 89. Note that the regional design is able to induce an improved response for this specific trajectory selection. However, due to the regional properties of the approach, there are no guarantees on the closed-loop behavior for larger references.

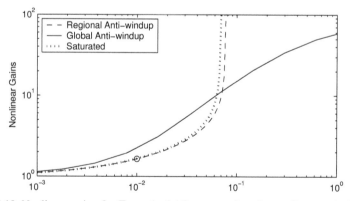

Figure 4.19  Nonlinear gains for Example 4.4.3 compared to the nonlinear gain from the global approach of Example 4.3.3.

The nonlinear gain characterization reported in Figure 4.19 shows that in the small signals range the regional approach (dashed curve) is more desirable than the global one (solid curve). It remains unclear why the nonlinear gain characterizing the closed loop before anti-windup (dotted curve) is so close to the regional one, despite the significant differences in the responses of Figure 4.18. This might be due to the limitations in the $\mathcal{L}_2$ gain characterization of performance or due to the conservativeness arising from the use of quadratic Lyapunov functions for the estimation of the nonlinear gains.                                                                    ☐

**Example 4.4.4**   (MIMO academic example no. 2)   Consider the problem statement considered first in Example 4.3.2 on page 85. One might wish that regional anti-windup would lead to improved nonlinear performance, at least for small values of $\|w\|_2$. However, when applying Algorithm 3 with $s = 1$ and $v = 0.01$, the following anti-windup gains are obtained:

$$D_{aw,1} = \begin{bmatrix} -0.65 & -0.6 \\ -0.8 & -1.2 \end{bmatrix}, \qquad D_{aw,2} = \begin{bmatrix} -2.4 & -4.32 \\ -2.65 & -2.4 \end{bmatrix},$$

which lead to a guaranteed regional performance level $\gamma = 1.1$. However, the

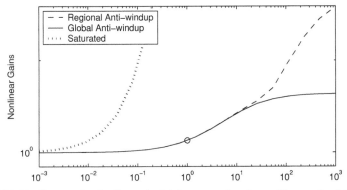

Figure 4.20   Nonlinear gains for Example 4.4.4 compared to the nonlinear gain obtained for the global design of Example 4.3.2.

simulations resulting from this regional gain selection are equivalent to the global ones reported in Example 4.3.2 and, most interestingly, the nonlinear $\mathcal{L}_2$ gain appears worse than that obtained for the global case. In particular, in Figure 4.20 the nonlinear gain obtained with regional anti-windup is compared to the one corresponding to global anti-windup (which comes from Figure 4.5 on page 87). In Figure 4.20, the small circle denotes the optimized regional gain, which in this case is just as good as the value of the nonlinear gain characterizing global anti-windup. However, since only regional guarantees are provided by Algorithm 3, the dashed curve for $s \gg 1$ is significantly deteriorated. For this specific example, global anti-windup appears to be more desirable than regional anti-windup.                                ☐

## 4.5  NOTES AND REFERENCES

Static anti-windup was for many years the predominant type of anti-windup employed. An important survey of how several anti-windup schemes could have been written within the static anti-windup framework is [SUR4]. Perhaps the first paper casting anti-windup as an LMI optimization is [MI2, MI1]. In [MI11], Mulder, Kothare, and Morari cast global, static full-authority anti-windup synthesis with algebraic loops as an LMI. In [MI18] (see also [MI19] for some case studies), Grimm et al. provided the system theoretic interpretation of the feasibility conditions, as reported at the end of Section 4.3.1. The algorithms reported here for regional performance using static anti-windup first appeared in [MI27] (which can be interpreted as the regional counterpart of [MI11]). Then, in [MI33] (which can be interpreted as the regional counterpart of [MI18]), the system theoretic interpretations of regional static anti-windup were given. The external anti-windup schemes here presented are taken from [MI22] which extends the full-authority results of [MI18] to the external case also giving system theoretic interpretation. Regional external anti-windup hasn't been published anywhere but is a straightforward extension of existing results. An alternative interesting approach is reported in [MI23], where an LMI-based procedure to minimize the URR gain instead of the $\mathcal{L}_2$ gain reported here is given. This approach is not reported in this book to keep the discussion simple.

All the algorithms given here rely on the application of the so-called "elimination lemma" (see [G17]) to the anti-windup problem. This type of approach was first proposed in [G5] and [G6] to address the $\mathcal{H}_\infty$ controller synthesis problem. The explicit formulas given in [G8] can also be used to bypass the last large LMI to be solved in the algorithms given in this chapter, which often causes numerical problems with large state-space dimensions. As an example, see [MI26], where those explicit formulas have been used for anti-windup design.

Discrete-time counterparts of the algorithms given here have been given in [MI24] (which can be seen as the discrete-time equivalent of the work in [MI11]) and in [MI30] (which can be seen as the discrete-time equivalent of the work in [MI18] and [MI33]).

Extensions of the direct linear LMI-based static anti-windup approaches to the case of systems with magnitude and rate saturations have been given in [MI32, MI29]. Those algorithms are not covered here, to keep the discussion simple.

The infeasible example for static anti-windup given in Example 4.3.1 was proposed in [MI18] where it was used to illustrate the feasibility conditions for static anti-windup. The MIMO academic example no. 2 treated in Examples 4.3.2, 4.3.6, and 4.4.4 is taken from [MI11] even though it had appeared before in [FC7] (it was also revisited several times later, e.g., in [MI19]). The passive electrical circuit described in Examples 4.3.3, 4.3.7, and 4.4.3 was introduced and used in [MI19] and later revisited in [MI22]. The SISO academic example no. 2 described and addressed in Examples 4.3.4 and 4.4.2 was introduced in [MI16], where it was noted that the quadratic construction of [MI11] was not always feasible (the system theoretic feasibility conditions appeared later in [MI18]). The F8 longitudinal dynamics example described in Examples 4.3.5 and 4.3.8 first appeared in [SAT1] and was later revisited in a large number of papers (e.g, [MI18]).

# Chapter Five

---

## Dynamic Linear Anti-windup Augmentation

### 5.1 OVERVIEW

In the previous chapter, linear static anti-windup designs inducing global and regional guarantees have been formulated and illustrated through several algorithms characterized by both full-authority and external architectures. It has been emphasized, however, that not all anti-windup problems admit such a simple compensation scheme.

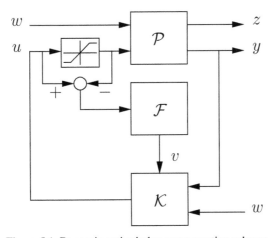

Figure 5.1 Dynamic anti-windup compensation scheme.

Motivated by this fact, this chapter discusses a generalization of the linear static scheme wherein the matrix gains are replaced by a linear dynamic anti-windup compensator $\mathcal{F}$ with internal states. The static case will be recovered by this dynamic generalization when the size of the anti-windup filter state is zero. The interconnection of the dynamic anti-windup filter resembles that of the static case as shown in Figure 5.1. In particular, the filter is driven by the difference between the input and the output of the saturation nonlinearity, and the anti-windup filter outputs affect the unconstrained controller dynamics through signals that have access to the state and output equations of the unconstrained controller for the full-authority architecture or to the input and output signals of the unconstrained controller for the external architecture.

It should be emphasized that not only the dynamic generalization of the linear static schemes allows one to address and solve problems that cannot be solved using static techniques. Indeed, the extra degrees of freedom residing in a dynamic anti-windup compensation scheme allow, in general, to achieve increased performance levels therefore leading, in general, to improved responses as compared to those obtained with static anti-windup. The price to be paid for these advantages is the complexity of the anti-windup filter, which has an internal state whose dimension sums up with the dimension of the unconstrained controller when evaluating the computational complexity of the anti-windup augmented control system.

Similar to the static case, several algorithms will be given in this chapter for dynamic direct linear anti-windup (DLAW) compensation. In particular, first algorithms with global performance and stability guarantees will be given. These algorithms are prone to similar applicability limitations as their static counterpart, namely, they are only applicable to problems related to exponentially stable linear plants. These algorithms will be formulated using both the full-authority architecture and the external architecture. Next, an algorithm for plant-order anti-windup design will be given, which only induces regional guarantees on the closed loop but that is applicable to any type of windup-prone linear control system. All the algorithms are aimed at guaranteeing optimal (regional or global) input-output performance of the type discussed in Section 2.4.2.

Within the context of dynamic anti-windup, an important parameter is the order $n_{aw}$ of the anti-windup filter. In this chapter, algorithms leading to compensators of the same order as the plant will be given first, because they are characterized by powerful feasibility conditions and convex formulations. Next, with the goal of reducing the complexity of the anti-windup compensators, algorithms for the design of reduced-order compensators will be given. These algorithms arise from approximate (convexified) solutions of nonconvex conditions, therefore they are not guaranteed to always provide a solution, but are often very effective for reduced-order anti-windup design.

## 5.2 KEY STATE-SPACE REPRESENTATIONS

Similar to the structure of Chapter 4, before introducing the anti-windup construction algorithms, it is useful to define some notation giving the equations for the plant and unconstrained controller under consideration and certain important matrices of the resulting closed loop.

The plant and unconstrained controller under consideration are the same as those in (4.1) and (4.2):

$$
\mathcal{P} \quad \begin{cases} \dot{x}_p &= A_p x_p + B_{p,u}\operatorname{sat}(u) + B_{p,w} w \\ z &= C_{p,z} x_p + D_{p,zu}\operatorname{sat}(u) + D_{p,zw} w \\ y &= C_{p,y} x_p + D_{p,yu}\operatorname{sat}(u) + D_{p,yw} w , \end{cases} \tag{5.1}
$$

$$
\mathcal{K} \quad \begin{cases} \dot{x}_c &= A_c x_c + B_{c,y} y + B_{c,w} w \\ u &= C_c x_c + D_{c,y} y + D_{c,w} w . \end{cases} \tag{5.2}
$$

The general structure of the dynamic anti-windup compensator is described as follows, where, depending on the adopted architecture (full-authority or external), the correction signals $v_1$ and $v_2$

$$\mathcal{F} \begin{cases} \dot{x}_{aw} &= A_{aw}x_{aw} + B_{aw}\left(u - \text{sat}(u)\right) \\ v_1 &= C_{aw,1}x_{aw} + D_{aw,1}\left(u - \text{sat}(u)\right) \\ v_2 &= C_{aw,2}x_{aw} + D_{aw,2}\left(u - \text{sat}(u)\right) \end{cases} \qquad (5.3)$$

are injected at different locations in the unconstrained controller, as clarified next.

The dimension $n_{aw}$ of the anti-windup state $x_{aw}$ is a design parameter, and $A_{aw}$, $B_{aw}, C_{aw} = \begin{bmatrix} C_{aw,1} \\ C_{aw,2} \end{bmatrix}$, and $D_{aw} = \begin{bmatrix} D_{aw,1} \\ D_{aw,2} \end{bmatrix}$ are matrices of appropriate dimensions that characterize the particular anti-windup solution adopted. In the special case where $n_{aw} = 0$, the size of the state $x_{aw}$ is zero and the compensation scheme reduces to $v_1 = D_{aw,1}\left(u - \text{sat}(u)\right)$ and $v_2 = D_{aw,2}\left(u - \text{sat}(u)\right)$, which corresponds to the static anti-windup scheme extensively discussed in the previous chapter.

The design approach for dynamic DLAW augmentation parallels that of the static DLAW augmentation. It is therefore necessary to rely on the same closed-loop representations that have been introduced and used in Section 4.2. They are reported next to make this chapter self-contained.

The *compact closed-loop representation* corresponds to the following system

$$\begin{aligned} \dot{x}_{cl} &= A_{cl}x_{cl} + B_{cl,q}q + B_{cl,w}w \\ z &= C_{cl,z}x_{cl} + D_{cl,zq}q + D_{cl,zw}w \\ u &= C_{cl,u}x_{cl} + D_{cl,uq}q + D_{cl,uw}w , \end{aligned} \qquad (5.4)$$

which describes the closed loop before anti-windup augmentation with $q = \text{dz}(u)$. The matrices in (5.4) are uniquely determined by the matrices in (5.1) and (5.2) and correspond to

$$\begin{bmatrix} A_{cl} \\ \hline C_{cl,z} \\ \hline C_{cl,u} \end{bmatrix} = \begin{bmatrix} A_p + B_{p,u}\Delta_u D_{c,y}C_{p,y} & B_{p,u}\Delta_u C_c \\ B_{c,y}\Delta_y C_{p,y} & A_c + B_{c,y}\Delta_y D_{p,yu}C_c \\ \hline D_{p,zu}\Delta_u D_{c,y}C_{p,y} + C_{p,z} & D_{p,zu}\Delta_u C_c \\ \hline \Delta_u D_{c,y}C_{p,y} & \Delta_u C_c \end{bmatrix} \qquad (5.5a)$$

$$\begin{bmatrix} B_{cl,q} & B_{cl,w} \\ \hline D_{cl,zq} & D_{cl,zw} \\ \hline D_{cl,uq} & D_{cl,uw} \end{bmatrix} = \begin{bmatrix} -B_{p,u}\Delta_u & B_{p,w} + B_{p,u}\Delta_u(D_{c,y}D_{p,yw} + D_{c,w}) \\ -B_{c,y}\Delta_y D_{p,yu} & B_{c,w} + B_{c,y}\Delta_y(D_{p,yu}D_{c,w} + D_{p,yw}) \\ \hline -D_{p,zu}\Delta_u & D_{p,zw} + D_{p,zu}\Delta_u(D_{c,y}D_{p,yw} + D_{c,w}) \\ \hline I - \Delta_u & \Delta_u(D_{c,w} + D_{c,y}D_{p,yw}) \end{bmatrix} , \qquad (5.5b)$$

where the matrices $\Delta_u := (I - D_{c,y}D_{p,yu})^{-1}$ and $\Delta_y := (I - D_{p,yu}D_{c,y})^{-1}$ are well-defined by the well-posedness of the unconstrained closed loop.

The closed-loop system (5.4) after anti-windup augmentation is equipped with an extra dynamical equation arising from the anti-windup compensator dynamics and by extra signals affecting the dynamics of the $x_{cl}$ state variable:

$$\begin{aligned} \dot{x}_{cl} &= A_{cl}x_{cl} + B_{cl,v}C_{aw}x_{aw} + (B_{cl,q} + B_{cl,v}D_{aw})q + B_{cl,w}w \\ \dot{x}_{aw} &= A_{aw}x_{aw} + B_{aw}q \\ z &= C_{cl,z}x_{cl} + D_{cl,zv}C_{aw}x_{aw} + (D_{cl,zq} + D_{cl,zv}D_{aw})q + D_{cl,zw}w \\ u &= C_{cl,u}x_{cl} + D_{cl,uv}C_{aw}x_{aw} + (D_{cl,uq} + D_{cl,uv}D_{aw})q + D_{cl,uw}w, \end{aligned} \qquad (5.6)$$

where the matrices $B_{cl,v}$, $D_{cl,uv}$, and $D_{cl,zv}$ depend on the architecture, full authority or external, of the compensation scheme under consideration. The values of these matrices are detailed separately for the two architectures in the following two sections.

### 5.2.1 Closed-loop representation with full-authority anti-windup

Full-authority dynamic linear anti-windup augmentation corresponds to modifying the controller equations (5.2) as follows:

$$\mathcal{K} \quad \begin{cases} \dot{x}_c = A_c x_c + B_{c,y} y + B_{c,w} w + v_1 \\ u = C_c x_c + D_{c,y} y + D_{c,w} w + v_2 , \end{cases} \tag{5.7}$$

where the two signals $v_1$ and $v_2$ are the two outputs of the anti-windup compensator (5.3).

When adopting the full-authority architecture, the matrices $B_{cl,v}$, $D_{cl,uv}$, and $D_{cl,zv}$ in (5.6) correspond to the following values:

$$\left[ \begin{array}{c} B_{cl,v} \\ \hline D_{cl,uv} \\ \hline D_{cl,zv} \end{array} \right] = \left[ \begin{array}{cc} 0 & B_{p,u}\Delta_u \\ I_{n_c} & B_{c,y}\Delta_y D_{p,yu} \\ 0 & \Delta_u \\ 0 & D_{p,zu}\Delta_u \end{array} \right] . \tag{5.8}$$

After designing the anti-windup compensator matrices using the full-authority algorithms reported in this chapter, the nonlinear performance of the resulting closed loop can be evaluated relying on the LMI-based tools introduced in Section 3.5.3 on page 66. To this end, the following equations will be useful to appropriately select the matrices in (3.54) to represent the dynamic full-authority linear anti-windup architecture with that notation:

$$\left[ \begin{array}{c|c|c} A & B & E \\ \hline C & D & F \\ \hline K & L & G \end{array} \right] = \left[ \begin{array}{cccc} A_{cl} & B_{cl,v}C_{aw} & B_{cl,q} + B_{cl,v}D_{aw} & B_{cl,w} \\ 0 & A_{aw} & B_{aw} & 0 \\ \hline C_{cl,z} & D_{cl,zv}C_{aw} & D_{cl,zq} + B_{cl,zv}D_{aw} & D_{cl,zw} \\ C_{cl,u} & D_{cl,uv}C_{aw} & D_{cl,uq} + B_{cl,uv}D_{aw} & D_{cl,uw} \end{array} \right] , \tag{5.9}$$

where $C_{aw} = \left[ \begin{array}{c} C_{aw,1} \\ C_{aw,2} \end{array} \right]$ and $D_{aw} = \left[ \begin{array}{c} D_{aw,1} \\ D_{aw,2} \end{array} \right]$. Most of the examples provided in this chapter will incorporate a nonlinear $\mathcal{L}_2$ gain analysis curve determined using transformation (5.9) and the tools of Section 3.5.3.

### 5.2.2 Closed-loop representation with external anti-windup

In the external anti-windup case, the modification signals denoted by $v$ and enforced on the controller by the anti-windup compensator are not allowed to access all the controller states but can only affect the external signals, namely, the controller input and output. The controller equations (5.2) then become:

$$\mathcal{K} \quad \begin{cases} \dot{x}_c = A_c x_c + B_{c,y} (y + v_1) + B_{c,w} w \\ u = C_c x_c + D_{c,y} (y + v_1) + D_{c,w} w + v_2 . \end{cases} \tag{5.10}$$

When adopting the external architecture, the matrices $B_{cl,v}$, $D_{cl,uv}$, and $D_{cl,zv}$ in (5.6) correspond to the following values:

$$\left[ \begin{array}{c} B_{cl,v} \\ \hline D_{cl,uv} \\ \hline D_{cl,zv} \end{array} \right] = \left[ \begin{array}{cc} -B_{p,u}\Delta_u D_{c,y} & B_{p,u}\Delta_u \\ B_{c,y}\Delta_y & B_{c,y}\Delta_y D_{p,yu} \\ \Delta_u D_{c,y} & \Delta_u \\ D_{p,zu}\Delta_u D_{c,y} & D_{p,zu}\Delta_u \end{array} \right] . \tag{5.11}$$

Similar to the full-authority case, external anti-windup schemes will also be analyzed by way of the nonlinear performance estimation tools of Section 3.5.3. To this end, the matrices in (3.54) should be selected using equation (5.9). However, for the external case, the matrices $B_{cl,v}$, $D_{cl,uv}$ and $D_{cl,zv}$ must be selected as in (5.11).

Paralleling the matrices given in Chapter 4, it is useful to introduce a last dynamical system called *compact open-loop representation*, which is described by the equations

$$\begin{aligned} \dot{x}_{ol} &= A_{ol}x_{ol} + B_{ol,w}w + B_{ol,v}v_1 \\ z_{ol} &= C_{ol,z}x_{ol} + D_{ol,zw}w, \end{aligned} \tag{5.12}$$

where the input $v_1$ resembles the first signal injected by the external anti-windup action and where the matrices correspond to the following selections:

$$\left[ \begin{array}{c|c|c} A_{ol} & B_{ol,w} & B_{ol,v} \\ \hline C_{ol,z} & D_{ol,zw} \end{array} \right] = \left[ \begin{array}{cc|c|c} A_p & 0 & B_{p,w} & 0 \\ 0 & A_c & B_{c,w} & B_{c,y} \\ \hline C_{p,z} & 0 & D_{p,zw} \end{array} \right]. \tag{5.13}$$

## 5.3 FACTORING RANK-DEFICIENT MATRICES

The goal of this section is to illustrate a simple mathematical tool which is required by the algorithms discussed in this chapter. In most cases, dynamic DLAW algorithms require an intermediate step where a symmetric rank-deficient matrix $\Xi \in \mathbb{R}^{n \times n}$ of rank $n_\xi < n$ is factored as $\Xi = NN^T$, where $N \in \mathbb{R}^{n \times n_\xi}$ is a rectangular matrix having full column rank. The next procedure allows one to construct one of the infinitely many solutions to the above mentioned factorization.

**Procedure 1** (Factorization of a rank-deficient matrix)

**Step a)** Define the symmetric matrix $\Xi \in \mathbb{R}^{n \times n}$ having rank $n_\xi < n$.

**Step b)** Compute the singular value decomposition of $\Xi$, namely, a diagonal matrix $S_\xi$ having nonnegative diagonal entries with the last $n - n_\xi$ diagonal entries equal to zero, and a square matrix $U_\xi$ such that $\Xi = U_\xi S_\xi U_\xi^T$. Numerically this can be done using, for example, the MATLAB command [U,S,V] = SVD(Xi), which returns U=V whenever its argument is symmetric.

**Step c)** Define $\bar{U}_\xi = U_\xi \left[ \begin{array}{c} I_{n_\xi} \\ 0 \end{array} \right]$ as the matrix having the first $n_\xi$ columns of $U_\xi$ and define $\bar{S}_\xi = \left[ \begin{array}{cc} I_{n_\xi} & 0 \end{array} \right] S_\xi \left[ \begin{array}{c} I_{n_\xi} \\ 0 \end{array} \right]$ as the upper left $n_\xi \times n_\xi$ square block of the matrix $S_\xi$. Since only the first $n_\xi$ diagonal entries of $S_\xi$ are nonzero, then $\bar{U}_\xi \bar{S}_\xi \bar{U}_\xi^T = U_\xi S_\xi U_\xi^T = \Xi$.

**Step d)** Define $\sqrt{\bar{S}_\xi}$ as a diagonal matrix having on its diagonal the square roots of the diagonal terms of $\bar{S}_\xi$, so that $\left( \sqrt{\bar{S}_\xi} \right)^2 = \bar{S}_\xi$. Numerically this

can be done using, for example, the MATLAB command $\mathtt{sqS = sqrtm(S)}$.
Select

$$N = \bar{U}_\xi \sqrt{\bar{S}_\xi},$$

and note that $N \in \mathbb{R}^{n_\xi \times n}$ and that $NN^T = \bar{U}_\xi \sqrt{\bar{S}_\xi} \sqrt{\bar{S}_\xi} \bar{U}_\xi^T = \bar{U}_\xi \bar{S}_\xi \bar{U}_\xi^T = \Xi$,
as desired.

$\star$

## 5.4 ALGORITHMS PROVIDING GLOBAL GUARANTEES

In this section constructive techniques for the selection of the dynamic anti-windup
compensator matrices in (5.3) are given, with the goal of guaranteeing global guarantees on the internal stability and input-output gain of the resulting closed-loop
system. This section should be then understood as the dynamic counterpart of
the previous Section 4.3 related to static DLAW. Similar to the static case, exponential stability of the plant is a necessary condition for dynamic DLAW with
global guarantees to be applicable, while to deal with non-exponentially stable
plants the reader should refer to the regional algorithms reported in the subsequent
Section 5.5. An appealing result that characterizes the dynamic DLAW techniques
is that whenever the plant is exponentially stable, there always exists a dynamic
anti-windup compensator of the same order as the plant that solves the global problem. Therefore, dynamic anti-windup compensation constitutes a step forward as
compared to static compensation techniques, where suitable feasibility conditions
were required to hold for the algorithms to be applicable.

### 5.4.1 Global full-authority plant-order augmentation

Full-authority linear dynamic anti-windup compensation synthesis amounts to selecting the matrices $(A_{aw}, B_{aw}, C_{aw,1}, D_{aw,1}, C_{aw,2}, D_{aw,2})$ of the anti-windup compensator (5.3) corresponding to the block $\mathcal{F}$ in Figure 5.1 when the signal $v$ acts
both on the state and on the output equations of the unconstrained controller. An
algorithm to construct a plant-order version of such a full-authority anti-windup
compensation is offered in this section. Reduced-order design will be addressed in
the following section.

**Algorithm 4** (Dynamic plant-order full-authority global DLAW)

| Applicability | | | | Architecture | | | Guarantee |
|---|---|---|---|---|---|---|---|
| Exp Stab | Marg Stab | Marg Unst | Exp Unst | Lin/ NonL | Dyn/ Static | Ext/ FullAu | Global/ Regional |
| $\checkmark$ | | | | L | D | FA | G |

> *Comments*: Dynamic anti-windup, with state dimension equal to that of the plant, overcomes the applicability limitations of Algorithm 1. Feasible for any loop containing an exponentially stable plant. Global input-output gain is optimized; it is never worse, and often better, than the input-output gain provided by Algorithm 1.

**Step 1.** Given a plant and a controller of the form (5.1), (5.2), construct the matrices of the compact closed-loop representation using (5.5) and (5.8). Select a constant $v \in [0, 1)$ used to enforce a strong well-posedness constraint (see Section 3.4.2 on page 60).

**Step 2.** Find the optimal solution $(R_{11}, S, \gamma) \in \mathbb{R}^{n_p \times n_p} \times \mathbb{R}^{n_{cl} \times n_{cl}} \times \mathbb{R}$ to the following LMI eigenvalue problem which is always feasible for exponentially stable plants:

$$\min_{R_{11}, S, \gamma} \gamma \ subject \ to:$$

*Positivity*:

$$R_{11} = R_{11}^T > 0,$$

$$S = S^T = \begin{bmatrix} S_{11} & S_{12} \\ S_{12}^T & S_{22} \end{bmatrix} > 0,$$

$$R_{11} - S_{11} > 0,$$

*Open-loop condition*:

$$\begin{bmatrix} R_{11}A_p^T + A_p R_{11} & B_{p,w} & R_{11}C_{p,z}^T \\ B_{p,w}^T & -\gamma I_{n_w} & D_{p,zw}^T \\ C_{p,z}R_{11} & D_{p,zw} & -\gamma I_{n_z} \end{bmatrix} < 0,$$

*Closed-loop condition*:

$$\begin{bmatrix} SA_{cl}^T + A_{cl}S & B_{cl,w} & SC_{cl,z}^T \\ B_{cl,w}^T & -\gamma I_{n_w} & D_{cl,zw}^T \\ C_{cl,z}S & D_{cl,zw} & -\gamma I_{n_z} \end{bmatrix} < 0.$$

**Step 3.** Based on the solution determined at the previous step, construct the symmetric matrix $R = \begin{bmatrix} R_{11} & S_{12} \\ S_{12}^T & S_{22} \end{bmatrix}$ and define $Q_{12} \in \mathbb{R}^{n_{cl} \times n_p}$ as any solution to the following equation:

$$Q_{12}Q_{12}^T = RS^{-1}R - R, \tag{5.14}$$

which is always solvable and admits infinitely many solutions because by construction, $\text{rank}(RS^{-1}R - R) = \text{rank}((RS^{-1}R - R)(R^{-1}S)) = \text{rank}(R - S) = n_p$. A solution can be determined following Procedure 1 on page 113 with $\Xi = RS^{-1}R - R$. Then define the following matrices:

$$Q_{22} = I_{n_p} + Q_{12}^T R^{-1} Q_{12}$$

$$Q = \begin{bmatrix} R & Q_{12} \\ Q_{12}^T & Q_{22} \end{bmatrix}.$$

**Step 4.** Based on the matrices determined at steps 1 and 3, define the following matrices (where the second row and column of zeros have size $n_p$):

$$
\left[
\begin{array}{c|c|c}
A_0 & B_{0,q} & B_{0,w} \\ \hline
C_{0,u} & D_{cl,yq} & D_{cl,yw} \\
C_{0,z} & D_{cl,zq} & D_{cl,zw}
\end{array}
\right]
=
\left[
\begin{array}{cc|cc}
A_{cl} & 0 & B_{cl,q} & B_{cl,w} \\
0 & 0_{n_p \times n_p} & 0 & 0 \\ \hline
C_{cl,u} & 0 & D_{cl,uq} & D_{cl,uw} \\
C_{cl,z} & 0 & D_{cl,zq} & D_{cl,zw}
\end{array}
\right]
$$

and construct the matrices $G_U \in \mathbb{R}^{(n_p+n_u)\times m}$, $H \in \mathbb{R}^{m\times(n_p+n_c+n_u)}$, and $\Psi \in \mathbb{R}^{m\times m}$, with $m := n_{cl}+n_p+n_u+n_w+n_z$, as follows:

$$
\Psi(U,\gamma) = \mathrm{He}
\begin{bmatrix}
A_0 Q & B_{0,q}U + Q C_{0,u}^T & B_{0,w} & Q C_{0,z}^T \\
0 & D_{cl,uq}U - U & D_{cl,uw} & U D_{cl,zq}^T \\
0 & 0 & -\frac{\gamma}{2}I & D_{cl,zw}^T \\
0 & 0 & 0 & -\frac{\gamma}{2}I
\end{bmatrix},
$$

$$
H =
\left[
\begin{array}{cc|cc|c}
0 & I_{n_p} & 0 & 0_{n_p \times n_w} & 0 \\
B_{cl,v}^T & 0 & D_{cl,uv}^T & 0 & D_{cl,zv}^T
\end{array}
\right]
$$

$$
G_U =
\left[
\begin{array}{cc|c|c|c}
Q_{12}^T & Q_{22} & 0 & 0 & 0 \\
0 & 0 & I_{n_u} & 0 & 0
\end{array}
\right],
$$

where for any square matrix $X$, $\mathrm{He}(X) = X + X^T$.

**Step 5.** Find the optimal solution $(\Lambda_U, U, \gamma)$ of the LMI eigenvalue problem:

$$
\min_{\Lambda_U, U, \gamma} \gamma \ \text{subject to:}
$$

*Strong well-posedness* (if $v = 0$, then redundant):

$$
-2(1-v)U + \mathrm{He}\left( D_{cl,uq}U + \left[\ 0_{n_u \times (n_p+n_c)} \ \ I_{n_u}\ \right]\Lambda_U \begin{bmatrix} 0_{n_u \times n_p} \\ I_{n_u} \end{bmatrix} \right) < 0,
$$

*Anti-windup gains LMI*:

$$
\Psi(U,\gamma) + G_U^T \Lambda_U^T H^T + H \Lambda_U G_U < 0.
$$

**Step 6.** Select the plant-order full-authority anti-windup compensator as:

$$
\dot{x}_{aw} = A_{aw} x_{aw} + B_{aw} q
$$
$$
v = C_{aw} x_{aw} + D_{aw} q
$$

with the selection $\left[\begin{array}{c|c} A_{aw} & B_{aw} \\ \hline C_{aw} & D_{aw} \end{array}\right] = \Lambda_U \left[\begin{array}{c|c} I_{n_p} & 0 \\ \hline 0 & U^{-1} \end{array}\right].$

$\star$

### 5.4.1.1 Case studies

The plant-order DLAW construction of Algorithm 4 will be now applied to the five examples used in Chapter 4 to illustrate global static DLAW designs. Since all of these examples are characterized by exponentially stable plants, plant-order anti-windup will be applicable to all of them. In some cases, the resulting scheme will lead to improved performance as compared to that induced by the static compensators. In some other cases, dynamic anti-windup will be mandatory because the static approach was not feasible. However, for a few of those examples, it will be evident that dynamic anti-windup leads to comparable responses to those obtained with the static compensation architecture. For these examples it will not be worth the extra computational effort required to implement a dynamic DLAW scheme and the static one will be the preferred solution.

The first example is the MIMO academic example no. 2. For this example, plant-order full-authority anti-windup behaves no better than the static anti-windup results illustrated in Example 4.3.2 for the static full-authority DLAW case and in Example 4.3.6 for the static external DLAW case.

**Example 5.4.1** (MIMO academic example no. 2) Consider the MIMO academic example whose data is given in Example 4.3.2 on page 85. Constructing a plant-order anti-windup compensator is of no use as highlighted above, but it is instructive to focus on the use of Algorithm 4 and the use of tricks to reduce the size of the resulting matrices, as discussed in Section 3.7.2. Indeed, when running Algorithm 4 as is, the achievable performance is $\gamma = 1.55$, which is roughly the same as what can be obtained with static anti-windup. However, the size of the entries in the anti-windup matrices obtained at step 6 are roughly $10^6$ due to some numerical issue with the LMI optimizer. More specifically, the matrices obtained are:

$$
\left[\begin{array}{c|c} A_{aw} & B_{aw} \\ \hline C_{aw} & D_{aw} \end{array}\right] = \left[\begin{array}{cc|cc} -2.5 \cdot 10^6 & 1.678 \cdot 10^6 & 3.611 & 4.429 \\ -1.65 \cdot 10^6 & -7.47 \cdot 10^6 & 2.246 & 3.218 \\ \hline -0.0138 & 0.0373 & -7.7 \cdot 10^{-6} & -9.26 \cdot 10^{-6} \\ -0.0183 & -0.0252 & -9.63 \cdot 10^{-6} & -1.23 \cdot 10^{-5} \\ 7.79 \cdot 10^5 & -5.537 \cdot 10^5 & -0.1256 & -1.379 \\ 6.2 \cdot 10^5 & -3.769 \cdot 10^5 & -0.8955 & -0.1005 \end{array}\right].
$$

The resulting simulation is unacceptably slow and reveals the unsuitability of this compensator. The numerical problem is easily solved by following the steps outlined in Section 3.7.2. In particular, the following two inequalities are added to the LMIs at step 5:

$$
U > I, \quad \left[\begin{array}{cc} \kappa I & \Lambda_U \\ \Lambda_U^T & \kappa I \end{array}\right] > 0, \tag{5.15}
$$

with $\kappa = 10$ (note that choosing $\kappa$ too small will result in infeasibility of the LMIs at step 5). The resulting compensator induces a performance level $\gamma = 1.56$ (slightly deteriorated) but the anti-windup compensator matrices become

$$
\left[\begin{array}{c|c} A_{aw} & B_{aw} \\ \hline C_{aw} & D_{aw} \end{array}\right] = \left[\begin{array}{cc|cc} -0.4724 & 0.03655 & -0.001020 & -0.005120 \\ 0.03636 & -9.741 & -0.007586 & 0.03723 \\ \hline -0.01129 & -0.02448 & -0.2185 & -0.9943 \\ -0.01481 & 0.02094 & -0.2731 & -1.326 \\ 0.03562 & -0.05718 & -0.1174 & -5.292 \\ 0.02746 & 0.04525 & -0.8721 & -3.135 \end{array}\right],
$$

much more reasonable from a numerical viewpoint. As anticipated above, the simulation results and the nonlinear $\mathcal{L}_2$ gain curve obtained from this plant-order compensator coincide with the static anti-windup solution of Example 4.3.2; therefore the static solution is preferred to this one due to its simpler architecture. ☐

The next example is the passive electrical circuit introduced in Example 4.3.3 and revisited in Example 4.3.7. For this example, the extra degrees of freedom available in the plant-order anti-windup scheme lead to improved performance and improved closed-loop responses.

**Example 5.4.2** (Electrical network) Consider the electrical network studied in Examples 4.3.3 and 4.4.3 and represented in Figure 4.6 on page 87 of the previous chapter.

For this example the global guarantees given by the static Algorithm 1 in Example 4.3.3 come at the price of a sluggish response (see Figure 4.3.3 on page 87). When moving into dynamic plant-order anti-windup, the response is much improved and the global guarantees are still at hand. Indeed, by selecting the parameters $\nu = 0.01$ and imposing the extra bound (5.15) with $\kappa = 1000$ on the size of the anti-windup matrices, Algorithm 4, provides the following anti-windup compensator:

$$\left[\begin{array}{c|c} A_{aw} & B_{aw} \\ \hline C_{aw} & D_{aw} \end{array}\right] = \left[\begin{array}{ccc|c} -10.22 & -6.257 & -0.101 & -0.05051 \\ 0.3395 & -0.3758 & -0.00833 & -0.02568 \\ -9.42\cdot 10^{-6} & -0.000463 & -577.3 & 0.000438 \\ -0.3003 & 0.8997 & -0.1456 & -0.01799 \\ -7.37\cdot 10^{-5} & -3.67\cdot 10^{-5} & 0.00179 & 7.47\cdot 10^{-5} \\ -24.50 & 71.77 & 1.263 & 0.04151 \end{array}\right],$$

which guarantees a global $\mathcal{L}_2$ performance level $\gamma = 58.428$ (much improved as compared to the value 85.22 obtained with static compensation). With this anti-windup compensator, the closed-loop response corresponds to the curves of Figure 5.2 where a significant improvement can be appreciated as compared to the static responses of Example 4.3.3.

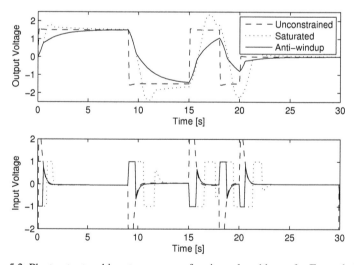

Figure 5.2 Plant output and input responses of various closed loops for Example 5.4.2.

It should be emphasized that an even more desirable response had been obtained when using regional static anti-windup compensation in Example 4.4.3 on page 105. However, the regional compensation did not provide guaranteed global $\mathcal{L}_2$ gain and the resulting global exponential stability property. This is guaranteed instead by the (possibly more conservative) compensation resulting from Algorithm 4.

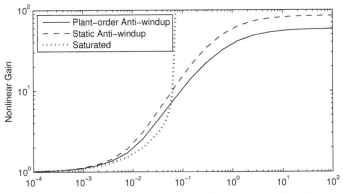

Figure 5.3 Nonlinear gains for Examples 4.3.3 and 5.4.2 characterizing the system before and after static and plant-order anti-windup augmentation.

Figure 5.3 shows the nonlinear gain curves corresponding to no anti-windup (dotted), static anti-windup (dashed), and plant-order anti-windup (solid). Clearly, the global $\mathcal{L}_2$ gain obtained from plant-order anti-windup is improved as compared to the static condition because the optimizer has extra degrees of freedom. For this example, the whole $\mathcal{L}_2$ gain curve associated with plant-order compensation is improved as compared to the static one. Therefore, improved responses on all signal ranges are to be expected (within the limits of the conservativeness of the quadratic $\mathcal{L}_2$ performance bound). □

The following example has been introduced in Example 4.3.4 where it was shown that static anti-windup design is not feasible (using either the full-authority or the external architecture). Plant-order anti-windup is feasible and provides a desirable solution to this windup problem.

**Example 5.4.3** (SISO academic example no. 2) Consider the SISO closed-loop system described in Example 4.3.4 on page 89. It was illustrated in that example that static anti-windup with global guarantees is not feasible. Conversely, Algorithm 4 always provides a feasible solution as long as the plant is exponentially stable and applying the algorithm with $v = 0.5$ and limiting the anti-windup matrices as in Example 5.4.1 (see equation (5.15)) with $\kappa = 5$, the following plant-order compensator is obtained:

$$\left[ \begin{array}{c|c} A_{aw} & B_{aw} \\ \hline C_{aw} & D_{aw} \end{array} \right] = \left[ \begin{array}{cc|c} -0.1962 & -0.009284 & -2.104 \\ 4.538 \cdot 10^{-5} & -2.838 & 0.007315 \\ \hline 0.6358 & -0.001821 & -0.9282 \\ 1.397 & 0.009778 & -0.01534 \end{array} \right],$$

which guarantees a global $\mathcal{L}_2$ gain $\gamma = 4.466$ and gives the desirable response represented in Figure 5.4.

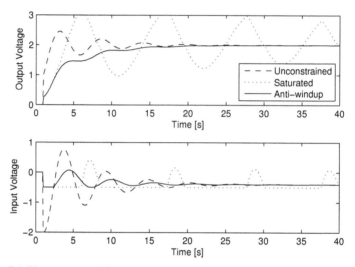

Figure 5.4 Plant output and input responses of various closed loops for Example 5.4.3.

Note that the anti-windup response of Figure 5.4 is pretty comparable to that one obtained using regional static anti-windup compensation as illustrated in Example 4.4.2 and in Figure 4.16 on page 105. However, the regional static design appears to be more aggressive than the global dynamic one, which is reasonable based on the fact that the aggressive regional design is not associated with any global asymptotic stability guarantee, while the less aggressive global one guarantees closed-loop global exponential stability.

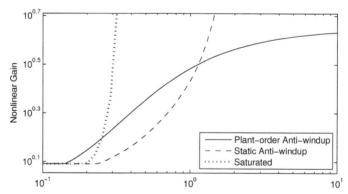

Figure 5.5 Nonlinear gains for Examples 4.4.2 and 5.4.3 characterizing the system before and after regional static and global plant-order anti-windup augmentation.

This fact is especially evident in Figure 5.5, where three nonlinear gains are compared to each other. The dotted curve represents the closed loop without anti-windup, the dashed one is the static regional anti-windup curve of Example 4.4.2 (see also Figure 4.17 on page 105), which enlarges the guaranteed stability region but still doesn't provide global properties, and the solid one is the plant-order anti-windup curve using the compensation described here. It is quite interesting also the fact that the nonlinear gains behavior for small signals suggests that the closed loop with plant-order anti-windup performs worse than no anti-windup. This is consistent with simulations involving smaller step references ($r = 0.2$) than that of Figure 5.4, where the saturated response appears closer to the unconstrained one.

□

In the next example, the longitudinal dynamics of an F8 aircraft introduced in Section 1.2.3 and revisited in Example 4.3.5 on page 90 is considered. In particular, for this system it is shown here that plant-order full-authority anti-windup can compensate for the unpleasant bias experienced in Example 4.3.5 when using static compensation.

**Example 5.4.4**   (F8 longitudinal dynamics) Consider the problem statement formally described and introduced in Example 4.3.5 on page 90. It was shown in that example (see in particular the simulations of Figure 4.9 on page 91) that static anti-windup could only lead to a partial solution of the windup problem. Indeed, in that simulation the anti-windup response exhibited a very slow convergence to the desired output value. It was also noted in Example 4.3.8 on page 97 later in that same chapter that external static anti-windup is not feasible for this study case.

When using plant-order anti-windup compensation, this problem can be overcome. In particular, by running Algorithm 4 with the same exact parameters as in Example 4.3.5 and enforcing the extra constraint (5.15) on the anti-windup compensator size with $\kappa = 8000$, the following anti-windup compensator is obtained:

$$A_{aw} = \begin{bmatrix} -0.7550 & 0.1290 & -10.39 & -0.1421 \\ 1.622 & 2.189 & -228.2 & 0.1980 \\ 0.2895 & 1.996 & -163.0 & -0.07277 \\ -0.0009871 & -0.0004719 & -0.002595 & -5774. \end{bmatrix}$$

$$B_{aw} = \begin{bmatrix} 34.16 & -20.85 \\ 547.6 & -493.6 \\ 404.2 & -349.5 \\ -0.0005585 & -0.001980 \end{bmatrix}$$

$$C_{aw} = \begin{bmatrix} 0.2514 & -0.1249 & -2.770 & 1.275 \\ -0.9223 & 0.2960 & 15.65 & -1.699 \\ -0.06782 & 0.2135 & 0.8155 & 0.6637 \\ -1266. & -1248. & 291.4 & 0.05931 \\ -0.3751 & 0.6115 & 0.5751 & -0.07910 \\ -1.624 & 0.6336 & -0.8472 & -0.01274 \\ 0.1635 & -0.06422 & -2.740 & 1.223 \\ -1.318 & 1.226 & 16.42 & -2.386 \\ -0.0002190 & -0.09643 & 11.29 & 0.005365 \\ -0.09764 & 1.327 & -91.86 & -0.2869 \end{bmatrix}$$

$$D_{aw} = \begin{bmatrix} -163.0 & 0.6179 \\ 38.27 & 5.611 \\ 5.787 & 0.7120 \\ 8481. & 105.9 \\ -6.192 & -1.218 \\ -9.293 & 0.3099 \\ -162.8 & 0.07452 \\ 42.33 & -3.657 \\ -27.54 & 24.51 \\ 31.6 & -197.9 \end{bmatrix}.$$

This compensator achieves global performance level $\gamma = 25.476$, clearly improved with respect to the one achievable by static anti-windup in Example 4.3.5, which was $\gamma = 26.44$.

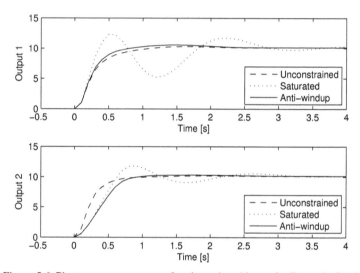

Figure 5.6  Plant output responses of various closed loops for Example 5.4.4.

Figure 5.6 represents the output response of the system with plant-order compensation. This should be compared to Figure 4.9 on page 91 representing the same response using static anti-windup. The new simulation is much improved because the slow convergence has been significantly suppressed by the new dynamic anti-windup solution. This increased quality of the dynamic anti-windup response is also confirmed by the comparison of the nonlinear $\mathcal{L}_2$ gain curves in Figure 5.7 where the plant-order solution (solid curve) lies below the static solution (dashed curve) on the whole disturbance input size range.                                                                    □

### 5.4.2 Global full-authority reduced-order augmentation

In Sections 4.3.1 and 5.4.1, respectively, design algorithms for static and plant-order full-authority DLAW augmentation with global guarantees have been given.

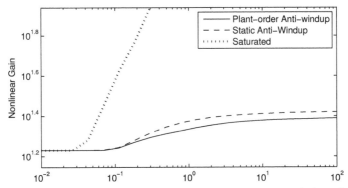

Figure 5.7 Nonlinear gains for Example 5.4.4 characterizing the system before (dotted) and after static (dashed) and plant-order (solid) anti-windup augmentation.

In particular, the construction procedures correspond to Algorithms 1 and 4, respectively. Is has also been emphasized that plant-order DLAW augmentation is always feasible for exponentially stable plants, whereas static augmentation is only possible provided certain feasibility conditions are satisfied by the problem data. It turns out that, as suggested by their similarity, both the abovementioned algorithms are special cases of a general result of linear full-authority anti-windup augmentation where the order of the compensator is any positive integer $n_{aw}$. However, for generic values of $n_{aw}$, the corresponding anti-windup construction cannot be formulated in terms of LMIs, as it is the case for Algorithms 1 and 4, and the corresponding anti-windup design becomes more complex. For a fixed anti-windup compensator order $n_{aw}$ (with $n_{aw} \neq 0$ and $n_{aw} \neq n_p$), what makes the generalized algorithm difficult to solve is determining the solution $R, S, \gamma$ at the first optimization step. All the remaining steps are roughly unchanged. The first optimization step generalizes the LMI optimization problem at step 3 of Algorithm 1 and the LMI optimization problem at step 2 of Algorithm 4. Given any integer value $n_{aw} \geq 0$, this generalization can be written as follows:

$$\min_{R_{11}, S, \gamma} \gamma \ subject \ to: \tag{5.16}$$

*Positivity*:

$$R_{11} = R_{11}^T > 0 ,$$

$$S = S^T = \begin{bmatrix} S_{11} & S_{12} \\ S_{12}^T & S_{22} \end{bmatrix} > 0 ,$$

$$R_{11} - S_{11} \geq 0 ,$$

*Nonlinear rank condition*:

$$\mathrm{rank}(R_{11} - S_{11}) \leq n_{aw},$$

*Open-loop condition*:

$$\begin{bmatrix} R_{11}A_p^T + A_p R_{11} & B_{p,w} & R_{11}C_{p,z}^T \\ B_{p,w}^T & -\gamma I_{n_w} & D_{p,zw}^T \\ C_{p,z}R_{11} & D_{p,zw} & -\gamma I_{n_z} \end{bmatrix} < 0 ,$$

*Closed-loop condition*:

$$\begin{bmatrix} SA_{cl}^T + A_{cl}S & B_{cl,w} & SC_{cl,z}^T \\ B_{cl,w}^T & -\gamma I_{n_w} & D_{cl,zw}^T \\ C_{cl,z}S & D_{cl,zw} & -\gamma I_{n_z} \end{bmatrix} < 0 \,.$$

The crucial constraint in the optimization problem (5.16) is the nonlinear rank constraint, $\text{rank}(R_{11} - S_{11}) \le n_{aw}$. It is easy to verify that for $n_{aw} = 0$ (static anti-windup), the constraint corresponds to requiring $R = S$, and (5.16) becomes linear and reduces to the LMI optimization reported at step 3 of Algorithm 1. Similarly, if $n_{aw} \ge n_p$, the nonlinear rank condition is automatically implied by the fact that the rank of the matrix $R_{11} - S_{11}$ cannot be larger than its size $n_p$. For any other selection of $n_{aw} \ge 0$, the problem (5.16) remains nonlinear and hard to solve.

Understanding the problem behind the optimization problem (5.16), however, allows one to introduce a new design tool which is useful whenever full-authority static anti-windup augmentation is not feasible or leads to unsatisfactory performance. In this case, it is of interest to look for "reduced-order" anti-windup augmentation with a certain performance guarantee $\gamma_G$, namely, to search for a minimized value of $n_{aw}$ such that the constraints (5.16) have a solution with $\gamma = \gamma_G$. The reduced-order DLAW problem can be addressed through an approximate approach arising from the observation that since $R_{11} - S_{11} \ge 0$, the trace of $R_{11} - S_{11}$ (namely, the sum of all the diagonal entries of $R_{11} - S_{11}$) is proportional to the size of $R - S$. Therefore, minimizing the trace of $R_{11} - S_{11}$ leads, in general, to a rank-deficient result whose rank will be the order of the anti-windup compensator. The optimization problem (5.16) can then be solved by removing the nonlinear rank constraint, replacing $\gamma$ by the constant $\gamma_G$ and minimizing the new cost variable $\text{trace}(R_{11} - S_{11})$. The value $\gamma_G$ can be selected as $\alpha \gamma_{n_p}$, where $\gamma_{n_p}$ is the achievable gain by plant-order anti-windup augmentation and $\alpha > 1$ is a constant parameter quantifying how much performance the designer is willing to give up to reduce the dimension $n_{aw}$ of the anti-windup compensator matrices. Note, however, that the final rank-deficient solution may need to give up a little on the pre-assigned guaranteed performance level $\gamma_G$ because a suitable correction on $S_{11}$ is required to make $R_{11} - S_{11}$ a rank-deficient matrix.

In the numerical implementation of this direct linear anti-windup algorithm, the constraint $R_{11} - S_{11} \ge 0$ will actually be replaced by the strict inequality constraint $R_{11} - S_{11} > 0$, and the solution $R_{11} - S_{11}$ will be very close to being rank deficient. Therefore, it will be useful to use singular value decomposition (SVD) to determine a new selection of $S$ for which $R_{11} - S_{11} \ge 0$ actually holds. Since the $S$ matrix is slightly changed by this modification, it is mandatory to verify that the new selection of $S$ still satisfies the closed-loop condition in (5.16). To this end, anticipating for such a perturbation, it is useful to add a small feasibility margin to the open closed-loop condition to make it robust to these perturbations (at least to a certain extent). The resulting constraint will be

$$\begin{bmatrix} SA_{cl}^T + A_{cl}S & B_{cl,w} & SC_{cl,z}^T \\ B_{cl,w}^T & -\gamma I_{n_w} & D_{cl,zw}^T \\ C_{cl,z}S & D_{cl,zw} & -\gamma I_{n_z} \end{bmatrix} + \varepsilon I < 0 \,,$$

where $\varepsilon > 0$ is a small constant. The complete procedure for the selection $R_{11}, S, \gamma$

characterizing a minimal order anti-windup compensator requires some tuning for $\varepsilon$ and $\gamma_G$ but is typically quite straightforward to apply. The main steps of the procedure are reported next.

**Procedure 2** (Computation of an $(R, S, \gamma)$ solution for reduced-order full-authority anti-windup with guaranteed performance)

**Step a)** A plant and a controller of the form (5.1), (5.2) and the matrices of the compact closed-loop representation constructed based on equations (5.5) and (5.8) are given. A small positive constant $\varepsilon$ to robustify the closed-loop LMI condition and a value $\gamma_G$ characterizing the desired guaranteed performance are also given. Note that the actual performance $\gamma^*$ achieved by the minimal-order $(R, S, \gamma)$ solution will be, in general, larger than $\gamma_G$ but often very close to it.

**Step b)** Find the optimal solution $(R_{11}, S) \in \mathbb{R}^{n_p \times n_p} \times \mathbb{R}^{n_{cl} \times n_{cl}}$ to the following LMI eigenvalue problem:

$$\min_{R_{11}, S} \text{trace}(R_{11} - S_{11}) \quad subject\ to:$$

*Positivity*:

$$R_{11} = R_{11}^T > 0 ,$$

$$S = S^T = \begin{bmatrix} S_{11} & S_{12} \\ S_{12}^T & S_{22} \end{bmatrix} > 0 ,$$

$$R_{11} - S_{11} > 0 ,$$

*Open-loop condition*:

$$\begin{bmatrix} R_{11}A_p^T + A_p R_{11} & B_{p,w} & R_{11}C_{p,z}^T \\ B_{p,w}^T & -\gamma_G I_{n_w} & D_{p,zw}^T \\ C_{p,z}R_{11} & D_{p,zw} & -\gamma_G I_{n_z} \end{bmatrix} < 0 ,$$

*Closed-loop condition*:

$$\begin{bmatrix} SA_{cl}^T + A_{cl}S & B_{cl,w} & SC_{cl,z}^T \\ B_{cl,w}^T & -\gamma_G I_{n_w} & D_{cl,zw}^T \\ C_{cl,z}S & D_{cl,zw} & -\gamma_G I_{n_z} \end{bmatrix} + \varepsilon I < 0 .$$

**Step c)** Compute the SVD of the symmetric matrix $R_{11} - S_{11}$ determined from the solution of the previous step, namely, compute $F, \Lambda$ such that $F\Lambda F^T = R_{11} - S_{11}$, where $\Lambda$ is a positive semidefinite diagonal matrix whose diagonal entries are nonincreasing (for example, use the MATLAB command [U,Lambda,V] = svd(R11-S11); F = (U+V)/2;). Set $n_{aw} = n_p$.

**Step d)** Call $\lambda_1, \ldots \lambda_{n_{aw}-1}$ the first $n_{aw} - 1$ entries on the diagonal of $\Lambda$ and define $\hat{\Lambda} := \text{diag}\{\lambda_1, \ldots, \lambda_{n_{aw}-1}, 0, \ldots, 0\}$, namely, $\hat{\Lambda}$ is a diagonal matrix having its first $n_{aw} - 1$ diagonal entries equal to the entries of $\Lambda$ and the remaining entries equal to zero. Also define $\hat{S}_{11} := S_{11} + F(\Lambda - \hat{\Lambda})F^T$.

**Step e)** Given the matrix $\hat{S} := \begin{bmatrix} \hat{S}_{11} & S_{12} \\ S_{12}^T & S_{22} \end{bmatrix}$, where $\hat{S}_{11}$ was determined at the previous step, verify the feasibility of the following LMI in the free variable $\gamma$:

$$
\begin{bmatrix}
\hat{S}A_{cl}^T + A_{cl}\hat{S} + \varepsilon I & B_{cl,w} & \hat{S}C_{cl,z}^T \\
B_{cl,w}^T & -\gamma I_{n_w} & D_{cl,zw}^T \\
C_{cl,z}\hat{S} & D_{cl,zw} & -\gamma I_{n_z}
\end{bmatrix} < 0.
$$

If the LMI is feasible, then set $(R_{11}^*, S^*, \gamma^*) := (R_{11}, \hat{S}, \gamma)$, set $n_{aw} = n_{aw} - 1$, and go to step d. If the LMI is not feasible, then go to step f.

**Step f)** Select the order of the anti-windup compensator as $n_{aw}$ and the $(R, S, \gamma)$ solution as $(R_{11}^*, S^*, \gamma^*)$.

$\star$

The performance level induced by the reduced-order anti-windup compensator is in general larger (namely, worse) than the performance $\gamma_{n_p}$ induced by the plant-order anti-windup design. This is true because of two reasons: first, the guaranteed performance $\gamma_G$ will have to be larger than $\gamma_{n_p}$, in general, to allow for satisfactory reductions of $n_{aw}$; second, due to the fact that the matrix $\hat{S}$ defined at step 5 is different from the original matrix $S$, the optimization at that same step might lead to a slightly deteriorated performance level. It has, however, been noted in several case studies that the value of $\gamma^*$ is not too different from $\gamma_G$.

Once a triple $(R_{11}^*, S^*, \gamma^*)$ satisfying (5.16) with a small enough $n_{aw}$ is determined from Procedure 2, the reduced-order anti-windup construction can be carried out following a simple generalization of the previous Algorithms 1 and 4, which is reported next.

**Algorithm 5** (Dynamic reduced-order full-authority global DLAW)

| Applicability | | | | Architecture | | | Guarantee |
|---|---|---|---|---|---|---|---|
| Exp Stab | Marg Stab | Marg Unst | Exp Unst | Lin/ NonL | Dyn/ Static | Ext/ FullAu | Global/ Regional |
| $\checkmark$ | | | | L | D* | FA | G |

*Comments*: Useful when static anti-windup is infeasible or performs poorly yet a low-order anti-windup augmentation is desired. Order reduction is carried out while maintaining a prescribed global input-output gain.
\* Generally of low order resulting from order reduction.

**Step 1.** Given a plant and a controller of the form (5.1), (5.2), construct the matrices of the compact closed-loop representation using (5.5) and (5.8). Select a small positive constant $\varepsilon$ to robustify the order minimization procedure and a constant $\alpha > 1$ characterizing the amount of performance to be traded in for order reduction. Select a constant $v \in [0, 1)$ used to enforce a strong well-posedness constraint (see Section 3.4.2 on page 60).

**Step 2.** Find the optimal solution $(R_{11}, S, \gamma_{n_p}) \in \mathbb{R}^{n_p \times n_p} \times \mathbb{R}^{n_{cl} \times n_{cl}} \times \mathbb{R}$ to the following LMI eigenvalue problem which is always feasible for exponentially stable plants:

$$\min_{R11,S,\gamma} \gamma \ \ subject \ to:$$

*Positivity*:

$$R_{11} = R_{11}^T > 0 \,,$$

$$S = S^T = \left[ \begin{array}{cc} S_{11} & S_{12} \\ S_{12}^T & S_{22} \end{array} \right] > 0 \,,$$

$$R_{11} - S_{11} > 0 \,,$$

*Open-loop condition*:

$$\left[ \begin{array}{ccc} R_{11}A_p^T + A_p R_{11} & B_{p,w} & R_{11}C_{p,z}^T \\ B_{p,w}^T & -\gamma I_{n_w} & D_{p,zw}^T \\ C_{p,z}R_{11} & D_{p,zw} & -\gamma I_{n_z} \end{array} \right] < 0 \,,$$

*Closed-loop condition*:

$$\left[ \begin{array}{ccc} SA_{cl}^T + A_{cl}S & B_{cl,w} & SC_{cl,z}^T \\ B_{cl,w}^T & -\gamma I_{n_w} & D_{cl,zw}^T \\ C_{cl,z}S & D_{cl,zw} & -\gamma I_{n_z} \end{array} \right] < 0 \,.$$

The resulting performance level $\gamma_{n_p}$ corresponds to the maximum performance achievable by plant-order anti-windup augmentation.

**Step 3.** Based on the matrices and parameters determined at step 1, select the desired guaranteed performance as $\gamma_G = \alpha \gamma_{n_p}$ and determine the reduced compensator order $n_{aw}$ and the triple $\left( R_{11}^*, S^* = \left[ \begin{array}{cc} S_{11}^* & S_{12}^* \\ (S_{12}^*)^T & S_{22}^* \end{array} \right], \gamma^* \right)$ from Procedure 2.

**Step 4.** Based on the solution determined at the previous step, construct the symmetric matrix $R = \left[ \begin{array}{cc} R_{11}^* & S_{12}^* \\ (S_{12}^*)^T & S_{22}^* \end{array} \right]$, and define $Q_{12} \in \mathbb{R}^{n_{cl} \times n_{aw}}$ as any solution to the following equation:

$$Q_{12}Q_{12}^T = RS^{-1}R - R,$$

which is always solvable and admits infinitely many solutions because by construction $\mathrm{rank}(RS^{-1}R - R) = \mathrm{rank}((RS^{-1}R - R)(R^{-1}S)) = \mathrm{rank}(R - S) = n_{aw}$. A solution can be determined following Procedure 1 on page 113 with $\Xi = RS^{-1}R - R$. Then define the following matrices:

$$Q_{22} = I_{n_{aw}} + Q_{12}^T R^{-1} Q_{12}$$

$$Q = \left[ \begin{array}{cc} R & Q_{12} \\ Q_{12}^T & Q_{22} \end{array} \right].$$

**Step 5.** Based on the matrices determined at steps 1 and 4, define the following matrices (where the second row and column of zeros have size $n_{aw}$):

$$\left[ \begin{array}{c|cc} A_0 & B_{0,q} & B_{0,w} \\ \hline C_{0,u} & D_{cl,yq} & D_{cl,yw} \\ C_{0,z} & D_{cl,zq} & D_{cl,zw} \end{array} \right] = \left[ \begin{array}{cc|cc} A_{cl} & 0 & B_{cl,q} & B_{cl,w} \\ 0 & 0_{n_{aw} \times n_{aw}} & 0 & 0 \\ \hline C_{cl,u} & 0 & D_{cl,uq} & D_{cl,uw} \\ C_{cl,z} & 0 & D_{cl,zq} & D_{cl,zw} \end{array} \right]$$

and construct the matrices $G_U \in \mathbb{R}^{(n_{aw}+n_u) \times m}$, $H \in \mathbb{R}^{m \times (n_{aw}+n_c+n_u)}$, and $\Psi \in \mathbb{R}^{m \times m}$, with $m := n_{cl} + n_{aw} + n_u + n_w + n_z$, as follows:

$$\Psi(U,\gamma) = \mathrm{He} \begin{bmatrix} A_0 Q & B_{0,q} U + Q C_{0,u}^T & B_{0,w} & Q C_{0,z}^T \\ 0 & D_{cl,uq} U - U & D_{cl,uw} & U D_{cl,zq}^T \\ 0 & 0 & -\frac{\gamma}{2} I & D_{cl,zw}^T \\ 0 & 0 & 0 & -\frac{\gamma}{2} I \end{bmatrix},$$

$$H = \begin{bmatrix} 0 & I_{n_{aw}} & 0 & O_{n_{aw} \times n_w} & 0 \\ B_{cl,v}^T & 0 & D_{cl,uv}^T & 0 & D_{cl,zv}^T \end{bmatrix}$$

$$G_U = \begin{bmatrix} Q_{12}^T & Q_{22} & 0 & 0 & 0 \\ 0 & 0 & I_{n_u} & 0 & 0 \end{bmatrix},$$

where for any square matrix $X$, $\mathrm{He}(X) = X + X^T$.

**Step 6.** Find the optimal solution $(\Lambda_U, U, \gamma)$ of the LMI eigenvalue problem:

$\min_{\Lambda_U, U, \gamma} \gamma$ *subject to*:

*Strong well-posedness* (if $v = 0$, then redundant):

$$-2(1-v)U + \mathrm{He}\left( D_{cl,uq} U + \begin{bmatrix} 0_{n_u \times (n_{aw}+n_c)} & I_{n_u} \end{bmatrix} \Lambda_U \begin{bmatrix} 0_{n_u \times n_{aw}} \\ I_{n_u} \end{bmatrix} \right) < 0,$$

*Anti-windup gains LMI*:

$$\Psi(U,\gamma) + G_U^T \Lambda_U^T H^T + H \Lambda_U G_U < 0.$$

**Step 7.** Select the plant-order full-authority anti-windup compensator as:

$$\dot{x}_{aw} = A_{aw} x_{aw} + B_{aw} q$$
$$v = C_{aw} x_{aw} + D_{aw} q$$

with the selection $\begin{bmatrix} A_{aw} & B_{aw} \\ \hline C_{aw} & D_{aw} \end{bmatrix} = \Lambda_U \begin{bmatrix} I_{n_{aw}} & 0 \\ \hline 0 & U^{-1} \end{bmatrix}.$

$\star$

Since the above reduced-order DLAW algorithm aims at reducing as much as possible the size of the anti-windup compensator state (while giving global stability and performance guarantees), its use is recommended in cases where the static anti-windup design of Algorithm 1 is not feasible or leads to unsatisfactory performance. Indeed, whenever Algorithm 1 is applicable, in the best case the reduced-order augmentation algorithm will lead to the same result, even though it will get there by a more complex route and, due to the approximate nature of the approach, it is not even guaranteed to provide the same level of performance. This is illustrated next in an example reconsidering the problem statement in Example 4.3.4 for which static full-authority anti-windup had been shown to be infeasible in Section 4.3.1. For this example, Algorithm 5 leads to a reduced-order linear anti-windup filter.

**Example 5.4.5** (SISO academic example no. 2) Consider the problem statement in Example 4.3.4 (on page 89) where no static full-authority anti-windup with global guarantees can be determined. That anti-windup problem has been revisited twice already: in Example 4.4.2 (on page 104) where a regional static anti-windup had been synthesized and in Example 5.4.3 (on page 119), where global plant-order anti-windup had been designed.

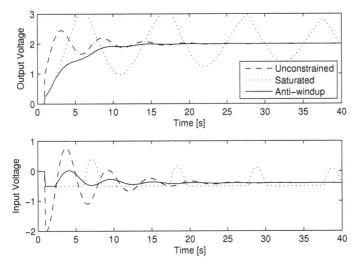

Figure 5.8  Plant output and input responses of various closed loops for Example 5.4.5.

The example is here revisited by running Algorithm 5 to obtain a reduced (or minimum)-order anti-windup guaranteeing a global performance almost as good as the global performance $\gamma = 4.466$ induced by the plant-order solution of Example 5.4.3. In particular, by choosing to give up by seventy percent on that global performance and selecting $\gamma_G = 1.7 \cdot 4.466 = 7.6$ and $\varepsilon = 10^{-4}$ in Procedure 2, a first-order compensator can be obtained guaranteeing a closed-loop performance $\gamma^* = 7.57$. The compensator matrices are obtained by following all the steps of Algorithm 5 as

$$\left[ \begin{array}{c|c} A_{aw} & B_{aw} \\ \hline C_{aw} & D_{aw} \end{array} \right] = \left[ \begin{array}{c|c} -0.3594 & -0.9334 \\ \hline 0.5792 & -0.1550 \\ -0.3631 & 0.9097 \end{array} \right].$$

To this end, no constraints have been enforced on the anti-windup matrices sizes and the parameter $\nu$ has been selected as $\nu = 0.1$.

Figure 5.8 represents the closed-loop responses of the saturated (dotted), unconstrained (dashed), and anti-windup (solid) closed loops. From comparing the plots to the other responses in the previous Examples 4.4.2 and 5.4.3, it appears that this reduced-order design provides a comparable type of response thus successfully solving the windup problem.

It is also instructive to inspect Figure 5.9 where the nonlinear $\mathcal{L}_2$ gains corresponding to the saturated closed loop (dotted) and to the various anti-windup closed loops synthesized in the previous Examples 4.4.2, 5.4.3 and in this one are compared. It appears from these curves that the reduced-order solution (solid) is expected to pretty much behave in the same way as the plant-order solution. Notice

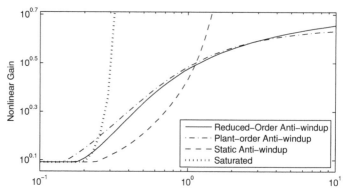

Figure 5.9 Nonlinear gains for Examples 4.4.2, 5.4.3, and 5.4.5 characterizing the system before (dotted) and after regional static (dashed), global plant-order (dashed-dotted), and global reduced-order (solid) anti-windup augmentation.

that in this case the nonlinear gain obtained from the reduced-order solution even improves upon the plant-order one on a wide range of input sizes. It is emphasized that the optimizers in all the algorithms providing global guarantees operate by only looking at the global $\mathcal{L}_2$ gain (namely, the far right of the nonlinear $\mathcal{L}_2$ gain curve) and therefore there is no a priori guarantee that one approach or the other will give better curves for values of the input size smaller than infinity. [1]                □

### 5.4.3 Global external plant-order augmentation

Paralleling the static case discussed in the previous chapter, external dynamic anti-windup compensation synthesis addresses the selection of the anti-windup compensator matrices ($A_{aw}$, $B_{aw}$, $C_{aw}$, and $D_{aw}$) when using the external architecture represented in Figure 5.10 where $\mathcal{F}$ represents the anti-windup compensator. In this and the following section, external anti-windup design algorithms will be given paralleling the full-authority approaches commented on in the previous sections.

When the anti-windup compensator order is equal to the size of the plant state, the algorithm provided here is always feasible for loops involving exponentially stable plants, thus overcoming certain limitations of the static external anti-windup approach of Algorithm 2 given on page 94.

**Algorithm 6** (Dynamic plant-order external global DLAW)

| Applicability | | | | Architecture | | | Guarantee |
|---|---|---|---|---|---|---|---|
| Exp Stab | Marg Stab | Marg Unst | Exp Unst | Lin/ NonL | Dyn/ Static | Ext/ FullAu | Global/ Regional |
| √ | | | | L | D | E | G |

---

[1] The value of the input size is referred to as $s$ in Section 3.5, where regional analysis tools are first introduced, and in Sections 4.4 and 5.5 where regional synthesis is discussed for the static and dynamic cases, respectively.

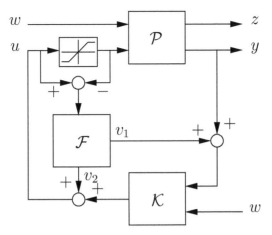

Figure 5.10  External anti-windup compensation scheme.

---

*Comments*: Useful when some internal states of the controller are inaccessible; otherwise not more effective than Algorithm 4. Feasible for any loop containing an exponentially stable plant. Global input-output gain is optimized; it is never worse, and often better, than the input-output gain provided by Algorithm 2.

---

**Step 1.** Given a plant and a controller of the form (5.1), (5.2), construct the matrices of the compact closed-loop representation using (5.5) and (5.11). Select a constant $v \in [0, 1)$ used to enforce a strong well-posedness constraint (see Section 3.4.2 on page 60). Moreover, based on the matrix $B_{ol,v}$ in equation (5.13) generate any matrix $B_{ol,v\perp}$ that spans the null space of its transpose, namely such that $B_{ol,v\perp}^T B_{ol,v} = 0$ and such that $\mathrm{rank}(B_{ol,v\perp}) + \mathrm{rank}(B_{ol,v}) = nc + np$. (Note that in general there can be infinite selections of $B_{ol,v\perp}$. As an example, using MATLAB, an orthonormal selection of $B_{ol,v\perp}$ is easily computed using `Bolperp = null(Bolv)`.)

**Step 2.** Define $M := \begin{bmatrix} B_{ol,v\perp} & 0 & 0 \\ 0 & I_{n_w} & 0 \\ 0 & 0 & I_{n_z} \end{bmatrix}$. Based on the matrices determined at step 1, find the optimal solution $(R, S, \gamma) \in \mathbb{R}^{n_{cl} \times n_{cl}} \times \mathbb{R}^{n_{cl} \times n_{cl}} \times \mathbb{R}$ of the following eigenvalue problem which is always feasible for exponentially stable plants:

$\min_{R,S,\gamma} \gamma$ *subject to*:

   *Positivity*:

$$R = R^T = \begin{bmatrix} R_{11} & S_{12} \\ S_{12}^T & S_{22} \end{bmatrix} > 0 ,$$

$$S = S^T = \begin{bmatrix} S_{11} & S_{12} \\ S_{12}^T & S_{22} \end{bmatrix} > 0 ,$$

$$R_{11} - S_{11} > 0,$$

*Open-loop condition*:

$$M^T \begin{bmatrix} RA_{ol}^T + A_{ol}R & B_{ol,w} & RC_{ol,z}^T \\ B_{ol,w}^T & -\gamma I_{n_w} & D_{ol,zw}^T \\ C_{ol,z}R & D_{ol,zw} & -\gamma I_{n_z} \end{bmatrix} M < 0,$$

*Closed-loop condition*:

$$\begin{bmatrix} SA_{cl}^T + A_{cl}S & B_{cl,w} & SC_{cl,z}^T \\ B_{cl,w}^T & -\gamma I_{n_w} & D_{cl,zw}^T \\ C_{cl,z}S & D_{cl,zw} & -\gamma I_{n_z} \end{bmatrix} < 0.$$

**Step 3.** Based on the solution determined at the previous step, define $Q_{12} \in \mathbb{R}^{n_{cl} \times n_p}$ as any solution to the following equation:

$$Q_{12}Q_{12}^T = RS^{-1}R - R,$$

which is always solvable and admits infinitely many solutions because by construction, $\mathrm{rank}(RS^{-1}R - R) = \mathrm{rank}((RS^{-1}R - R)(R^{-1}S)) = \mathrm{rank}(R - S) = n_p$. A solution can be determined following Procedure 1 on page 113 with $\Xi = RS^{-1}R - R$. Then define the following matrices:

$$Q_{22} = I_{n_p} + Q_{12}^T R^{-1} Q_{12}$$

$$Q = \begin{bmatrix} R & Q_{12} \\ Q_{12}^T & Q_{22} \end{bmatrix}.$$

**Step 4.** Based on the matrices determined at steps 1 and 3 define the following matrices (where the second row and column of zeros has size $n_p$):

$$\left[ \begin{array}{c|cc} A_0 & B_{0,q} & B_{0,w} \\ \hline C_{0,u} & D_{cl,yq} & D_{cl,yw} \\ C_{0,z} & D_{cl,zq} & D_{cl,zw} \end{array} \right] = \left[ \begin{array}{cc|cc} A_{cl} & 0 & B_{cl,q} & B_{cl,w} \\ 0 & 0_{n_p \times n_p} & 0 & 0 \\ \hline C_{cl,u} & 0 & D_{cl,uq} & D_{cl,uw} \\ C_{cl,z} & 0 & D_{cl,zq} & D_{cl,zw} \end{array} \right]$$

and construct the matrices $G_U \in \mathbb{R}^{(n_p+n_u) \times m}$, and $H \in \mathbb{R}^{m \times (n_p+n_y+n_u)}$, and $\Psi \in \mathbb{R}^{m \times m}$, with $m := n_{cl} + n_p + n_u + n_w + n_z$, as follows:

$$\Psi(U,\gamma) = \mathrm{He} \begin{bmatrix} A_0 Q & B_{0,q}U + QC_{0,u}^T & B_{0,w} & QC_{0,z}^T \\ 0 & D_{cl,uq}U - U & D_{cl,uw} & UD_{cl,zq}^T \\ 0 & 0 & -\frac{\gamma}{2}I & D_{cl,zw}^T \\ 0 & 0 & 0 & -\frac{\gamma}{2}I \end{bmatrix},$$

$$H = \left[ \begin{array}{cc|c|c|c} 0 & I_{n_p} & 0 & 0_{n_p \times n_w} & 0 \\ B_{cl,v}^T & 0 & D_{cl,uv}^T & 0 & D_{cl,zv}^T \end{array} \right]$$

$$G_U = \left[ \begin{array}{cc|c|c|c} Q_{12}^T & Q_{22} & 0 & 0 & 0 \\ 0 & 0 & I_{n_u} & 0 & 0 \end{array} \right],$$

where for any square matrix $X$, $\mathrm{He}(X) = X + X^T$.

**Step 5.** Find the optimal solution $(\Lambda_U, U, \gamma)$ of the LMI eigenvalue problem:

$$\min_{\Lambda_U, U, \gamma} \gamma \ \text{subject to:}$$

Strong well-posedness (if $v = 0$, then redundant):

$$-2(1-v)U + \text{He}\left(D_{cl,uq}U + \begin{bmatrix} 0_{n_u \times (n_p + n_y)} & I_{n_u} \end{bmatrix} \Lambda_U \right) < 0,$$

Anti-windup gains LMI:

$$\Psi(U, \gamma) + G_U^T \Lambda_U^T H^T + H \Lambda_U G_U < 0 .$$

**Step 6.** Select the plant-order external anti-windup compensator as:

$$\dot{x}_{aw} = A_{aw} x_{aw} + B_{aw} q$$
$$v = C_{aw} x_{aw} + D_{aw} q$$

with the selection $\begin{bmatrix} A_{aw} & B_{aw} \\ C_{aw} & D_{aw} \end{bmatrix} = \Lambda_U \begin{bmatrix} I_{n_p} & 0 \\ 0 & U^{-1} \end{bmatrix}.$

★

**Example 5.4.6** (Electrical network) Consider again the electrical network problem studied in Examples 4.3.3 and 4.4.3 of the previous chapter and revisited in Example 5.4.2 of this chapter.

For this example, the full-authority plant-order anti-windup in Example 5.4.2 already showed an improvement compared to the static solutions of the previous chapter. Using external compensation allows practically the same performance to be achieved but with a simpler architecture. In particular, following Algorithm 6 and selecting the same parameters as those of the full-authority case in Example 5.4.2 (namely, $v = 0.01$ and $\kappa = 1000$), the following compensator is determined:

$$\begin{bmatrix} A_{aw} & B_{aw} \\ C_{aw} & D_{aw} \end{bmatrix} = \begin{bmatrix} -14.60 & -46.07 & -21.20 & 0.02451 \\ -39.86 & -435.6 & -245.6 & -0.01175 \\ -23.65 & -245.0 & -138.3 & -0.02755 \\ 0.05428 & -0.05083 & 0.1095 & -0.001282 \\ -0.9075 & -10.06 & -5.611 & 1.003 \end{bmatrix},$$

which induces the global $\mathcal{L}_2$ gain $\gamma = 58.7$ (almost the same as the gain $\gamma = 58.428$ induced by the full-authority solution of Example 5.4.2). Moreover, the simulation responses and the nonlinear $\mathcal{L}_2$ gain curve achieved using this external anti-windup compensator are perfectly matched to those ones achieved using the full-authority one and reported in Figures 5.2 and 5.3. For this example, external anti-windup is therefore more desirable than full-authority anti-windup because of its simpler architecture for an identical performance. □

In contrast to the above example, in the next example it is evident that the lesser degrees of freedom available to external anti-windup design lead to a solution which is not as desirable as the one obtained using the full-authority design.

**Example 5.4.7** (F8 longitudinal dynamics) Consider the problem statement formally described and introduced in Example 4.3.5 on page 90. This example has been revisited a few pages earlier as Example 5.4.4 on page 121. In particular, it was shown in this last example that full-authority plant-order anti-windup is able to overcome some undesirable slow convergence experienced with static full-authority anti-windup in Example 4.3.5 (compare Figure 4.9 on page 91 with Figure 5.6 on page 122).

For this same case study, it was also discussed in Example 4.3.8 on page 97 that external static anti-windup is not feasible. Thus, already with static anti-windup designs, the extra degrees of freedom available to full-authority design in this case do make a difference in the feasibility conditions and on the achievable performance. This is further illustrated here, where it is shown that external plant-order anti-windup is feasible but the performance that it leads to is not as good as the one obtained from the full-authority design. When running Algorithm 6 for this problem statement (also condition (5.15) on the anti-windup compensator size is used, with $\kappa = 100$, but this doesn't impact the achievable performance), the achievable closed-loop performance is $\gamma = 51.18$, highly deteriorated as compared to the $\gamma = 25.476$ achieved by full-authority compensation.

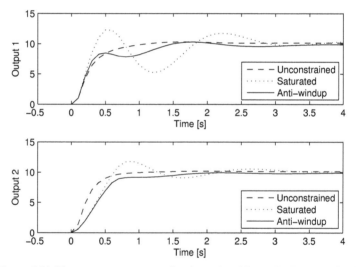

Figure 5.11 Plant output responses of various closed loops for Example 5.4.7.

The anti-windup compensator matrices are the following:

$$A_{aw} = \begin{bmatrix} -32.08 & 0.2592 & -2.576 & -17.20 \\ -17.97 & -0.07452 & -8.052 & -4.463 \\ 3.278 & 0.1544 & -36.41 & 26.02 \\ -7.271 & 0.05788 & 27.23 & -24.78 \end{bmatrix}$$

$$B_{aw} = \begin{bmatrix} -0.3225 & -0.2322 & -0.156 & 0.08838 \\ 0.7652 & 0.507 & 0.2578 & -0.08209 \end{bmatrix}^T$$

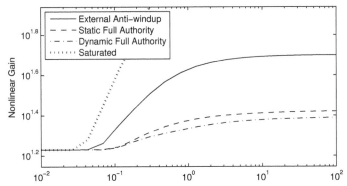

Figure 5.12 Nonlinear gains for Example 5.4.7 characterizing the system before (dotted) and after static full-authority (dashed), plant-order full-authority (dashed-dotted), and plant-order external anti-windup augmentation.

$$C_{aw} = \begin{bmatrix} -0.1364 & 0.3025 & -0.1216 & -0.08926 \\ 0.04730 & 0.2345 & 0.04293 & 0.002186 \\ -0.8590 & 0.007851 & -1.073 & 0.2361 \\ 13.05 & -0.08613 & 8.141 & 1.644 \end{bmatrix}$$

$$D_{aw} = \begin{bmatrix} -0.005001 & -0.004227 & 0.974 & 0.2207 \\ 0.0001178 & 0.004450 & 0.02915 & 0.6027 \end{bmatrix}^{T},$$

clearly less computationally intensive than the full-authority matrices in Example 5.4.4. However, the increased simplicity of the anti-windup architecture comes at the cost of a deteriorated output response. Figure 5.11 shows this response which should be compared to the more desirable response in Figure 5.6 on page 122.

This performance deterioration is also confirmed by the comparison of the nonlinear $\mathcal{L}_2$ gain curves, which is reported in Figure 5.12. In particular, in Figure 5.12 four nonlinear $\mathcal{L}_2$ gains are compared to each other: the saturated system without anti-windup (dotted), the system with the static full-authority anti-windup compensation of Example 4.3.5 (dashed), the plant-order full-authority anti-windup compensation of Example 5.4.4 (dashed-dotted), and the plant-order external compensation of this example (solid). Surprisingly, external dynamic compensation is not even able to do better than static full-authority compensation (the solid curve is all above the dashed curve), which means that the extra control directions available to the full-authority compensation are even more useful than having the extra dynamics available to the external plant-order approach. Indeed, the performance comparison suggested by the curves in Figure 5.12 is in agreement with the output responses experienced in Figures 4.9, 5.6, and 5.11 which refer to the three Examples 4.3.5, 5.4.4, and 5.4.7 under consideration. □

### 5.4.4 Global external reduced-order augmentation

Similar to the full-authority construction introduced in Section 5.4.2 on page 122, reduced-order external anti-windup can also be achieved by suitably generalizing the static and plant-order constructions of Algorithms 2 and 4 on pages 94 and 114, respectively. In particular, in the external case, when seeking for reduced-order anti-windup augmentation with a guaranteed performance level, the conditions at step 3 of Algorithm 2 and at step 2 of Algorithm 4 generalize to the following nonlinear conditions involving a nonlinear rank constraint and paralleling the full-authority conditions (5.16) seen in Section 5.4.2 on page 122:

$$\min_{R,S,\gamma} \gamma \quad subject\ to: \tag{5.17}$$

*Positivity*:

$$R = R^T > 0\ ,$$

$$S = S^T > 0\ ,$$

$$R - S \geq 0\ ,$$

*Nonlinear rank condition*:

$$\mathrm{rank}(R - S) \leq n_{aw},$$

*Open-loop condition*:

$$M^T \begin{bmatrix} RA_{ol}^T + A_{ol}R & B_{ol,w} & RC_{ol,z}^T \\ B_{ol,w}^T & -\gamma I_{n_w} & D_{ol,zw}^T \\ C_{ol,z}R & D_{ol,zw} & -\gamma I_{n_z} \end{bmatrix} M < 0\ ,$$

*Closed-loop condition*:

$$\begin{bmatrix} SA_{cl}^T + A_{cl}S & B_{cl,w} & SC_{cl,z}^T \\ B_{cl,w}^T & -\gamma I_{n_w} & D_{cl,zw}^T \\ C_{cl,z}S & D_{cl,zw} & -\gamma I_{n_z} \end{bmatrix} < 0\ ,$$

where $M := \begin{bmatrix} B_{ol,v\perp} & 0 & 0 \\ 0 & I_{n_w} & 0 \\ 0 & 0 & I_{n_z} \end{bmatrix}$ and $B_{ol,v\perp}$ is the usual full column rank matrix

spanning the null space of $B_{ol,v}^T$.

Similar to the full-authority case on page 123, conditions (5.17) are not easily solvable because of the presence of the nonlinear rank constraint. It is, however, possible to solve a simplified problem whereby minimizing the trace of $R - S$ results in generating a rank-deficient matrix, thereby reducing to a certain extent the size $n_{aw}$ of the compensator. The approximate approach already adopted in the full-authority case can then be used also here to construct a triple $(R^*, S^*, \gamma^*)$ which serves as a starting point for reduced-order anti-windup design, where $\gamma^*$ is close to a desired guaranteed performance level $\gamma_G$.

A procedure is reported next which parallels (for the external case) the full-authority Procedure 2. An important difference from Procedure 2 is that in the external case it is necessary to concentrate on the whole matrix $R$, rather than only on its upper left block, because of the generalized condition given by the open-loop condition in (5.17) which involves the whole matrix $R$.

**Procedure 3** (Computation of an $(R, S, \gamma)$ solution for reduced-order external anti-windup with guaranteed performance)

**Step a)** A plant and a controller of the form (5.1), (5.2), and the matrices of the compact closed-loop representation constructed based on equations (5.5) and (5.11) are given. A small positive constant $\varepsilon$ to robustify the closed-loop LMI condition and a value $\gamma_G$ characterizing the desired guaranteed performance are also given. Note that the actual performance $\gamma^*$ achieved by the minimal-order $(R, S, \gamma)$ solution will be, in general, larger than $\gamma_G$ but often very close to it.

**Step b)** Find the optimal solution $(R, S) \in \mathbb{R}^{n_{cl} \times n_{cl}} \times \mathbb{R}^{n_{cl} \times n_{cl}}$ to the following LMI eigenvalue problem:

$$\min_{R,S} \text{ trace}(R - S) \text{ subject to:}$$

   *Positivity*:
$$R = R^T > 0,$$
$$S = S^T > 0,$$
$$R - S > 0,$$

   *Open-loop condition*:
$$M^T \begin{bmatrix} RA_{ol}^T + A_{ol}R & B_{ol,w} & RC_{ol,z}^T \\ B_{ol,w}^T & -\gamma_G I_{n_w} & D_{ol,zw}^T \\ C_{ol,z}R & D_{ol,zw} & -\gamma_G I_{n_z} \end{bmatrix} M < 0,$$

   *Closed-loop condition*:
$$\begin{bmatrix} SA_{cl}^T + A_{cl}S & B_{cl,w} & SC_{cl,z}^T \\ B_{cl,w}^T & -\gamma_G I_{n_w} & D_{cl,zw}^T \\ C_{cl,z}S & D_{cl,zw} & -\gamma_G I_{n_z} \end{bmatrix} + \varepsilon I < 0.$$

**Step c)** Compute the singular value decomposition (SVD) of the symmetric matrix $R - S$ determined from the solution of the previous step. That is, compute $F, \Lambda$ such that $F \Lambda F^T = R - S$, where $\Lambda$ is a positive semidefinite diagonal matrix whose diagonal entries are nonincreasing (for example, use the MATLAB command [U,Lambda,V] = svd(R-S); F = (U+V)/2;). Set $n_{aw} = n_p + n_c$.

**Step d)** Call $\lambda_1, \dots \lambda_{n_{aw}-1}$ the first $n_{aw} - 1$ entries on the diagonal of $\Lambda$ and define $\hat{\Lambda} := \text{diag}\{\lambda_1, \dots \lambda_{n_{aw}-1}, 0, \dots, 0\}$. That is, $\hat{\Lambda}$ is a diagonal matrix having its first $n_{aw} - 1$ diagonal entries equal to the entries of $\Lambda$ and the remaining entries equal to zero. Also define $\hat{S} := S + F(\Lambda - \hat{\Lambda})F^T$.

**Step e)** Given the matrix $\hat{S}$ determined at the previous step, verify the feasibility of the following LMI in the variable $\gamma$:

$$\begin{bmatrix} \hat{S}A_{cl}^T + A_{cl}\hat{S} + \varepsilon I & B_{cl,w} & \hat{S}C_{cl,z}^T \\ B_{cl,w}^T & -\gamma I_{n_w} & D_{cl,zw}^T \\ C_{cl,z}\hat{S} & D_{cl,zw} & -\gamma I_{n_z} \end{bmatrix} < 0.$$

If the LMI is feasible, then set $(R^*, S^*) := (R, \hat{S})$, set $n_{aw} = n_{aw} - 1$ and go to step d. If the LMI is not feasible, then go to step f.

**Step f)** Select the order of the anti-windup compensator as $n_{aw}$ and the $(R, S, \gamma)$ solution as $(R^*, S^*, \gamma^*)$.

                                                                    ★

In the worst case, Procedure 3 will give a solution $(R, S, \gamma)$ characterizing an anti-windup compensation of order $n_{aw} = n_p + n_c$, which is unacceptably large (indeed, in that worst case the direct construction for an $n_p$-order anti-windup compensation provided by Algorithm 2 would be easier to apply and more desirable). However, several case studies confirmed that the reduced order obtained with this algorithm is very likely to be smaller than or equal to $n_p$.

Once a triple $(R^*, S^*, \gamma^*)$ satisfying (5.17) with a small enough $n_{aw}$ is determined from Procedure 3, the reduced-order anti-windup construction can be carried out following a simple generalization of the previous Algorithms 2 and 4, which is reported next.

**Algorithm 7** (Dynamic reduced-order external global DLAW)

| Applicability | | | | Architecture | | | Guarantee |
|:---:|:---:|:---:|:---:|:---:|:---:|:---:|:---:|
| Exp Stab | Marg Stab | Marg Unst | Exp Unst | Lin/ NonL | Dyn/ Static | Ext/ FullAu | Global/ Regional |
| √ | | | | L | D* | E | G |

| Comments: Useful when Algorithm 2 is not feasible or performs poorly yet a low-order external anti-windup augmentation is desired. Order reduction is carried out while maintaining a prescribed global input-output gain.<br>* Generally of low order resulting from order reduction. |
|:---|

**Step 1.** Given a plant and a controller of the form (5.1), (5.2), construct the matrices of the compact closed-loop representation using (5.5) and (5.11). Select a small positive constant $\varepsilon$ to robustify the order minimization procedure and a constant $\alpha > 1$ characterizing the amount of performance to be traded in for order reduction. Select a constant $v \in [0, 1)$ used to enforce a strong well-posedness constraint (see Section 3.4.2 on page 60). Moreover, based on the matrix $B_{ol,v}$ in equation (5.13), generate any matrix $B_{ol,v\perp}$ that spans the null space of its transpose, namely, such that $B_{ol,v\perp}^T B_{ol,v} = 0$ and such that $\text{rank}(B_{ol,v\perp}) + \text{rank}(B_{ol,v}) = nc + np$. (Note that in general there can be infinite selections of $B_{ol,v\perp}$. As an example, using MATLAB, an orthonormal selection of $B_{ol,v\perp}$ is easily computed using `Bolperp = null(Bolv)`.)

**Step 2.** Find the optimal solution $(R_{11}, S, \gamma_{n_p}) \in \mathbb{R}^{n_p \times n_p} \times \mathbb{R}^{n_{cl} \times n_{cl}} \times \mathbb{R}$ to the following LMI eigenvalue problem which is always feasible for exponentially stable plants:

$$\min_{R11, S, \gamma} \gamma \text{ subject to:}$$

    Positivity:

$$R_{11} = R_{11}^T > 0,$$

$$S = S^T = \begin{bmatrix} S_{11} & S_{12} \\ S_{12}^T & S_{22} \end{bmatrix} > 0 ,$$

$$R_{11} - S_{11} > 0 ,$$

*Open-loop condition*:

$$\begin{bmatrix} R_{11}A_p^T + A_pR_{11} & B_{p,w} & R_{11}C_{p,z}^T \\ B_{p,w}^T & -\gamma I_{n_w} & D_{p,zw}^T \\ C_{p,z}R_{11} & D_{p,zw} & -\gamma I_{n_z} \end{bmatrix} < 0 ,$$

*Closed-loop condition*:

$$\begin{bmatrix} SA_{cl}^T + A_{cl}S & B_{cl,w} & SC_{cl,z}^T \\ B_{cl,w}^T & -\gamma I_{n_w} & D_{cl,zw}^T \\ C_{cl,z}S & D_{cl,zw} & -\gamma I_{n_z} \end{bmatrix} < 0 .$$

The resulting performance level $\gamma_{n_p}$ corresponds to the maximum performance achievable by plant-order full-authority anti-windup augmentation.

**Step 3.** Based on the matrices and parameters determined at step 1, select the desired guaranteed performance as $\gamma_G = \alpha \gamma_{n_p}$ and determine the reduced compensator order $n_{aw}$ and the triple $(R^*, S^*, \gamma^*)$ from Procedure 3.

**Step 4.** Based on the solution determined at the previous step, define $Q_{12} \in \mathbb{R}^{n_{cl} \times n_{aw}}$ as any solution to the following equation:

$$Q_{12}Q_{12}^T = RS^{-1}R - R,$$

which is always solvable and admits infinitely many solutions because by construction, $\text{rank}(RS^{-1}R - R) = \text{rank}((RS^{-1}R - R)(R^{-1}S)) = \text{rank}(R - S) = n_{aw}$. A solution can be determined following Procedure 1 on page 113 with $\Xi = RS^{-1}R - R$. Then define the following matrices:

$$Q_{22} = I_{n_{aw}} + Q_{12}^T R^{-1} Q_{12}$$

$$Q = \begin{bmatrix} R & Q_{12} \\ Q_{12}^T & Q_{22} \end{bmatrix} .$$

**Step 5.** Based on the matrices determined at steps 3 and 4, define the following matrices (where the second row and column of zeros has size $n_{aw}$):

$$\left[ \begin{array}{c|cc} A_0 & B_{0,q} & B_{0,w} \\ \hline C_{0,u} & D_{cl,yq} & D_{cl,yw} \\ C_{0,z} & D_{cl,zq} & D_{cl,zw} \end{array} \right] = \left[ \begin{array}{cc|cc} A_{cl} & 0 & B_{cl,q} & B_{cl,w} \\ 0 & 0_{n_{aw} \times n_{aw}} & 0 & 0 \\ \hline C_{cl,u} & 0 & D_{cl,uq} & D_{cl,uw} \\ C_{cl,z} & 0 & D_{cl,zq} & D_{cl,zw} \end{array} \right]$$

and construct the matrices $G_U \in \mathbb{R}^{(n_{aw}+n_u) \times m}$, $H \in \mathbb{R}^{m \times (n_{aw}+n_y+n_u)}$, and $\Psi \in \mathbb{R}^{m \times m}$, with $m := n_{cl} + n_{aw} + n_u + n_w + n_z$, as follows:

$$\Psi(U, \gamma) = \text{He} \begin{bmatrix} A_0 Q & B_{0,q}U + QC_{0,u}^T & B_{0,w} & QC_{0,z}^T \\ 0 & D_{cl,uq}U - U & D_{cl,uw} & UD_{cl,zq}^T \\ 0 & 0 & -\frac{\gamma}{2}I & D_{cl,zw}^T \\ 0 & 0 & 0 & -\frac{\gamma}{2}I \end{bmatrix} ,$$

$$H = \left[\begin{array}{cc|c|c|c} 0 & I_{naw} & 0 & 0_{naw \times nw} & 0 \\ B_{cl,v}^T & 0 & D_{cl,uv}^T & 0 & D_{cl,zv}^T \end{array}\right]$$

$$G_U = \left[\begin{array}{cc|c|c|c} Q_{12}^T & Q_{22} & 0 & 0 & 0 \\ 0 & 0 & I_{nu} & 0 & 0 \end{array}\right],$$

where for any square matrix $X$, $\mathrm{He}(X) = X + X^T$.

**Step 6.** Find the optimal solution $(\Lambda_U, U, \gamma)$ of the LMI eigenvalue problem:

$$\min_{\Lambda_U, U, \gamma} \gamma \ \textit{subject to}:$$

*Strong well-posedness (if $v = 0$, then redundant):*

$$-2(1-v)U + \mathrm{He}\left(D_{cl,uq}U + \left[\begin{array}{cc} 0_{nu \times (naw+ny)} & I_{nu} \end{array}\right]\Lambda_U\right) < 0,$$

*Anti-windup gains LMI:*

$$\Psi(U, \gamma) + G_U^T \Lambda_U^T H^T + H \Lambda_U G_U < 0.$$

**Step 7.** Select the plant-order external anti-windup compensator as:

$$\dot{x}_{aw} = A_{aw} x_{aw} + B_{aw} q$$

$$v = C_{aw} x_{aw} + D_{aw} q$$

with the selection $\left[\begin{array}{c|c} A_{aw} & B_{aw} \\ \hline C_{aw} & D_{aw} \end{array}\right] = \Lambda_U \left[\begin{array}{c|c} I_{naw} & 0 \\ \hline 0 & U^{-1} \end{array}\right].$

$\star$

Similar to the full-authority case, it is only reasonable to apply Algorithm 7 when the static external anti-windup construction of Algorithm 2 is not feasible or does not lead to satisfactory performance. As a matter of fact, if the feasibility conditions for Algorithm 2 are satisfied with a desirable performance level, the reduced-order design of the above Algorithm 7 would lead to the same compensator following a more involved design construction. Conversely, in the cases where static external anti-windup is not feasible or provides an unsatisfactory performance level, low-order anti-windup augmentation is typically obtained from Algorithm 7 as clarified from the following examples.

The reduced-order external anti-windup construction is first illustrated by giving an external solution to the windup problem first introduced in Example 4.3.4 on page 89, where it was shown that full-authority static anti-windup is not feasible. Static external anti-windup is also not feasible because it is a less powerful architecture. However, the reduced-order design leads to a one-dimensional external compensator.

**Example 5.4.8** (SISO academic example no. 2) Consider the example study first introduced in Example 4.3.4 on page 89. Among other solutions, a reduced-order full-authority anti-windup compensator has been designed in Example 5.4.5. This compensator had been shown to perform as desirably as the full-order solution thereby being more desirable due to its increased numerical simplicity.

For the same example when running Algorithm 7 with the same parameters as in Example 5.4.5, the following compensator is obtained:

$$\left[\begin{array}{c|c} A_{aw} & B_{aw} \\ \hline C_{aw} & D_{aw} \end{array}\right] = \left[\begin{array}{c|c} -0.1938 & -1.935 \\ \hline 0.2717 & -0.3655 \\ 0.05351 & 1.330 \end{array}\right].$$

This induces the same performance level as the full-authority compensator in Example 5.4.5. In addition, both the simulation curves and the nonlinear gain curves overlap with those obtained from Example 5.4.5. There are thus two alternative solutions to the same problem with the latter, synthesized here, having the advantage of being numerically simpler due to its external nature.                                           □

Reduced-order anti-windup design is illustrated below for the F8 longitudinal dynamics control system. For this system, it was shown in Example 4.3.8 that static external anti-windup is not feasible (although, as shown in Example 4.3.5, full-authority static anti-windup is feasible for this same example). Reduced-order anti-windup design leads once again to a low-order external anti-windup compensator which behaves pretty much the same as the plant-order compensator synthesized in Example 5.4.7.

**Example 5.4.9** (F8 longitudinal dynamics) Consider the model of the F8 longitudinal dynamics first addressed and described in Example 4.3.5 on page 90. This case study was addressed using plant-order external anti-windup augmentation a few pages back, in Example 5.4.7 on page 133.

For this example, Algorithm 7 can be followed using the same parameters selected in Example 5.4.7, with $\gamma_G = 70$ and $\varepsilon = 10^{-4}$ in Procedure 3. The solution obtained reduces the anti-windup compensator order from four to two states and guarantees a global performance $\gamma^* = 70.336$. The corresponding anti-windup compensator matrices are

$$
\left[ \begin{array}{c|c} A_{aw} & B_{aw} \\ \hline C_{aw} & D_{aw} \end{array} \right] = \left[ \begin{array}{cc|cc}
-159.9 & -73.35 & -6.341 & 0.4484 \\
-18.61 & -12.00 & -0.7000 & 0.05463 \\
-1.368 & -0.7770 & -0.07716 & 0.004313 \\
-0.2337 & -0.3300 & -0.02819 & 0.002176 \\
-2.775 & -1.151 & 0.8826 & 0.007642 \\
8.641 & 5.346 & 0.2688 & 0.9702
\end{array} \right].
$$

The simulations carried out with this compensator are not shown because they are equivalent to the ones obtained using the plant-order solution in Example 5.4.7. Similarly, the nonlinear gain curve is only slightly deteriorated as compared to the one obtained in Example 5.4.7 but substantially predicts the same type of behavior.                                                    □

## 5.5 ALGORITHMS PROVIDING REGIONAL GUARANTEES

All the algorithms given so far in this chapter provide anti-windup compensators inducing global exponential stability properties on the compensated closed loop. It was pointed out in Chapter 4 that this property is sometimes desirable but may be overkill because of the intrinsic limitations of plants stabilized by a bounded signal.

In particular, an intrinsic limitation characterizing all the dynamic anti-windup algorithms listed above is that the plant needs to be exponentially stable for the LMI-based constructions to go through. In other words, all the plant poles need to lie strictly in the negative half plane, so that even a simple integrator will not be good for this. This fact is actually not surprising if one keeps in mind that it is impossible, via a bounded input, to stabilize exponentially and globally any plant which is not already exponentially stable.

Paralleling Section 4.4 on regional *static* anti-windup designs, this section then fills in the missing bricks within the anti-windup constructions proposed in the past two chapters: it deals with dynamic regional anti-windup augmentation. When looking at regional properties, the importance of using dynamics in anti-windup is not as clear as for the global case. Indeed, for the global case, introducing plant-order compensation made it possible to overcome the feasibility conditions discussed at length in Chapter 4. For the regional case these feasibility conditions actually depend on the size of the guaranteed stability region, therefore the necessity for dynamics in the anti-windup compensator should be seen from the point of view of wanting to enlarge as much as possible the stability region. Hence the effectiveness of the anti-windup solution.

Similar to the static case, from the nonlinear performance viewpoint of Section 2.4.4, the regional algorithms seek for anti-windup gains optimizing the nonlinear gain curve at a certain horizontal coordinate, corresponding to a prescribed bound on $\|w\|_2$. Generally this also leads to a finite gain for larger values of $\|w\|_2$ and even global stability guarantees in some cases, however, this result is not a priori guaranteed, and it should be checked by analyzing the nonlinear $\mathcal{L}_2$ gain curve after the compensator matrices have been synthesized.

In the next sections, algorithms for constructing full-authority plant-order and reduced-order anti-windup compensators will be given. These two algorithms generalize the algorithms in Sections 5.4.1 and 5.4.2 by throwing in the generalized sector concepts discussed in Section 3.5. Therefore, their structures completely parallel those of their global counterpart except for a few extra terms in the LMIs which are responsible for the regional properties and for the increased feasibility properties.

External regional anti-windup constructions are not discussed here but can be determined by extending the approaches in the external algorithms of this chapter along similar lines.

As already discussed in Section 4.4 for the static anti-windup case, when only regional properties are guaranteed, three closed-loop properties become important: 1) the size of the stability region, 2) the size of the reachable region from inputs with bounded $\mathcal{L}_2$ norm, and 3) the regional $\mathcal{L}_2$ gain. The following algorithms are all stated with the goal in mind to minimize the regional $\mathcal{L}_2$ gain under certain constraints on the stability region size. However, alternative similar formulations could be also derived to maximize the stability region or minimize the reachable region from bounded inputs under certain constraints on the other two performance parameters. To keep the discussion simple, these generalizations are not included here.

Finally, both the algorithms given next depend on the saturation levels $\bar{u}_i > 0$, $i = 1, \ldots, n_u$, which may all be different from each other. The algorithms lead to solutions with guaranteed stability and performance also for time-varying saturation limits, as long as the vector $\bar{u} = [\bar{u}_1, \ldots, \bar{u}_{n_u}]^T$ is a lower bound for the saturation values at all times. As a consequence, the algorithm is also applicable to nonsymmetric saturations as long as for each input channel the value $u_i$, $i = \in \{1, \ldots, n_u\}$ is selected as the most stringent saturation limit for that channel. In this case, the stability and performance estimates provided by the algorithm increase in conservativeness.

### 5.5.1 Regional full-authority plant-order augmentation

Regional plant-order full-authority anti-windup augmentation shares the same architecture as the global counterpart addressed earlier in this chapter and corresponds to selecting the anti-windup matrices in equation (5.3). The following algorithm corresponds to the regional counterpart of Algorithm 4 on page 114.

**Algorithm 8** (Dynamic plant-order full-authority regional DLAW)

| Applicability | | | | Architecture | | | Guarantee |
|:---:|:---:|:---:|:---:|:---:|:---:|:---:|:---:|
| Exp Stab | Marg Stab | Marg Unst | Exp Unst | Lin/ NonL | Dyn/ Static | Ext/ FullAu | Global/ Regional |
| √ | √ | √ | √ | L | D | FA | R |

*Comments*: Extends the applicability of Algorithm 4 by only requiring regional properties. Applicable to any loop. Regional input-output gain is optimized.

**Step 1.** Given a plant and a controller of the form (5.1), (5.2), construct the matrices of the compact closed-loop representation using (5.5) and (5.8). Select a constant $v \in [0, 1)$ used to enforce a strong well-posedness constraint (see Section 3.4.2 on page 60) and select the maximum size $s$ of allowable external inputs $\|w\|_2 < s$ over which the gain minimization has to be performed.

Possibly select two matrices $S_p$ and $R_p$ characterizing a desirable guaranteed reachable set contained in $\mathcal{R}_p = \{x_p : x_p^T R_p^{-1} x_p\}$ and a desirable stability region containing $\mathcal{S}_p = \{x_p : x_p^T S_p^{-1} x_p\}$ in the plant state directions; otherwise ignore the conditions involving $S_p$ and $R_p$ below.

**Step 2.** Find the optimal solution $(R_{11}, S, Z, \gamma^2) \in \mathbb{R}^{n_p \times n_p} \times \mathbb{R}^{n_{cl} \times n_{cl}} \times \mathbb{R}^{n_u \times n_p} \times \mathbb{R}$ to the following LMI eigenvalue problem which is always feasible for exponentially stable plants:

$$\min_{R_{11}, S, Z, \gamma^2} \gamma^2 \ \ subject \ to:$$

*Positivity*:

$$R_{11} = R_{11}^T > 0 ,$$

$$S = S^T = \begin{bmatrix} S_{11} & S_{12} \\ S_{12}^T & S_{22} \end{bmatrix} > 0 ,$$

$$R_{11} - S_{11} > 0 ,$$

*Open-loop condition:*

$$\text{He} \begin{bmatrix} A_p R_{11} + B_{p,u} Z & B_{p,w} & 0 \\ 0 & -\frac{I_{n_w}}{2} & 0 \\ C_{p,z} R_{11} + D_{p,zu} Z & D_{p,zw} & -\frac{\gamma^2 I_{n_z}}{2} \end{bmatrix} < 0,$$

*Closed-loop condition:*

$$\begin{bmatrix} SA_{cl}^T + A_{cl} S & B_{cl,w} & SC_{cl,z}^T \\ B_{cl,w}^T & -I_{n_w} & D_{cl,zw}^T \\ C_{cl,z} S & D_{cl,zw} & -\gamma^2 I_{n_z} \end{bmatrix} < 0,$$

*Regional conditions:*

$$\begin{bmatrix} \bar{u}_i^2/s^2 & Z_i \\ Z_i^T & R_{11} \end{bmatrix} \geq 0, \quad i = 1, \dots, n_u,$$

*Guaranteed reachable set and stability region:*

$$R_p \leq R_{11} \leq S_p.$$

**Step 3.** Based on the solution determined at the previous step, construct the symmetric matrix $R = \begin{bmatrix} R_{11} & S_{12} \\ S_{12}^T & S_{22} \end{bmatrix}$ and define $Q_{12} \in \mathbb{R}^{n_{cl} \times n_p}$ as any solution to the following equation:

$$Q_{12} Q_{12}^T = RS^{-1} R - R, \tag{5.18}$$

which is always solvable and admits infinitely many solutions because by construction, $\text{rank}(RS^{-1}R - R) = \text{rank}((RS^{-1}R - R)(R^{-1}S)) = \text{rank}(R - S) = n_p$. A solution can be determined following Procedure 1 on page 113, with $\Xi = RS^{-1}R - R$. Then define the following matrices:

$$Q_{22} = I_{n_p} + Q_{12}^T R^{-1} Q_{12}$$

$$Q = \begin{bmatrix} R & Q_{12} \\ Q_{12}^T & Q_{22} \end{bmatrix}.$$

**Step 4.** Based on the matrices determined at steps 1 and 3, define the following matrices (where the second row and column of zeros have size $n_p$):

$$\begin{bmatrix} A_0 & B_{0,q} & B_{0,w} \\ C_{0,u} & D_{cl,yq} & D_{cl,yw} \\ C_{0,z} & D_{cl,zq} & D_{cl,zw} \end{bmatrix} = \begin{bmatrix} A_{cl} & 0 & B_{cl,q} & B_{cl,w} \\ 0 & 0_{n_p \times n_p} & 0 & 0 \\ C_{cl,u} & 0 & D_{cl,uq} & D_{cl,uw} \\ C_{cl,z} & 0 & D_{cl,zq} & D_{cl,zw} \end{bmatrix}$$

and construct the matrices $G_U \in \mathbb{R}^{(n_p + n_u) \times m}$, $H \in \mathbb{R}^{m \times (n_p + n_c + n_u)}$, and $\Psi \in \mathbb{R}^{m \times m}$, with $m := n_{cl} + n_p + n_u + n_w + n_z$, as follows:

$$\Psi(U, \gamma^2) = \text{He} \begin{bmatrix} A_0 Q & B_{0,q} U + Q C_{0,u}^T & B_{0,w} & Q C_{0,z}^T \\ -Y & D_{cl,uq} U - U & D_{cl,uw} & U D_{cl,zq}^T \\ 0 & 0 & -\frac{1}{2}I & D_{cl,zw}^T \\ 0 & 0 & 0 & -\frac{\gamma^2}{2}I \end{bmatrix},$$

$$H = \begin{bmatrix} 0 & I_{n_p} & 0 & 0_{n_p \times n_w} & 0 \\ B_{cl,v}^T & 0 & D_{cl,uv}^T & 0 & D_{cl,zv}^T \end{bmatrix}$$

$$G_U = \begin{bmatrix} Q_{12}^T & Q_{22} & 0 & 0 & 0 \\ 0 & 0 & I_{n_u} & 0 & 0 \end{bmatrix},$$

where for any square matrix $X$, $He(X) = X + X^T$, and where $Y = \begin{bmatrix} Z & ZR_{11}^{-1}R_{12} \end{bmatrix}$ $ZR_{11}^{-1}\bar{Q}$ (with $\bar{Q}$ being the matrix formed by the first $n_p$ rows of $Q_{12}$).

**Step 5.** Find the optimal solution $(\Lambda_U, U, \gamma^2)$ of the LMI eigenvalue problem:

$$\min_{\Lambda_U, U, \gamma^2} \gamma^2 \text{ subject to:}$$

*Strong well-posedness* (if $v = 0$, then redundant):

$$-2(1-v)U + He\left(D_{cl,uq}U + \begin{bmatrix} 0_{n_u \times (n_p + n_c)} & I_{n_u} \end{bmatrix} \Lambda_U \begin{bmatrix} 0_{n_u \times n_p} \\ I_{n_u} \end{bmatrix}\right) < 0,$$

*Anti-windup gains LMI*:

$$\Psi(U, \gamma^2) + G_U^T \Lambda_U^T H^T + H \Lambda_U G_U < 0 .$$

**Step 6.** Select the plant-order full-authority anti-windup compensator as:

$$\dot{x}_{aw} = A_{aw}x_{aw} + B_{aw}q$$
$$v = C_{aw}x_{aw} + D_{aw}q$$

with the selection $\begin{bmatrix} A_{aw} & B_{aw} \\ C_{aw} & D_{aw} \end{bmatrix} = \Lambda_U \begin{bmatrix} I_{n_p} & 0 \\ 0 & U^{-1} \end{bmatrix}.$

$\star$

**Example 5.5.1** (Electrical network) Consider the electrical network problem whose parameters are given in Example 4.3.3 on page 87 of the previous chapter. This example has been revisited several times and has been last studied in Example 5.4.6 on page 133. It has been noted in Examples 4.3.3 and 5.4.2 that requiring global properties from the anti-windup compensation leads to a slow closed-loop response.

Performing a regional plant-order design following Algorithm 8 with $s = 0.1$ and selecting the other parameters as in Example 5.4.2 yields the following anti-windup compensator:

$$\begin{bmatrix} A_{aw} & B_{aw} \\ C_{aw} & D_{aw} \end{bmatrix} = \begin{bmatrix} -10.84 & -6.521 & -3.816 & -0.05033 \\ 1.457 & -0.7155 & -0.7416 & -0.02048 \\ -0.005009 & -0.1858 & -562.9 & 0.01726 \\ -1.950 & 0.1510 & -8.114 & -0.02413 \\ 0.02099 & 0.003800 & 0.1015 & 0.0001427 \\ -22.21 & 37.35 & 48.98 & -0.001113 \end{bmatrix},$$

which induces a regional performance level $\gamma = 6.067$ at $s = 0.1$.

Figure 5.13 illustrates the closed-loop response obtained when using this anti-windup compensation. It is quite evident that this is the best response obtained so far for this example. The reason for this is that giving up the global stability guarantees allows the optimizer to select a more aggressive control solution. This fact is partially seen from Figure 5.14, where several nonlinear gain curves are compared. For small signals, the regional approach of this example behaves better than

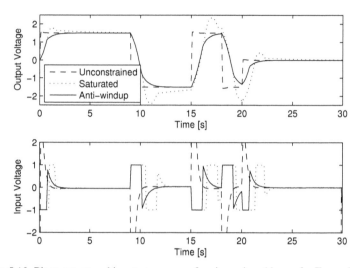

Figure 5.13 Plant output and input responses of various closed loops for Example 5.5.1.

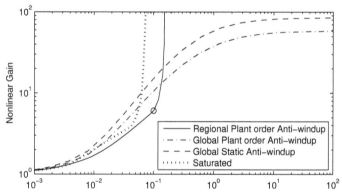

Figure 5.14 Nonlinear gains for Examples 4.3.3, 5.4.2, and 5.5.1 characterizing the system
before and after static global (dotted), plant-order global (dashed-dotted), and
plant-order regional (solid) anti-windup augmentation.

the other approaches (the little circle indicates the optimized value of the gain at
$s = 0.1$). However, the nonlinear gain associated with this regional compensation
blows up for larger values of $\|w\|_2$, which indicates that no global stability guaran-
tee can be given with this compensation solution.                                  $\square$

### 5.5.2 Regional full-authority reduced-order augmentation

Paralleling the full-authority construction introduced in Section 5.4.2 on page 122,
reduced-order regional full-authority anti-windup can be achieved by suitably generalizing the static and plant-order constructions of Algorithms 3 and 8 on pages 99
and 143, respectively. In particular, in the regional full-authority case, when seek-

ing for reduced-order anti-windup augmentation with a guaranteed performance level, the conditions at step 3 of Algorithm 3 and at step 2 of Algorithm 8 generalize to the following nonlinear conditions involving a nonlinear rank constraint and paralleling the full-authority conditions (5.16) seen in Section 5.4.2 on page 122:

$$\min_{R_{11},S,Z,\gamma^2} \gamma^2 \ \text{subject to:} \tag{5.19}$$

*Positivity*:

$$R_{11} = R_{11}^T > 0,$$

$$S = S^T = \begin{bmatrix} S_{11} & S_{12} \\ S_{12}^T & S_{22} \end{bmatrix} > 0,$$

$$R_{11} - S_{11} \geq 0,$$

*Nonlinear rank condition*:

$$\text{rank}(R_{11} - S_{11}) \leq n_{aw},$$

*Open-loop condition*:

$$\text{He} \begin{bmatrix} A_p R_{11} + B_{p,u} Z & B_{p,w} & 0 \\ 0 & -\frac{I_{n_w}}{2} & 0 \\ C_{p,z} R_{11} + D_{p,zu} Z & D_{p,zw} & -\frac{\gamma^2 I_{n_z}}{2} \end{bmatrix} < 0,$$

*Closed-loop condition*:

$$\begin{bmatrix} SA_{cl}^T + A_{cl}S & B_{cl,w} & SC_{cl,z}^T \\ B_{cl,w}^T & -I_{n_w} & D_{cl,zw}^T \\ C_{cl,z}S & D_{cl,zw} & -\gamma^2 I_{n_z} \end{bmatrix} < 0,$$

*Regional conditions*:

$$\begin{bmatrix} \bar{u}_i^2/s^2 & Z_i \\ Z_i^T & R_{11} \end{bmatrix} \geq 0, \quad i = 1,\ldots,n_u,$$

*Guaranteed reachable set and stability region*:

$$R_p \leq R_{11} \leq S_p.$$

Similar to the full-authority case on page 123, conditions (5.17) are not easily solvable because of the presence of the nonlinear rank constraint. It is, however, possible to solve a simplified problem whereby minimizing the trace of $R - S$ results in generating a rank-deficient matrix, thereby reducing to a certain extent the size $n_{aw}$ of the compensator. The approximate approach already adopted in the full-authority case can then be used also here to construct a triple $(R^*, S^*, \gamma^*)$ which serves as a starting point for reduced-order anti-windup design, where $\gamma^*$ is close to a desired guaranteed performance level $\gamma_G$.

A procedure is reported next which parallels (for the regional case) the global Procedure 2.

**Procedure 4** (Computation of an $(R, S, \gamma^2)$ solution for reduced-order full-authority anti-windup with guaranteed performance)

   **Step a)** A plant and a controller of the form (5.1), (5.2), and the matrices of the compact closed-loop representation constructed based on equations (5.5) and (5.8) are given. A maximum size $s$ of allowable external inputs $\|w\|_2 < s$ over which the gain minimization has to be performed is given. Possibly, two matrices $S_p$ and $R_p$ are given, characterizing a desirable guaranteed reachable set contained in $\mathcal{R}_p = \{x_p : x_p^T R_p^{-1} x_p\}$ and a desirable stability region containing $\mathcal{S}_p = \{x_p : x_p^T S_p^{-1} x_p\}$ in the plant state directions, otherwise ignore the conditions involving $S_p$ and $R_p$ below. A small positive constant $\varepsilon$ to robustify the closed-loop LMI condition and a value $\gamma_G$ characterizing the desired guaranteed performance are also given. Note that the actual performance $\gamma^*$ achieved by the minimal-order $(R, S, \gamma)$ solution will be, in general, larger than $\gamma_G$ but often very close to it.

**Step b)** Find the optimal solution $(R_{11}, S, Z) \in \mathbb{R}^{n_p \times n_p} \times \mathbb{R}^{n_{cl} \times n_{cl}} \times \mathbb{R}^{n_u \times n_p}$ to the following LMI eigenvalue problem:

$$\min_{R_{11}, S, Z} \; \text{trace}(R_{11} - S_{11}) \quad subject \; to:$$

*Positivity*:

$$R_{11} = R_{11}^T > 0 \,,$$

$$S = S^T = \begin{bmatrix} S_{11} & S_{12} \\ S_{12}^T & S_{22} \end{bmatrix} > 0 \,,$$

$$R_{11} - S_{11} > 0 \,,$$

*Open-loop condition*:

$$\text{He} \begin{bmatrix} A_p R_{11} + B_{p,u} Z & B_{p,w} & 0 \\ 0 & -\frac{I_{n_w}}{2} & 0 \\ C_{p,z} R_{11} + D_{p,zu} Z & D_{p,zw} & -\frac{\gamma_G^2 I_{n_z}}{2} \end{bmatrix} < 0 \,,$$

*Closed-loop condition*:

$$\begin{bmatrix} SA_{cl}^T + A_{cl}S & B_{cl,w} & SC_{cl,z}^T \\ B_{cl,w}^T & -I_{n_w} & D_{cl,zw}^T \\ C_{cl,z}S & D_{cl,zw} & -\gamma_G^2 I_{n_z} \end{bmatrix} + \varepsilon I < 0 \,,$$

*Regional conditions*:

$$\begin{bmatrix} \bar{u}_i^2/s^2 & Z_i \\ Z_i^T & R_{11} \end{bmatrix} \geq 0, \quad i = 1, \ldots, n_u,$$

*Guaranteed reachable set and stability region*:

$$R_p \leq R_{11} \leq S_p.$$

**Step c)** Compute the SVD of the symmetric matrix $R_{11} - S_{11}$ determined from the solution of the previous step. That is, compute $F, \Lambda$ such that $F \Lambda F^T = R_{11} - S_{11}$, where $\Lambda$ is a positive semidefinite diagonal matrix whose diagonal entries are nonincreasing (for example, use the MATLAB command [U,Lambda,V] = svd(R11-S11); F = (U+V)/2;). Set $n_{aw} = n_p$.

**Step d)** Call $\lambda_1, \ldots \lambda_{n_{aw}-1}$ the first $n_{aw} - 1$ entries on the diagonal of $\Lambda$ and define $\hat{\Lambda} := \text{diag}\{\lambda_1, \ldots, \lambda_{n_{aw}-1}, 0, \ldots, 0\}$. That is, $\hat{\Lambda}$ is a diagonal matrix having its first $n_{aw} - 1$ diagonal entries equal to the entries of $\Lambda$ and the remaining entries equal to zero. Also define $\hat{S}_{11} := S_{11} + F(\Lambda - \hat{\Lambda})F^T$.

**Step e)** Given the matrix $\hat{S} := \begin{bmatrix} \hat{S}_{11} & S_{12} \\ S_{12}^T & S_{22} \end{bmatrix}$, where $\hat{S}_{11}$ was determined at the previous step, verify the feasibility of the following LMI in the variable $\gamma^2$:

$$\begin{bmatrix} \hat{S}A_{cl}^T + A_{cl}\hat{S} + \varepsilon I & B_{cl,w} & \hat{S}C_{cl,z}^T \\ B_{cl,w}^T & -I_{n_w} & D_{cl,zw}^T \\ C_{cl,z}\hat{S} & D_{cl,zw} & -\gamma^2 I_{n_z} \end{bmatrix} < 0.$$

If the LMI is feasible, then set $(R_{11}^*, S^*, \gamma^*) := (R_{11}, \hat{S}, \sqrt{\gamma^2})$, set $n_{aw} = n_{aw} - 1$, and go to step d. If the LMI is not feasible, then go to step f.

**Step f)** Select the order of the anti-windup compensator as $n_{aw}$ and the $(R, S, \gamma)$ solution as $(R_{11}^*, S^*, \gamma^*)$.

$\star$

In the worst case, Procedure 4 will give a solution $(R, S, \gamma)$ characterizing an anti-windup compensation of order $n_{aw} = n_p$. However, several case studies confirmed that the reduced order obtained with this algorithm is very likely to be smaller than or equal to $n_p$. Clearly, its value depends on the level of guaranteed performance $\gamma_G$ requested from the construction.

Once a triple $(R^*, S^*, \gamma^*)$ satisfying (5.19) with a small enough $n_{aw}$ is determined from Procedure 4, the reduced-order anti-windup construction can be carried out following a simple generalization of the previous Algorithms 3 and 8, which is reported next.

**Algorithm 9** (Dynamic reduced-order full-authority regional DLAW)

| Applicability | | | | Architecture | | | Guarantee |
|---|---|---|---|---|---|---|---|
| Exp Stab | Marg Stab | Marg Unst | Exp Unst | Lin/ NonL | Dyn/ Static | Ext/ FullAu | Global/ Regional |
| $\checkmark$ | $\checkmark$ | $\checkmark$ | $\checkmark$ | L | D* | FA | R |

*Comments*: Useful when Algorithm 3 is not feasible or performs poorly yet a low-order anti-windup augmentation is desired. Order reduction is carried out while maintaining a prescribed global input-output gain.
* Generally of low order resulting from order reduction.

**Step 1.** Given a plant and a controller of the form (5.1), (5.2), construct the matrices of the compact closed-loop representation using (5.5) and (5.8). Select a small positive constant $\varepsilon$ to robustify the order minimization procedure and a constant $\alpha > 1$ characterizing the amount of performance to be traded in for order reduction. Select a constant $v \in [0,1)$ used to enforce a strong well-posedness constraint (see Section 3.4.2 on page 60) and select the maximum size $s$ of allowable external inputs $\|w\|_2 < s$ over which the gain minimization has to be performed.

Possibly select two matrices $S_p$ and $R_p$ characterizing a desirable guaranteed reachable set contained in $\mathcal{R}_p = \{x_p : x_p^T R_p^{-1} x_p\}$ and a desirable stability region containing $\mathcal{S}_p = \{x_p : x_p^T S_p^{-1} x_p\}$ in the plant state directions, otherwise ignore the conditions involving $\mathcal{S}_p$ and $R_p$ below.

**Step 2.** Find the optimal solution $(R_{11}, S, Z, \gamma_{n_p}^2) \in \mathbb{R}^{n_p \times n_p} \times \mathbb{R}^{n_{cl} \times n_{cl}} \times \mathbb{R}^{n_u \times n_p}$ $\times \mathbb{R}$ to the following LMI eigenvalue problem which is always feasible for small enough values of $s$:

$$\min_{R11, S, Z, \gamma^2} \gamma \text{ subject to:}$$

*Positivity*:

$$R_{11} = R_{11}^T > 0,$$

$$S = S^T = \begin{bmatrix} S_{11} & S_{12} \\ S_{12}^T & S_{22} \end{bmatrix} > 0,$$

$$R_{11} - S_{11} > 0,$$

*Open-loop condition*:

$$\text{He} \begin{bmatrix} A_p R_{11} + B_{p,u} Z & B_{p,w} & 0 \\ 0 & -\frac{I_{n_w}}{2} & 0 \\ C_{p,z} R_{11} + D_{p,zu} Z & D_{p,zw} & -\frac{\gamma^2 I_{n_z}}{2} \end{bmatrix} < 0,$$

*Closed-loop condition*:

$$\begin{bmatrix} SA_{cl}^T + A_{cl} S & B_{cl,w} & SC_{cl,z}^T \\ B_{cl,w}^T & -\gamma I_{n_w} & D_{cl,zw}^T \\ C_{cl,z} S & D_{cl,zw} & -\gamma I_{n_z} \end{bmatrix} < 0,$$

*Regional conditions*:

$$\begin{bmatrix} \bar{u}_i^2 / s^2 & Z_i \\ Z_i^T & R_{11} \end{bmatrix} \geq 0, \quad i = 1, \dots, n_u,$$

*Guaranteed reachable set and stability region*:

$$R_p \leq R_{11} \leq S_p.$$

The resulting performance level $\gamma_{n_p}$ corresponds to the maximum performance achievable by plant-order anti-windup augmentation.

**Step 3.** Based on the matrices and parameters determined at step 1, select the desired guaranteed performance as $\gamma_G = \alpha \gamma_{n_p}$ and determine the reduced compensator order $n_{aw}$ and the triple $\left( R_{11}^*, S^* = \begin{bmatrix} S_{11}^* & S_{12}^* \\ (S_{12}^*)^T & S_{22}^* \end{bmatrix}, \gamma^* \right)$ from Procedure 4.

**Step 4.** Based on the solution determined at the previous step, construct the symmetric matrix $R = \begin{bmatrix} R_{11}^* & S_{12}^* \\ (S_{12}^*)^T & S_{22}^* \end{bmatrix}$ and define $Q_{12} \in \mathbb{R}^{n_{cl} \times n_{aw}}$ as any solution to the following equation:

$$Q_{12} Q_{12}^T = RS^{-1} R - R,$$

which is always solvable and admits infinitely many solutions because by construction, $\text{rank}(RS^{-1}R - R) = \text{rank}((RS^{-1}R - R)(R^{-1}S)) = \text{rank}(R - S) =$

$n_{aw}$. A solution can be determined following Procedure 1 on page 113 with $\Xi = RS^{-1}R - R$. Then define the following matrices:

$$Q_{22} = I_{n_{aw}} + Q_{12}^T R^{-1} Q_{12}$$

$$Q = \begin{bmatrix} R & Q_{12} \\ Q_{12}^T & Q_{22} \end{bmatrix}.$$

**Step 5.** Based on the matrices determined at steps 1 and 4 define the following matrices (where the second row and column of zeros have size $n_{aw}$):

$$\left[ \begin{array}{c|cc} A_0 & B_{0,q} & B_{0,w} \\ \hline C_{0,u} & D_{cl,yq} & D_{cl,yw} \\ C_{0,z} & D_{cl,zq} & D_{cl,zw} \end{array} \right] = \left[ \begin{array}{cc|cc} A_{cl} & 0 & B_{cl,q} & B_{cl,w} \\ 0 & 0_{n_{aw} \times n_{aw}} & 0 & 0 \\ \hline C_{cl,u} & 0 & D_{cl,uq} & D_{cl,uw} \\ C_{cl,z} & 0 & D_{cl,zq} & D_{cl,zw} \end{array} \right]$$

and construct the matrices $G_U \in \mathbb{R}^{(n_{aw}+n_u) \times m}$, $H \in \mathbb{R}^{m \times (n_{aw}+n_c+n_u)}$, and $\Psi \in \mathbb{R}^{m \times m}$, with $m := n_{cl} + n_{aw} + n_u + n_w + n_z$, as follows:

$$\Psi(U, \gamma^2) = \mathrm{He} \begin{bmatrix} A_0 Q & B_{0,q} U + Q C_{0,u}^T & B_{0,w} & Q C_{0,z}^T \\ -Y & D_{cl,uq} U - U & D_{cl,uw} & U D_{cl,zq}^T \\ 0 & 0 & -\frac{1}{2}I & D_{cl,zw}^T \\ 0 & 0 & 0 & -\frac{\gamma^2}{2}I \end{bmatrix},$$

$$H = \left[ \begin{array}{cc|c|c|c} 0 & I_{n_{aw}} & 0 & 0_{n_{aw} \times n_w} & 0 \\ B_{cl,v}^T & 0 & D_{cl,uv}^T & 0 & D_{cl,zv}^T \end{array} \right]$$

$$G_U = \left[ \begin{array}{cc|c|c|c} Q_{12}^T & Q_{22} & 0 & 0 & 0 \\ 0 & 0 & I_{n_u} & 0 & 0 \end{array} \right],$$

where for any square matrix $X$, $\mathrm{He}(X) = X + X^T$, and where $Y = \begin{bmatrix} Z & Z R_{11}^{-1} R_{12} & Z R_{11}^{-1} \bar{Q} \end{bmatrix}$ (with $\bar{Q}$ being the matrix formed by the first $n_p$ rows of $Q_{12}$).

**Step 6.** Find the optimal solution $(\Lambda_U, U, \gamma^2)$ of the LMI eigenvalue problem:

$$\min_{\Lambda_U, U, \gamma^2} \gamma^2 \text{ subject to:}$$

*Strong well-posedness (if $\nu = 0$, then redundant):*

$$-2(1-\nu)U + \mathrm{He}\left( D_{cl,uq}U + \begin{bmatrix} 0_{n_u \times (n_{aw}+n_c)} & I_{n_u} \end{bmatrix} \Lambda_U \begin{bmatrix} 0_{n_u \times n_{aw}} \\ I_{n_u} \end{bmatrix} \right) < 0,$$

*Anti-windup gains LMI:*
$$\Psi(U, \gamma^2) + G_U^T \Lambda_U^T H^T + H \Lambda_U G_U < 0.$$

**Step 7.** Select the plant-order full-authority anti-windup compensator as:

$$\dot{x}_{aw} = A_{aw} x_{aw} + B_{aw} q$$
$$v = C_{aw} x_{aw} + D_{aw} q$$

with the selection $\left[ \begin{array}{c|c} A_{aw} & B_{aw} \\ \hline C_{aw} & D_{aw} \end{array} \right] = \Lambda_U \left[ \begin{array}{c|c} I_{n_{aw}} & 0 \\ \hline 0 & U^{-1} \end{array} \right].$

$\star$

**Example 5.5.2** (Electrical network) Consider the same example for which regional plant-order anti-windup has been designed in Example 5.5.1. The regional performance obtained there can be given up for an increased simplicity in the anti-windup architecture. Keeping the value $s = 0.1$ used in Example 5.5.1 and fixing $\alpha = 1.3$ (so that $\gamma_G = 7.88$ is obtained), Algorithm 9 provides the following first-order anti-windup compensator:

$$
\left[ \begin{array}{c|c} A_{aw} & B_{aw} \\ \hline C_{aw} & D_{aw} \end{array} \right] = \left[ \begin{array}{c|c} -1.056 & -0.001306 \\ 157.1 & 2.431 \\ -1.955 & -0.03047 \\ 161.4 & 0.8000 \end{array} \right],
$$

inducing a regional performance $\gamma = 7.35$ (note that due to numerical reasons, this performance is slightly better than $\gamma_G$).

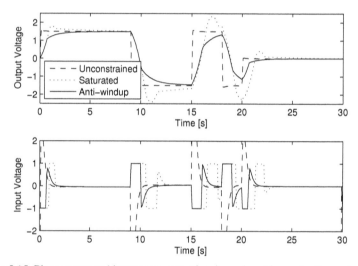

Figure 5.15 Plant output and input responses of various closed loops for Example 5.5.2.

Figure 5.15 represents a simulation showing how the reduced-order solution leads to a slightly slower response, as compared to the equivalent plant-order regional solution of Example 5.5.1. This slight deterioration of the closed-loop performance is also confirmed by the curves in Figure 5.16, where it is shown that the nonlinear gain curve of Example 5.5.1 (dashed) lies below the nonlinear gain curve of this example.                                                                    □

## 5.6 NOTES AND REFERENCES

Dynamic direct linear anti-windup designs are definitely more recent than the static anti-windup schemes discussed in the previous chapters. To the best of the authors' knowledge, the first systematic LMI-based approach for dynamic anti-windup design appeared in in [MI18] (see also [MI19] for some case studies). Note, however, that alternative dynamic approaches to anti-windup had already appeared in

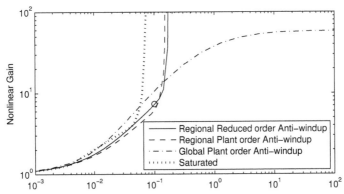

Figure 5.16 Nonlinear gains for Examples 5.4.2, 5.5.1, and 5.5.1 characterizing the system before and after plant-order global (dashed-dotted), plant-order regional (dashed), and reduced-order regional (solid) anti-windup augmentation.

[MR1, MR3, H1, H3] and other works. The full-authority work in [MI18] was later extended to the external case in [MI22]. It was also extended to the full-authority regional anti-windup case in [MI33]. Regional external anti-windup hasn't been published anywhere but is a straightforward extension of existing results. An alternative interesting approach is reported in [MI23], where an LMI-based procedure to minimize the URR gain instead of the $\mathcal{L}_2$ gain reported here is given. This approach is not reported in this book to keep the discussion simple. The reduced-order anti-windup designs reported here are taken from [MI28]. Alternative reduced-order designs can be found in [MI25] as well as [MI31] and [MI34] and references therein.

Discrete-time counterparts of the algorithms given here have been given in [MI30] (which can be seen as the discrete-time equivalent of the work in [MI18] and [MI33]).

The MIMO academic example no. 2 treated in Example 5.4.1 is taken from [MI11] even though it had appeared before in [FC7] (it was also revisited several times later, e.g., in [MI19]). The passive electrical circuit used in Examples 5.4.2, 5.4.6, 5.5.1, and 5.5.2 was introduced and used in [MI19] and later revisited in [MI22]. The SISO academic example no. 2 addressed in Examples 5.4.3, 5.4.5 and 5.4.8 was introduced in [MI16], where it was discussed that the quadratic construction of [MI11] was not always feasible (the system theoretic feasibility conditions appeared later in [MI18]). The F8 longitudinal dynamics example addressed in Examples 5.4.4, 5.4.7, and 5.4.9 first appeared in [SAT1] and was later revisited in a large number of papers (e.g., [MI18]).

PART 3

Model Recovery Anti-windup Augmentation

# Chapter Six

## The MRAW Framework

### 6.1 INTRODUCTION

All of the anti-windup algorithms presented in this part of the book share a common, external anti-windup structure. The term "model recovery anti-windup" (MRAW) will be used for these algorithms because the structure they use recovers the unconstrained plant model as seen from the viewpoint of the unconstrained controller. This feature keeps the unconstrained controller from misbehaving in the constrained closed loop with anti-windup augmentation, and it gives this closed loop some information about how it would be performing in the ideal case without constraints. Equipped with this information, the degrees of freedom that remain within the MRAW structure can be used to address recovering this unconstrained response, which turns out to be the most natural anti-windup performance measure to use for MRAW design. The MRAW structure reduces the anti-windup synthesis problem to a problem of direct control design for input saturation. In other words, the degrees of freedom can be constructed ignoring the existence of the predesigned linear controller. Consequently, the degrees of freedom in MRAW can be synthesized using many existing control algorithms, including simple LMI-based algorithms. More advanced algorithms will be introduced to achieve nearly maximal stability domains for unstable plants and nonlinear systems in general.

Most of this book focuses on anti-windup design for linear systems with input constraints. However, some attention will be given to more general nonlinear systems hereafter. These results permit tackling linear systems with simultaneous input magnitude and input rate constraints, as well as general fully actuated Euler-Lagrange systems. Later on, systems with input time delays will be addressed using the MRAW structure. Moreover, the different but related problems of bumpless transfer and reliable control will be considered. For all of these problems, the literature on constrained control for general nonlinear systems is relevant and provides algorithms that can be used within the MRAW structure.

In the next section, the MRAW structure will be described through block diagrams and transfer functions first, then via state-space representations. The transfer function description may be easier to understand, while the state-space representation reveals that the idea can also be applied to nonlinear systems.

## 6.2 A BLOCK DIAGRAM/TRANSFER FUNCTION DESCRIPTION

### 6.2.1 The augmentation structure

In terms of block diagrams, the MRAW structure appears as shown in Figure 6.1, where the upper diagram represents the unconstrained closed-loop system and the lower diagram represents the anti-windup augmented closed-loop system. In the lower diagram, the saturation nonlinearity has been included, and an extra dynamical system denoted $\widehat{\mathcal{P}}$ has been introduced. This notation is used because $\widehat{\mathcal{P}}$ is chosen to be a model of the plant $\mathcal{P}$. The degrees of freedom in the design of the anti-windup algorithm reside in the selection of the signal $v$, the source of which has been left unspecified in the diagram. Notice that the input of the unconstrained controller has been changed from $y$ to $y - y_{aw}$, where $y$ is the output of the plant $\mathcal{P}$ and $y_{aw}$ is the corresponding output of the extra dynamics $\widehat{\mathcal{P}}$. The input to the saturation nonlinearity has been changed from $y_c$ to $y_c + v$, where $y_c$ is the output of the unconstrained controller and $v$ is to be designed.

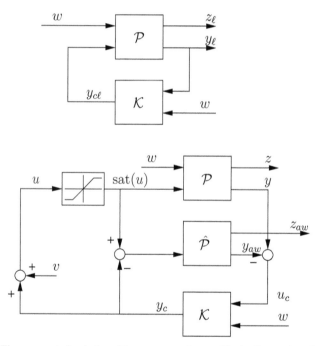

Figure 6.1 The unconstrained closed-loop system (upper block diagram) and the MRAW compensation scheme (lower block diagram).

### 6.2.2 Model recovery

When $\widehat{\mathcal{P}} = \mathcal{P}$ and these systems are linear, the effect of the input $\mathrm{sat}(u)$ on the real plant is equal to its effect on the copy of the plant introduced for anti-windup purposes. Therefore, the effect of $\mathrm{sat}(u)$ on the outputs of the two plants is equal and the corresponding responses cancel each other before reaching the controller input $u_c$. This happens regardless of the selection of the signal $v$. Indeed, the only signals that affect the measurement input of the controller are the $w$ signal, through the upper plant, and the $y_c$ signal, through the lower plant.

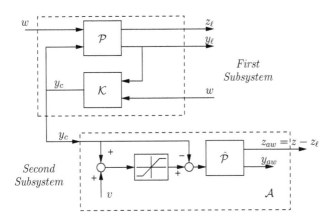

Figure 6.2  The cascade structure of the MRAW compensation scheme.

Still using the assumption that $\widehat{P} = P$, the anti-windup augmented system can be transformed into the cascaded structure of Figure 6.2, where the upper subsystem exactly reproduces the unconstrained dynamics and the lower subsystem captures information about the deviation of the anti-windup response from the desirable unconstrained one. Indeed, the signal $z_{aw}$ captures the mismatch between the cost variable $z_\ell$ in the unconstrained closed-loop system and the cost variable $z$ in the constrained closed-loop system with anti-windup augmentation.

### 6.2.3  Connections to the anti-windup filter $\mathcal{F}$

The signal $v$ is selected with the goal of driving to zero the state of the second subsystem, which is the state of the anti-windup filter. Since this state captures the mismatch between the actual and the desired responses when $\widehat{P} = P$, driving it to zero corresponds to recovering the unconstrained response. It should be clear from the lower diagram in Figure 6.2 that, even when $\widetilde{P} \neq P$, driving to zero the state of $\widehat{P}$ results in restoring the unconstrained closed-loop system. With this aim, the signal $v$ is generally determined as a feedback signal from the state and input variables of the anti-windup filter, as shown in Figure 6.3. The feedback loop generated by this choice of $v$ generates the overall anti-windup filter $\mathcal{F}$, which is driven by $\text{sat}(u) - u$ and produces signals that are fed to the unconstrained controller input and output. This corresponds to the "external" anti-windup structure described in Section 2.3.6. As discussed there, this structure guarantees the small signal preservation property regardless of how $v$ is chosen.

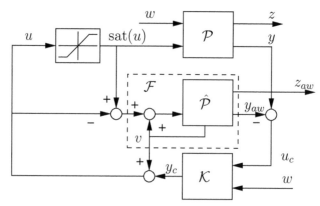

Figure 6.3 A typical block diagram representation of MRAW augmentation. The degrees of freedom in the design reside in determining the signal $v$, which is indicated as a supplemental output from the plant model $\widehat{\mathcal{P}}$.

### 6.2.4 An equivalent transfer function description

In terms of transfer functions, the MRAW structure for constrained linear systems is described as follows:

Given are a linear plant with constrained inputs

$$y = \mathcal{P} \begin{bmatrix} \mathrm{sat}(u) \\ d \end{bmatrix} \tag{6.1}$$

and a controller that was constructed for the linear model

$$y = \mathcal{P} \begin{bmatrix} u \\ d \end{bmatrix}. \tag{6.2}$$

MRAW augmentation then begins by:

1. building a filter, with state $x_{aw}$ that has dimension equal to that of the plant, satisfying

$$y_{aw} = \widehat{\mathcal{P}} \begin{bmatrix} \mathrm{sat}(u) - y_c \\ 0 \end{bmatrix}, \tag{6.3a}$$

   where $y_c$ is the output of the unconstrained controller;
2. changing the input of the unconstrained controller to

$$u_c = y - y_{aw} ; \tag{6.3b}$$

3. changing the input of the plant to

$$u = y_c + v, \tag{6.3c}$$

where again $y_c$ is the output of the unconstrained controller and $v$ represents the degrees of freedom within the MRAW structure used to achieve good anti-windup performance.

In the case where $\widehat{\mathcal{P}} = \mathcal{P}$, the structure given above recovers the unconstrained plant model as seen from the viewpoint of the unconstrained controller, thus motivating the "model recovery anti-windup" terminology. This is noted as follows.

Using the equations above and exploiting the linearity of $\mathcal{P}$, it follows that the input to the unconstrained controller satisfies

$$u_c = y - y_{aw} = \mathcal{P}\begin{bmatrix} y_c \\ d \end{bmatrix}, \tag{6.4}$$

where $y_c$ is the output of the unconstrained controller. Thus, the dynamical system that is connected in feedback around the unconstrained controller has the model $\mathcal{P}$ for which the unconstrained controller was constructed. Because of this, the unconstrained controller will not misbehave in the constrained closed loop with anti-windup augmentation. Moreover, the response of the interconnection of the unconstrained controller with $\mathcal{P}$ gives the unconstrained response. In particular, $y_c$ will match the output of the unconstrained controller in the unconstrained closed loop. In many reasonable cases this signal is guaranteed to be converging to a region strictly inside the set where the saturation function equals the identity mapping. The effect of $v$ on the anti-windup augmented closed loop is through the anti-windup filter

$$y_{aw} = \mathcal{P}\begin{bmatrix} \text{sat}(y_c + v) - y_c \\ 0 \end{bmatrix}. \tag{6.5}$$

The task of $v$ is then to keep the state of this filter small and drive it to zero in the presence of signals $y_c$ converging to a region inside the set where the saturation function equals the identity mapping.

## 6.3 A STATE-SPACE DESCRIPTION (LINEARITY NOT NEEDED)

### 6.3.1 The nonlinear augmentation structure

In state-space form, which permits extensions to nonlinear systems, the MRAW structure is described as follows.

Given is a plant

$$\begin{aligned} \dot{x} &= f(x,u) + d_1 \\ y &= h(x,u) + d_2, \end{aligned} \tag{6.6}$$

which, for constrained linear systems, would correspond to

$$\begin{aligned} f(x,u) &= A_p x + B_{p,u}\text{sat}(u) \\ h(x,u) &= C_{p,y} x + D_{p,yu}\text{sat}(u). \end{aligned}$$

Also given is a controller, which won't be specified, that was constructed for a model related to the plant (6.6) and given by

$$\begin{aligned} \dot{x} &= F(x,u) + d_1 \\ y &= H(x,u) + d_2. \end{aligned} \tag{6.7}$$

For linear systems, typically it would be

$$\begin{aligned} F(x,u) &= A_p x + B_{p,u} u \\ H(x,u) &= C_{p,y} x + D_{p,yu} u. \end{aligned}$$

MRAW augmentation then begins by

1. building a filter, with state $x_{aw}$ that has dimension equal to that of the plant, satisfying

$$
\begin{aligned}
\dot{x}_{aw} &= f(x,u) - F(x - x_{aw}, y_c) \\
y_{aw} &= h(x,u) - H(x - x_{aw}, y_c),
\end{aligned}
\tag{6.8}
$$

where $y_c$ is the output of the unconstrained controller;

2. changing the input of the unconstrained controller from $u_c = y$ to

$$
u_c = y - y_{aw} ;
\tag{6.9}
$$

3. changing the input of plant from $u = y_c$ to

$$
u = y_c + v,
\tag{6.10}
$$

where again $y_c$ is the output of the unconstrained controller and where the function $v$ is to be designed in order to achieve good anti-windup performance.

In the general nonlinear case, where the anti-windup filter is given by the equation

$$
\dot{x}_{aw} = f(x,u) - F(x - x_{aw}, u - v),
\tag{6.11}
$$

the small signal behavior is preserved if, near the operating point of interest, all functions used to describe the system are locally Lipschitz[1] $f(x,u) = F(x,u)$, $h(x,u) = H(x,u)$ and $v$ evaluates to zero when $x_{aw} = 0$. Indeed, in this case, if the unconstrained response remains near the operating point and $x_{aw}(0)$, then the unique solution for $x_{aw}$ is $x_{aw}(t) \equiv 0$.

### 6.3.2 Model recovery

As noted now, the structure given above recovers the unconstrained plant model as seen from the viewpoint of the unconstrained controller, thus motivating the "model recovery anti-windup" terminology.

Using the equations above, it follows that the input to the unconstrained controller satisfies

$$
\begin{aligned}
u_c = y - y_{aw} &= h(x,u) + d_2 - [h(x,u) - H(x - x_{aw}, y_c)] \\
&= H(x - x_{aw}, y_c) + d_2 .
\end{aligned}
\tag{6.12}
$$

This input depends on the state $x - x_{aw}$, which satisfies the equation

$$
\begin{aligned}
\overset{\cdot}{\overbrace{x - x_{aw}}} &= f(x,u) + d_1 - [f(x,u) - F(x - x_{aw}, y_c)] \\
&= F(x - x_{aw}, y_c) + d_1 .
\end{aligned}
\tag{6.13}
$$

Thus, the system that is connected in feedback around the unconstrained controller evolves according to the model, expressed in terms of the functions $F$ and $H$, for which the unconstrained controller was designed. Because of this, the unconstrained controller will not misbehave in the constrained closed loop with anti-windup augmentation. Moreover, since $x - x_{aw}$ will match the unconstrained plant

---

[1]It is well known in nonlinear systems analysis that the Lipschitz continuity assumption ensures (existence and) uniqueness of solutions.

response, the filter state $x_{aw}$ is the difference, or mismatch, between the unconstrained plant response and the constrained closed-loop response with anti-windup
augmentation. So, the selection of $v$ should be aimed at making an appropriate
measure of $x_{aw}$ small.

It is instructive to consider how $v$ affects $x_{aw}$ in the case of constrained linear
systems where $f(x,u) = A_p x + B_{p,u}\text{sat}(u)$ and $F(x,u) = A_p x + B_{p,u} u$. In this case,
the filter equation becomes the following mismatch equation

$$\dot{x}_{aw} = A_p x_{aw} + B_{p,u}\left[\text{sat}(v + y_c) - y_c\right], \tag{6.14}$$

the signal $y_c$ being the output of the unconstrained controller. According to the discussion above, the signal $y_c$ is exactly the response that would occur at the output of
the unconstrained controller in the unconstrained closed loop. In many reasonable
cases this signal is guaranteed to be converging to a region strictly inside the set
where the saturation function equals the identity mapping. The task of $v$ is then to
keep $x_{aw}$ small and drive it to zero in the presence of such signals $y_c$. This control
problem, both in continuous time and from a sampled-data point of view, has received considerable attention in the control theory literature. This part of the book
discusses the use of such results within the MRAW structure to achieve suitable
anti-windup performance.

### 6.3.3 Connections to the anti-windup filter $\mathcal{F}$

In the special case where $v$ is chosen to satisfy

$$v = K x_{aw} + L\left[\text{sat}(v + y_c) - y_c\right] \tag{6.15}$$

for some suitable matrices $K$ and $L$, it is possible to be explicit about the anti-
windup filter $\mathcal{F}$ that is generated, in accordance with the description of anti-windup
given in Section 2.3.6. Indeed, using that $u = v + y_c$, equation (6.15) becomes

$$v = K x_{aw} + Lv + L\left[\text{sat}(u) - u\right], \tag{6.16}$$

i.e.,

$$v = (I - L)^{-1} K x_{aw} + (I - L)^{-1} L\left[\text{sat}(u) - u\right], \tag{6.17}$$

and then equation (6.14) becomes

$$\begin{aligned}
\dot{x}_{aw} &= A_p x_{aw} + B_{p,u} v + B_{p,u}\left[\text{sat}(u) - u\right] \\
&= (A_p + B_{p,u}(I - L)^{-1} K) x_{aw} + B_{p,u}(I - L)^{-1}\left[\text{sat}(u) - u\right].
\end{aligned} \tag{6.18}$$

Similarly, if in (6.6) the output equation is linear, namely, if $h(x,u) = C_{p,y} x + D_{p,yu}\text{sat}(u)$, then the output equation providing $y_{aw}$ in (6.8) becomes

$$y_{aw} = (I - L)^{-1} C_{p,y} x_{aw} + (I - L)^{-1} D_{p,yu}\left[\text{sat}(u) - u\right]. \tag{6.19}$$

Thus, the anti-windup filter $\mathcal{F}$ corresponds to a linear system with state-space
representation

$$\left[\begin{array}{c|c} A_{aw} & B_{aw} \\ \hline C_{aw} & D_{aw} \end{array}\right] = \left[\begin{array}{c|c} A_p + B_{p,u}(I-L)^{-1}K & B_{p,u}(I-L)^{-1} \\ (I-L)^{-1}C_{p,y} & (I-L)^{-1}D_{p,yu} \\ \hline (I-L)^{-1}K & (I-L)^{-1}L \\ (I-L)^{-1}C_{p,z} & (I-L)^{-1}D_{p,zu} \end{array}\right], \tag{6.20}$$

whose output has three stacked components $\begin{bmatrix} y_{aw} \\ v \\ z_{aw} \end{bmatrix}$. Within the description of Section 2.3.6, this anti-windup compensation is external and the first anti-windup output $y_{aw}$ should be added to the controller input $u_c$, the second anti-windup output $v$ should be added to the controller output $y_c$ and the third anti-windup output $z_{aw}$ is a measure of performance representing the mismatch between the unconstrained performance output $z_\ell$ and the actual performance output $z$. This third output is not necessarily computed in the anti-windup implementation but plays a major role in the anti-windup design phase because it captures the information on the effectiveness of the anti-windup compensation.

### 6.3.4 Measurements used in nonlinear MRAW augmentation

For the general nonlinear case, the anti-windup filter used in MRAW requires measurements of the components of $x$ that appear explicitly in the function

$$f(x, y_c + v) - F(x - x_{aw}, y_c) .$$

In the case of constrained linear systems,

$$f(x, y_c + v) - F(x - x_{aw}, y_c) = A_p x_{aw} + B_{p,u} \left[ \text{sat}(y_c + v) - v \right].$$

Thus, no components of $x$ show up in this function. The filter depends only on its states $x_{aw}$, the output of the unconstrained controller $y_c$, and the control commanded to the constrained plant $y_c + v$.

More generally, for nonlinear systems and models that are linear up to input and output injection, the only measurements that are needed are the original inputs and outputs. For the application of MRAW to general Euler-Lagrange systems, the structure as described requires measurement of both positions and generalized velocities. On the other hand, it is always possible to estimate velocities using an observer and to use those estimates in place of the actual state values.

In some, but not all, of the algorithms for exponentially unstable linear systems, the function $v$ will be chosen to depend on the exponentially unstable modes of the plant in addition to the states of the anti-windup filter. In this case, those extra measurements are needed in principle. Again, an observer can be used to avoid this requirement.

## 6.4 ROBUST, FRAGILE, OR BOTH?

### 6.4.1 Plant model recovery

The ability of the MRAW structure to recover the unconstrained plant model as seen from the viewpoint of the unconstrained controller depends crucially on having an accurate constrained plant model. This is because the constrained plant model is used to synthesize the anti-windup filter and cancellation is used to recover the unconstrained plant model as seen from the viewpoint of the unconstrained controller. This cancellation leads to a cascade structure for the anti-windup augmented closed loop (see Figure 6.2): one system corresponds to the unconstrained closed loop and one of its signals drives the anti-windup filter which is to be stabilized via the function $v$ to be designed. Obviously, cancellation and the ensuing cascade structure are not robust. These properties do not persist even in the presence of an arbitrarily small mismatch between the actual constrained plant and the constrained plant model.

### 6.4.2  High-performance anti-windup

While the plant model recovery feature is not robust, that is not the goal. The goal is high-performance anti-windup. For this purpose, the function $v$ is designed to stabilize the dynamics of the anti-windup filter, often with the best stability margins possible. Because of this, the anti-windup augmented closed loop can tolerate some mismatch between the actual constrained plant and the constrained plant model that would lead to an interacting structure between the anti-windup filter and a perturbation of the unconstrained closed loop. To appreciate why this is the case consider, for cascaded exponentially stable linear systems, the robustness of exponentially stability to perturbations that destroy the cascade structure. The situation for nonlinear systems, while more complicated, is quite similar. The amount of mismatch that can be tolerated is problem specific. Sometimes the amount is quite large while other times it is small. In the case where the control engineer has very little confidence in the accuracy of the constrained plant model, the MRAW structure is probably not the best anti-windup structure to use.

### 6.4.3  Mathematical explanation of robustness*

In particular, in cases where the unconstrained controller is linear,[2] much can be said on the robustness of the scheme to perturbations on the plant dynamics of the following type:

$$\mathcal{P} \begin{cases} \dot{x}_p &= A_p x_p + B_p u + B_{pw} w + \psi_x \\ y &= C_y x_p + D_{uy} u + D_{wy} w + \psi_y \\ z &= C_z x_p + D_{uz} u + D_{wz} w + \psi_z, \end{cases} \tag{6.21}$$

where $\psi_x$, $\psi_y$, and $\psi_z$ are the outputs of a linear dynamic system representing the unmodeled dynamics, which can be written as

$$\Psi := \begin{bmatrix} \psi_x \\ \psi_y \\ \psi_z \end{bmatrix} = \Delta(s) \begin{bmatrix} x_p \\ u \\ w \end{bmatrix} = \begin{bmatrix} \Delta_x(s) \\ \Delta_y(s) \\ \Delta_z(s) \end{bmatrix} \begin{bmatrix} x_p \\ u \\ w \end{bmatrix}, \tag{6.22}$$

where $\Delta(s)$ is a linear time-invariant exponentially stable system.

According to the representation (6.22) for the unmodeled dynamics and the notation in (6.21) for the plant dynamics, the MRAW compensation scheme with the approximate dynamics (7.2) will in general guarantee robust properties, in the sense that the performance and stability properties of the closed loop will be robust with respect to the unmodeled dynamics (6.22), as long as the input-output $\mathcal{L}_2$ gain of (6.22) is sufficiently small.

---

[2]The linearity assumption for the unconstrained controller could actually be relaxed to a weaker incremental stability assumption for the unconstrained closed-loop system.

Note that the representation (6.22) for the unmodeled dynamics is a fairly general one when the actual plant is assumed to be linear. The following example describes the description of two frequent cases using the notation in (6.22).

**Example 6.4.1**   When uncertainties affect the entries of the matrices of the plant model, the actual plant dynamics (6.21) correspond to:

$$\dot{x}_p = (A_p + \varepsilon \Delta A_p) x_p + (B_p + \varepsilon \Delta B_p) \, \mathrm{sat}(u) + (B_{pw} + \varepsilon \Delta B_{pw}) w$$
$$y = (C_y + \varepsilon \Delta C_y) x_p + (D_{uy} + \varepsilon \Delta D_{uy}) \, \mathrm{sat}(u) + (D_{wy} + \varepsilon \Delta D_{wy}) w$$
$$z = (C_z + \varepsilon \Delta C_z) x_p + (D_{uz} + \varepsilon \Delta D_{uz}) \, \mathrm{sat}(u) + (D_{wz} + \varepsilon \Delta D_{wz}) w,$$

which can be written in the form (6.21), (6.22) by selecting the unmodeled dynamics $\Delta(s)$ as the following static system:

$$\Delta \left\{ \begin{bmatrix} \psi_x \\ \psi_y \\ \psi_z \end{bmatrix} = \varepsilon \begin{bmatrix} \Delta A_p & \Delta B_p & \Delta B_{pw} \\ \Delta C_y & \Delta D_{uy} & \Delta D_{wy} \\ \Delta C_z & \Delta D_{uz} & \Delta D_{wz} \end{bmatrix} \begin{bmatrix} x_p \\ \mathrm{sat}(u) \\ w \end{bmatrix} \right. .$$

Note that $\varepsilon$ measures the size of the perturbation of each matrix in the plant equations and, as expected, the $\mathcal{L}_2$ gain of the perturbation $\Delta$ is proportional to $\varepsilon$.

Consider also the case where a reduced-order model of the plant is used for antiwindup design. Assume in particular that the state variables $x_p$ are a subset of the overall plant states $(x_p, x_\delta)$, and that the dynamics of $x_\delta$ are weakly observable and/or controllable. Then the actual plant dynamics can be written as

$$\begin{bmatrix} \dot{x}_p \\ \dot{x}_\delta \end{bmatrix} = \begin{bmatrix} A_p & A_{p12} \\ A_{p21} & A_{p22} \end{bmatrix} \begin{bmatrix} x_p \\ x_\delta \end{bmatrix} + \begin{bmatrix} B_p \\ B_{p\delta} \end{bmatrix} \mathrm{sat}(u) + \begin{bmatrix} B_{pw} \\ B_{pw\delta} \end{bmatrix} w$$

$$y = \begin{bmatrix} C_y & C_{y\delta} \end{bmatrix} \begin{bmatrix} x_p \\ x_\delta \end{bmatrix} + D_{uy} \, \mathrm{sat}(u) + D_{wy} w$$

$$z = \begin{bmatrix} C_z & C_{z\delta} \end{bmatrix} \begin{bmatrix} x_p \\ x_\delta \end{bmatrix} + D_{uz} \, \mathrm{sat}(u) + D_{wz} w,$$

which can be also written in the form (6.21), (6.22) by selecting the unmodeled dynamics $\Delta(s)$ as the following dynamical system:

$$\Delta \left\{ \begin{array}{rcl} \dot{x}_\delta & = & A_{p22} x_\delta + A_{p21} x_p + B_{p\delta} \, \mathrm{sat}(u) + B_{pw\delta} w \\ \begin{bmatrix} \psi_x \\ \psi_y \\ \psi_z \end{bmatrix} & = & \begin{bmatrix} A_{p12} \\ C_{y\delta} \\ C_{z\delta} \end{bmatrix} x_\delta . \end{array} \right.$$

As a last case, consider the addition of a strictly proper fast linear filter at each channel of the measured output $y$ of the plant. This situation resembles the presence of possibly very fast unmodeled sensor dynamics. This can be modeled by the transfer function

$$y(s) = \frac{1}{\varepsilon s + 1} y_U(s),$$

where $y_U(s)$ denotes the unperturbed output signal $y_U = C_y x_p + D_{uy} u + D_{wy} w$, and where the constant $\varepsilon$ is inversely proportional to the speed of the linear filter, so that

$\varepsilon = 0$ corresponds to the unperturbed system. This situation can be represented with the notation (6.22) by choosing

$$
\Delta \quad \left\{
\begin{array}{rcl}
\dot{x}_\delta & = & -\dfrac{\delta}{\varepsilon} - (C_y x_p + D_{uy} u + D_{wy} w) \\[2ex]
\left[
\begin{array}{c}
\psi_x \\
\psi_y \\
\psi_z
\end{array}
\right]
& = &
\left[
\begin{array}{c}
0 \\
I \\
0
\end{array}
\right] x_\delta.
\end{array}
\right.
\tag{6.23}
$$

□

## 6.5 NOTES AND REFERENCES

The presentation in this chapter is based on the conference papers [MR4] and [MR3]. In that work, various linear and nonlinear selections for the signal $v$ were proposed, all of them not inducing any algebraic loop. In later work [MR15], linear selections for $v$ were proposed involving a deadzone injection, thus inducing an algebraic loop in the compensation scheme. The corresponding gains were selected there based on the solution of suitable LMIs. Within the same framework, nonlinear selections for $v$ were also considered later, both when seeking for guaranteed stability regions in the case of exponentially unstable plants [MR5, AP6, MR30] and when seeking for extreme anti-windup performance with exponentially stable plants, via scheduled designs with hybrid switching [MR21, MR28], scheduling [MR29, MR27], and via sampled-data implementations of MPC-based control laws [MR17]. MRAW was also applied to fully actuated Euler-Lagrange systems (a special class of feedback linearizable nonlinear systems) in [MR19]. Finally, the MRAW ideas have been extended to different problems: MRAW anti-windup design for magnitude- and rate-saturated systems has been given in [AP2, AP6, MR12]; anti-windup design for dead-time plants, namely, plants with a pure delay at the plant input or output, has been given in [MR24]; and a formal definition of the bumpless transfer problem together with the MRAW-based solution has been given in [MR25]. The robustness aspects of MRAW anti-windup compensation are discussed in [MR26] (see also [H11]) in terms of suitable tradeoffs that should be carried out between performance and robustness. Discrete-time counterparts of some of the above MRAW techniques have been given in [MR32]. Most of these selections and extensions will be the topic of subsequent chapters.

A large but tractable number of schemes that have been introduced in the anti-windup literature make use of a copy of the plant dynamics in some way to synthesize a dynamic anti-windup filter. Many of these schemes can be interpreted in the framework discussed in this chapter, and correspond to a particular linear choice of the signal $v$. Several of these schemes are discussed below.

### 6.5.0.1 The IMC scheme

One of the first successful attempts to employ dynamic anti-windup design for guaranteed nonlinear closed-loop stability was carried out in [CS4] (see also [FC7]). In this approach, a model of the plant is employed to suitably design the dynamic anti-windup filter $\mathcal{F}$ so that the compensated closed-loop system is globally exponentially stable whenever the plant $\mathcal{P}$ is exponentially stable too. This model-based solution guarantees global exponential stability if (and only if) the plant is exponentially stable, yet it does not explicitly address performance. This augmentation scheme results in very sluggish responses when the plant contains slow modes.

In its original formulation, the IMC anti-windup construction has been formalized as follows. The unconstrained controller $\mathcal{K}$ is factored on the basis of the model of the plant $\mathcal{P}$ as

$$
\mathcal{K} = \left( I + Q(s) \mathcal{P} \right)^{-1} Q(s),
\tag{6.24}
$$

where $Q(s) := (I - \mathcal{K}\mathcal{P})^{-1} \mathcal{K}$. The resulting scheme is represented in Figure 6.4. From the figure, it is clear that the effect of this anti-windup construction is to assign to zero the transfer function seen by the saturation block. Reinterpreting this anti-windup construction within the MRAW framework of Figures 6.1 and 6.2 will clarify in a very intuitive manner the reason for the stability and performance properties of the IMC-based anti-windup scheme.

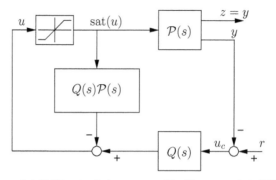

Figure 6.4  IMC anti-windup proposed by Campo et al. in [CS4].

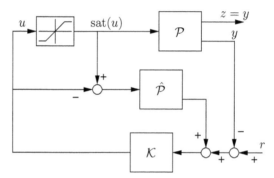

Figure 6.5  Model-based anti-windup proposed by Irving and reported in [SUR1].

To transform the scheme of Figure 6.4 into the MRAW scheme, consider the following transformations:

$$u = Q(s)u_c - Q(s)\mathcal{P}\text{sat}(u) + (Q(s)\mathcal{P}u - Q(s)\mathcal{P}u)$$
$$= Q(s)(u_c - \mathcal{P}(\text{sat}(u) - u)) - Q(s)\mathcal{P}u.$$

Solving the above equation for $u$ and recalling the definition (6.24), the following holds:

$$u = (I + Q(s)\mathcal{P})^{-1}Q(s)(u_c - \mathcal{P}(\text{sat}(u) - u))$$
$$= \mathcal{K}(u_c + \mathcal{P}(\text{sat}(u) - u)),$$

which corresponds to the block diagram in Figure 6.5. Note that this equivalent representation of the IMC-based anti-windup scheme corresponds to the MRAW compensator (6.3) with the particular selection $v = 0$. [3]

Based on the cascaded structure of Figure 6.2 it then becomes clear why the IMC scheme induces stability whenever the plant is exponentially stable and poor performance whenever the plant has slow modes. Defining the deadzone nonlinearity as $\text{dz}(u) := u - \text{sat}(u)$, the effect of choosing $v = 0$ is that the nonlinear closed-loop system is equivalent to the cascade of the unconstrained closed loop, a deadzone nonlinearity, and a model of the open-loop plant. Hence, exponential stability of the cascade is guaranteed by the linearity of the two systems and by their exponential stability, while the bad performance experienced when the plant has slow modes is related to the slow convergence to zero of the slow modes in the plant model. Indeed, as already discussed before, the states of the plant model correspond to the mismatch between the unconstrained and the actual trajectories, which eventually converges to zero, but

---

[3] Interestingly, the equivalent scheme in Figure 6.5 coincides with a model-based scheme reported by Hanus in the survey paper [SUR1], and therein attributed to Irving.

exhibits a transient response as slow as the slowest plant time constant. Nevertheless, every time the plant dynamics is fast enough to induce a satisfactory transient response, the IMC solution guarantees global exponential stability and good performance of the anti-windup closed-loop system, in addition to being the simplest possible MRAW construction.

### 6.5.0.2 Anti-windup proposed by Horowitz

An interesting anti-windup scheme based on dynamic compensation was introduced by Horowitz in 1983 [FC4]. The scheme was actually introduced in [FC4] where it was not really intended to address anti-windup but it has interpreted as such later in [CS1]. In this construction, represented in Figure 6.6, the author proposed the following factorization for the controller $\mathcal{K}$:

$$\mathcal{K}(s) = \left(I + \frac{L}{s} + Q(s)\mathcal{P}\right)^{-1} Q(s), \tag{6.25}$$

where $L > 0$ is a constant design parameter and $Q(s)$ is defined as

$$Q(s) := \left(I + \frac{L}{s}\right)(I - \mathcal{K}\mathcal{P})^{-1}\mathcal{K}.$$

It is interesting to observe the similarities between the IMC scheme and the scheme in Figure 6.6. Indeed, taking the limit as $L$ converges to zero, the compensation scheme converges to IMC anti-windup.

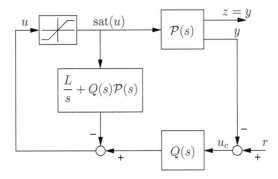

Figure 6.6  Dynamic anti-windup compensation proposed by Horowitz in [FC4].

Similar to the IMC case, by a simple loop transformation, the anti-windup scheme of Figure 6.6 can be viewed in terms of the $\mathcal{L}_2$ anti-windup scheme of Figure 6.1 and of equation (6.3), which are equivalent to the cascade structure of Figure 6.2. In particular, consider the following transformation:

$$u = Q(s)u_c - \left(\frac{L}{s} + Q(s)\mathcal{P}\right)\text{sat}(u) + (Q(s)\mathcal{P}y_c - Q(s)\mathcal{P}y_c)$$

$$= Q(s)(u_c - \mathcal{P}(\text{sat}(u) - y_c)) - Q(s)\mathcal{P}y_c - \frac{L}{s}y_c - \frac{L}{s}(\text{sat}(u) - y_c)$$

$$= Q(s)(u_c - \mathcal{P}(\text{sat}(u) - y_c)) - \left(Q(s)\mathcal{P} + \frac{L}{s}\right)y_c - \frac{L}{s}(\text{sat}(u) - y_c).$$

Select $v = -\frac{L}{s}(\text{sat}(u) - y_c)$. Then, recalling that $u = y_c + v$, the previous equation transforms into

$$u = Q(s)(u_c - \mathcal{P}(\text{sat}(u) - y_c)) - \left(Q(s)\mathcal{P} + \frac{L}{s}\right)(y_c + v - v) + v$$

$$= Q(s)(u_c - \mathcal{P}(\text{sat}(u) - y_c)) - \left(Q(s)\mathcal{P} + \frac{L}{s}\right)u + \left(Q(s)\mathcal{P} + \frac{L}{s} + I\right)v.$$

Finally, by solving the equation for $u$ and substituting (6.25) the following holds:

$$u = \left(I + \frac{L}{s} + Q(s)\mathcal{P}\right)^{-1} Q(s)(u_c - \mathcal{P}(\mathrm{sat}(u) - y_c)) + v$$

$$= \mathcal{K}(s)(u_c + \mathcal{P}(\mathrm{sat}(u) - y_c)) - \frac{L}{s}(\mathrm{sat}(u) - y_c),$$

which corresponds to the block diagram in Figure 6.7.

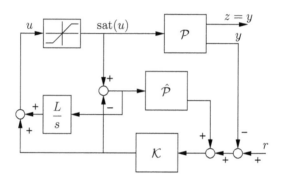

Figure 6.7 Equivalent representation of the anti-windup compensation proposed by Horowitz.

Note that the equivalent representation of Figure 6.7 corresponds to a generalization of the MRAW equations in (6.3), where the signal $v$ depends dynamically on the anti-windup compensator input. Therefore, the scheme of Figure 6.7 can be once again viewed as the cascade structure in Figure 6.2. From that figure, the exponential stability of the resulting scheme is once again evident in the case where the plant is exponentially stable, because the negative feedback around the saturation nonlinearity effected by $v$ only changes the input nonlinearity to the second subsystem, but preserves the cascade structure for the two linear exponentially stable systems.

It is also evident from the scheme of Figure 6.7 that improved performance is expected from this scheme as compared to IMC-based anti-windup. As a matter of fact, if $L$ is large enough, the effect of the signal $v$ is to compensate for large inputs affecting the plant model dynamics, and the $x_{aw}$ state (which measures the mismatch between the actual and the unconstrained response) is, to a certain extent, not allowed to grow too large. Nevertheless, the main limitation of this compensation scheme is that the knowledge of the state $x_{aw}$ is not at all exploited for performance improvement. Therefore, it is evident that this scheme leaves space for performance improvement by suitably designing selections for $v$ based on the value of the mismatch dynamics $x_{aw}$.

### 6.5.0.3 Anti-windup proposed by Park and Choi

In [MR1], Park and Choi proposed a model-based anti-windup scheme where the controller state equation was modified by an optimality-based selection of a signal produced by a linear filter, whose input is the deviation of the excess of saturation at the plant input (namely, the plant input passed through a deadzone $\mathrm{dz}(u) = u - \mathrm{sat}(u)$). The scheme is represented in Figure 6.8, where $D_{cp}$ denotes the input-output matrix related to the unconstrained controller $\mathcal{K}$ and $\overline{\mathcal{K}}$ is the strictly proper part of that controller.

Interestingly, the authors proved in [MR1] that the selection for the transfer function of the filter feeding the controller input is optimal with the goal of minimizing the performance index

$$J = \int_0^\infty |x_c(t) - x_{c\ell}(t)|^2 dt,$$

where $x_c(t)$ denotes the controller state related to the anti-windup closed-loop response and $x_{c\ell}(t)$ denotes the controller state related to the unconstrained closed-loop response. In particular, the authors observe a posteriori that the effect of the compensation scheme in Figure 6.8 is to impose $x_c(t) - x_{c\ell}(t) = 0$

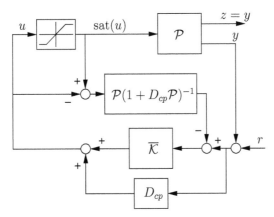

Figure 6.8 Model-based anti-windup proposed by Park and Choi in [MR1].

at all times, so that $J = 0$ regardless of the saturation activation. This property is indeed shared by all the MRAW schemes that fit into the general framework represented in Figure 6.1.

Once again, suitable transformations can be carried out on the scheme of Figure 6.8 to determine its representation within the MRAW framework of Figure 6.1 and equation (6.3). First consider the following equality:

$$
\begin{aligned}
u &= \overline{\mathcal{K}}u_c + D_{cp}u_c + \overline{\mathcal{K}}\mathcal{P}(I + D_{cp}\mathcal{P})^{-1}(\mathrm{sat}(u) - u) \\
&= \mathcal{K}u_c + \overline{\mathcal{K}}\mathcal{P}(I + D_{cp}\mathcal{P})^{-1}(\mathrm{sat}(u) - u).
\end{aligned}
\tag{6.26}
$$

Similar to what has been done in the previous section, set $u = y_c + v$, with the selection $v = -D_{cp}\mathcal{P}(\mathrm{sat}(u) - y_c)$ and consider the last term on the right side of equation (6.26):

$$
\begin{aligned}
u - \mathcal{K}u_c &= \overline{\mathcal{K}}\mathcal{P}(I + D_{cp}\mathcal{P})^{-1}(\mathrm{sat}(u) - y_c - v) \\
&= \overline{\mathcal{K}}\mathcal{P}(I + D_{cp}\mathcal{P})^{-1}((\mathrm{sat}(u) - y_c) + D_{cp}\mathcal{P}(\mathrm{sat}(u) - y_c)) \\
&= \overline{\mathcal{K}}\mathcal{P}(I + D_{cp}\mathcal{P})^{-1}(I + D_{cp})(\mathrm{sat}(u) - y_c)) \\
&= \overline{\mathcal{K}}\mathcal{P}(\mathrm{sat}(u) - y_c)).
\end{aligned}
$$

Substituting $\overline{\mathcal{K}} = \mathcal{K} - D_{cp}$ and recalling the definition of $v$, the following holds

$$
\begin{aligned}
u &= \mathcal{K}u_c + \overline{\mathcal{K}}\mathcal{P}(\mathrm{sat}(u) - y_c)) \\
&= \mathcal{K}(u_c + \mathcal{P}(\mathrm{sat}(u) - y_c))) - D_{cp}\mathcal{P}(\mathrm{sat}(u) - y_c) \\
&= \mathcal{K}(u_c + v_2) + v,
\end{aligned}
$$

which corresponds to the block diagram in Figure 6.9, which fits into the framework of Figure 6.1 and can be written as in equation (6.3) with the selection $v = -D_{cp}C_y x_{aw} - D_{cp}D_{py}(\mathrm{sat}(y_c) - u)$.

Based on the equivalent representation of Figure 6.9, interesting interpretations of the properties of this model-based anti-windup scheme can be given, based on the cascaded representation of Figure 6.2. Indeed, when writing this compensation scheme in that cascaded form, the mismatch dynamics can be written as:

$$
\begin{aligned}
\dot{x}_{aw} &= A_p x_{aw} + B_p(\mathrm{sat}(y_c + v) - y_c) \\
v_1 &= -D_{cp}C_y x_{aw} - D_{cp}D_{py}(\mathrm{sat}(y_c + v) - y_c),
\end{aligned}
\tag{6.27}
$$

and the stability of these dynamics imply the stability of the overall scheme. A necessary condition for the stability of (6.27) is the exponential stability of the linear system arising in all the cases when the input saturation is not active. In particular, it is necessary that $S := A_p - B_p(I - D_{cp}D_{py})^{-1}D_{cp}C_y$ be a Hurwitz matrix, as it is required in Assumption A3) of [MR1]. The cascade representation Figure 6.2 gives a straightforward interpretation of that assumption. Finally, a (conservative) sufficient condition for the exponential stability of the anti-windup scheme can be drawn based on the $[0, I]$ sector properties of the saturation function, by requiring that the transfer function from $(\mathrm{sat}(y_c + v_1) - y_c)$ to $v_1$ corresponding to system (6.27) belong to the sector $[-I, I]$. This corresponds to what is proven in Theorem 3 of [MR1], where it is assumed that the infinity norm of $D_{cp}\mathcal{P}$ is smaller than 1.

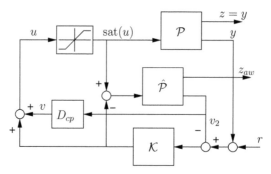

Figure 6.9  Equivalent representation of the anti-windup scheme of [MR1].

### 6.5.0.4 *Anti-windup proposed by Weston and Postlethwaite*

In [H2, H5], Weston and Postlethwaite proposed a linear anti-windup scheme based on coprime factorizations. The resulting anti-windup construction is based on the insertion of the following filter in the closed loop:

$$\mathcal{F} \begin{cases} \dot{x}_F &= (A_p + B_p F)x_{aw} + B_p(u - \mathrm{sat}(u)) \\ v_1 &= F x_{aw} \\ v_2 &= (C_y + D_{py} F)x_{aw} + D_{py}(u - \mathrm{sat}(u)), \end{cases} \tag{6.28}$$

where the matrix $F$ has to be chosen to make $A_p + B_p F$ Hurwitz and the signals $w_1$ and $w_2$ modify, respectively, the output and the input of the unconstrained controller, as illustrated in Figure 6.10. In Figure 6.10, it is also emphasized by means of block diagrams that the filter (6.28) has an implicit state feedback structure effected through the matrix gain $F$.

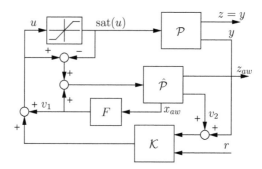

Figure 6.10  Model-based anti-windup proposed by Weston and Postlethwaite in [H5].

Transforming the anti-windup scheme of Figure 6.10 into the general MRAW framework of Figure 6.1 amounts to effecting a simple block transformation on the figure. Indeed, after inverting the signs of the summing block below the saturation and moving the position where the signal $v$ affects the nonlinearity input, the equivalent scheme of Figure 6.9 is obtained, which, once again, can be interpreted on the basis of the general framework of Figure 6.1 and of equation (6.14) and of its equivalent cascade representation of Figure 6.2.

Based on the scheme of Figure 6.11, it is easy to determine that the introduction of the filter (6.28) corresponds to the MRAW compensation (6.14) with the selection $v = -F x_{aw}$. Hence, according to the cascaded structure in Figure 6.2, the stability of the anti-windup closed-loop system corresponds to the stability of the mismatch system

$$\begin{aligned} \dot{x}_{aw} &= A_p x_{aw} + B_p(\mathrm{sat}(y_c + v) - y_c) \\ v &= -F x_{aw}. \end{aligned} \tag{6.29}$$

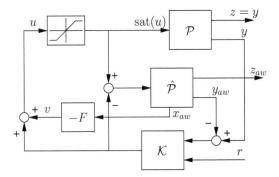

Figure 6.11  Equivalent representation of the anti-windup scheme of [H5].

Note that a necessary (but not sufficient) condition for exponential stability of (6.29) is that whenever the saturation is not active, the corresponding linear dynamics are exponentially stable. This property is ensured by guaranteeing that $A_p + B_p F$ is Hurwitz, which corresponds to the design requirement imposed in [H5] for the selection of the matrix $F$. Note, however, that by the presence of saturation, not all selections of $F$ that satisfy this requirement will guarantee global exponential stability of the mismatch dynamics (6.29). Hence, alternative (possibly nonlinear) stability tests need to be carried out to verify the stability of the resulting anti-windup closed-loop system. The work in [H2, H5] was later extended in [H9, H10] to incorporate a new free parameter which added extra degrees of freedom in the design. When reinterpreted in the MRAW framework, the goal of this parameter is to allow for a more general selection of the signal $v$ in (6.29), namely,

$$v = -F_1 x_{aw} - F_2(\text{sat}(y_c + v) - y_c), \qquad (6.30)$$

where $F_1$ and $F_2$ are both free parameters only constrained by the need of a well-posed nonlinear algebraic loop in (6.30). Quite interestingly, seen in the MRAW framework, the new selection of [H9, H10] has the same degrees of freedom as the linear selection for $v$ commented on in equation (6.15).

## Chapter Seven

## Linear MRAW Synthesis

### 7.1 INTRODUCTION

Figure 7.1 represents the model recovery anti-windup (MRAW) scheme introduced in Chapter 6. In that scheme, the upper block $\mathcal{P}$ represents the plant, the lower block $\mathcal{K}$ represents the unconstrained controller and the intermediate block $\widehat{\mathcal{P}}$, which corresponds to a model of the plant, contains the dynamical states of the anti-windup augmentation. This structure permits recovering information about the unconstrained response, so that unconstrained response recovery is possible through the extra degree of freedom represented by the unspecified compensation signal $v$.

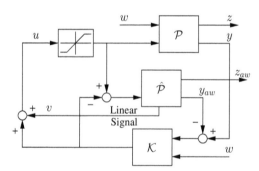

Figure 7.1 The model recovery anti-windup compensation scheme.

Throughout this chapter, it will be useful to refer to suitable state-space representations of the plant and of the anti-windup compensator dynamics. In particular, assuming that the plant is linear, the following state-space representation will be used:

$$\mathcal{P} \quad \begin{cases} \dot{x}_p &= A_p x_p + B_{p,u}\, \sigma + B_{p,w}\, w \\ y &= C_{p,y} x_p + D_{p,yu}\, \sigma + D_{p,yw}\, w \\ z &= C_{p,z} x_p + D_{p,zu}\, \sigma + D_{p,zw}\, w\,, \end{cases} \tag{7.1}$$

where $x_p \in \mathbb{R}^{n_p}$ is the state, $y \in \mathbb{R}^{n_y}$ is the measured output, $z \in \mathbb{R}^{n_z}$ is the performance output, $\sigma \in \mathbb{R}^{n_u}$ is the control input, and $w$ is the external input gathering references and disturbances. Moreover, since the anti-windup filter $\mathcal{F}$ is based on a model of the plant from the input $u$ to the output $y$, it will have the following state-space representation:

$$\mathcal{F} \begin{cases} \dot{x}_{aw} &=& A_p x_{aw} + B_{p,u}\left(\sigma - y_c\right) \\ z_{aw} &=& C_{p,z} x_{aw} + D_{pz}\left(\sigma - y_c\right) \\ y_{aw} &=& C_{p,y} x_{aw} + D_{py}\left(\sigma - y_c\right) \\ v &=& \text{to be designed.} \end{cases} \tag{7.2}$$

The anti-windup closed loop then corresponds to the interconnection of (7.1), (7.2) and the unconstrained controller via the anti-windup interconnection equations

$$u_c = y - y_{aw}, \qquad u = y_c + v, \qquad \sigma = \text{sat}(u), \tag{7.3}$$

where $u_c$ is the measurement input and $y_c$ is the output of the unconstrained controller. It follows that the anti-windup compensator (7.2) can be written in the equivalent form

$$\mathcal{F} \begin{cases} \dot{x}_{aw} &=& A_p x_{aw} + B_p v + B_{p,u}\left(\text{sat}(u) - u\right) \\ z_{aw} &=& C_{p,z} x_{aw} + D_{p,zu} v + D_{p,zu}\left(\text{sat}(u) - u\right) \\ y_{aw} &=& C_{p,y} x_{aw} + D_{p,yu} v + D_{p,yu}\left(\text{sat}(u) - u\right) \\ v &=& \text{to be designed.} \end{cases} \tag{7.4}$$

The stability and performance properties induced by MRAW on the closed loop depend on the choice of the feedback signal $v$. Because of the cascaded structure described in Figure 6.2 on page 159, this signal performs a stabilizing action aimed at driving the actual plant response toward the (ideal) unconstrained responses whose information is captured through the extra dynamics of the anti-windup filter. The MRAW scheme of Figure 7.1 allows for very general selections of the compensation signal $v$, ranging from linear to nonlinear and sampled-data ones. The natural starting point for the selection of $v$ is the linear feedback

$$v = K x_{aw} + L(\text{sat}(y_c + v) - y_c), \tag{7.5}$$

which can be written equivalently as

$$v = K x_{aw} + Lv + L(\text{sat}(u) - u) \tag{7.6}$$

or

$$v = (I - L)^{-1} K x_{aw} + (I - L)^{-1} L(\text{sat}(u) - u). \tag{7.7}$$

All the linear MRAW algorithms given in this chapter correspond to selecting the linear gains $K$ and $L$ in (7.5) according to different procedures with certain applicability and expected performance. All of them will be specified by giving the final selection for $K$ and $L$ in the compensation scheme given by (7.2), (7.3), (7.5). However, it is useful to emphasize up front that the scheme is fully equivalent to the implementation (7.4), (7.3), (7.5) and also fully equivalent to the use of an external linear anti-windup compensation filter $\mathcal{F}$ driven by $\text{sat}(u) - u$ as in Section 2.3.6 on page 34 and having state-space representation

$$\left[ \begin{array}{c|c} A_{aw} & B_{aw} \\ \hline C_{aw} & D_{aw} \end{array} \right] = \left[ \begin{array}{c|c} A_p + B_p(I-L)^{-1}K & B_p(I-L)^{-1} \\ (I-L)^{-1}C_{p,y} & (I-L)^{-1}D_{p,yu} \\ (I-L)^{-1}K & (I-L)^{-1}L \end{array} \right] \tag{7.8}$$

whose output is partitioned in two signals, the first one to be added to the un-constrained controller input and the second one to be added to the unconstrained controller output as indicated in Section 2.3.6. These expressions were given in Section 6.3.3 as well.

The rest of this chapter is dedicated to prescriptive algorithms for generating the feedback matrices $K$ and $L$. When $L \neq 0$, the equation (7.5), or any of its equivalent forms, contains an algebraic loop that must be solved during implementation. All of the algorithms in this chapter guarantee that this algebraic loop is well-posed.

In Section 7.2, LMI-based synthesis algorithms will be provide for which global stability and performance are guaranteed. These algorithms are not applicable to closed-loop systems with unstable plants. For exponentially stable plants, they do not require a priori information about the size of the disturbances acting on the system. The performance induced by these algorithms can be too conservative for small to medium disturbances. On the other hand, sometimes these algorithms still perform quite well over all disturbance levels.

In Section 7.3, LMI-based synthesis algorithms for regional stability and perfor-mance will be provided. These algorithms are applicable to any closed-loop system, but the region over which the algorithms may be applied is limited when the plant is exponentially unstable. Moreover, information about the size of the disturbances acting on the system is required if guarantees on performance are required.


## 7.2  GLOBAL STABILITY-BASED ALGORITHMS

### 7.2.1  Global exponential stability

When seeking for global guarantees, a straightforward strategy for selecting the compensation signal $v$ is given by the following Lyapunov-based procedure, where the degrees of freedom available in the design process can often lead to good perfor-mance, even though this should be tuned by means of trial-and-error approaches. First the case where the plant is exponentially stable will be addressed and then a generalization of the algorithm will be introduced that also applies to the case when the plant is stable, but not exponentially (namely, it has single poles on the imaginary axis).

The following algorithm requires two parameters to be chosen before being ap-plied. One of them is the positive definite matrix $Q$, which has an impact on the way the different states of the plant are weighted within the control action. The sec-ond one is the positive scalar factor $\rho$, which is proportional to the aggressiveness of the stabilizing action.

**Algorithm 10** (Stability-based MRAW for exponentially stable plants)

| Applicability | | | | Architecture | | | Guarantee |
|---|---|---|---|---|---|---|---|
| Exp Stab | Marg Stab | Marg Unst | Exp Unst | Lin/ NonL | Dyn/ Static | Ext/ FullAu | Global/ Regional |
| $\checkmark$ | | | | L | D | E | G |

> *Comments*: Special cases include IMC anti-windup and Lyapunov-based anti-windup, neither of which creates an algebraic loop. Global exponential stability is guaranteed but no performance measure is optimized.

**Step 1.** Select a constant $v > 0$ used to enforce a strong well-posedness condition.

**Step 2.** Solve the following LMI feasibility problem in the free variables $Q = Q^T > 0$, $U > 0$ diagonal, $X_1$ and $X_2$, where $X_1$ has the same dimension as $K$ and $X_2$ has the same dimension as $L$:

*Strong well-posedness*:
$$X_2^T + X_2 - 2(1-v)U < 0.$$

*Global stability*:
$$\begin{bmatrix} QA_p^T + A_pQ & B_{p,u}U + X_1^T \\ UB_{p,u}^T + X_1 & X_2^T + X_2 - 2U \end{bmatrix} < 0.$$

**Step 3.** Using a feasible solution from the previous step, compute the compensation gains as follows:
$$K = X_1 Q^{-1}, \quad L = X_2 U^{-1}. \tag{7.9}$$

**Step 4.** Construct the anti-windup compensation as (7.2), (7.3), (7.5) with the selection (7.9). Equivalently, use the filter (7.8) in an external linear anti-windup compensation scheme.

$\star$

### 7.2.1.1 Special classes of feasible solutions

In Algorithm 10, there is no hope of the synthesis LMIs being feasible unless $A_p$ is exponentially stable. When $A_p$ is exponentially stable, there are some obvious solutions to the synthesis LMI.

IMC: One can take $X_1 = 0$, $X_2 = 0$, $U = I$, and $Q = \rho P^{-1}$, where $\rho > 0$ is sufficiently large and $P > 0$ satisfies $A_p^T P + P A_p < 0$. Such a matrix $P$ will always exist when $A_p$ is exponentially stable. These choices lead to picking $K = 0$ and $L = 0$ in (7.9). In this case, $\mathcal{F}$ corresponds to a copy of the plant $\mathcal{P}$. This fact leads to the label "internal model control," or IMC for short.

Lyapunov-based synthesis: One can take $X_1 = -UB_p^T$, $X_2 = 0$, $U > 0$ diagonal, and $Q = P^{-1}$, where $P > 0$ satisfies $A_p^T P + P A_p < 0$. These choices lead to picking $K = -UB_p^T P$, where $U$ is an arbitrary positive definite, diagonal matrix and $L = 0$ in (7.9). Despite its simplicity, in many cases Algorithm 10 with these particular choices is sufficiently good to induce highly desirable responses on the resulting closed-loop system.

The following examples illustrate this technique and compare it to the IMC approach.

**Example 7.2.1** (MIMO academic example) Consider a two-dimensional MIMO plant modeled by the following set of equations:

$$\dot{x}_1 = -0.1x_1 + 0.5\text{sat}_1(u_1) + 0.4\text{sat}_2(u_2)$$
$$\dot{x}_2 = -0.1x_2 + 0.4\text{sat}_1(u_1) + 0.3\text{sat}_2(u_2)$$
$$z = x$$
$$y = x,$$

where $\text{sat}_1(\cdot)$ and $\text{sat}_2(\cdot)$ range between $-3$ and $3$ and $-10$ and $10$, respectively. At time $t = 0$, the outputs $x_1$ and $x_2$ are subject to pulse setpoint changes of duration 5 seconds and magnitudes 0.6 and 0.4 respectively. The following PI controller can be used to guarantee zero steady-state error to step references:

$$\begin{aligned}
\dot{x}_{c1} &= y_{sp1} - x_1 \\
\dot{x}_{c2} &= y_{sp2} - x_2 \\
u_1 &= 10(y_{sp1} - x_1) + x_{c1} \\
u_2 &= -10(y_{sp2} - x_2) + x_{c2},
\end{aligned} \tag{7.10}$$

where $y_{sp1}$ and $y_{sp2}$ represent the setpoints for the outputs.

The two variants of Algorithm 10 can be applied to this academic example:

1. For the IMC approach, nothing needs to be computed: since $A_p$ is exponentially stable (all the eigenvalues have a negative real part), the anti-windup compensation is feasible and corresponds to the selection $K = 0$ and $L = 0$ in (7.2), (7.3), (7.5).
2. For the Lyapunov-based synthesis, the matrix $P$ is selected as the solution of the Lyapunov equation $A_p^T P + P A_p = -Q$, with $Q = 0.2I < 0$. Picking $U = I$, the solution becomes:

$$K = \begin{bmatrix} -1.500 & -1.200 \\ -4.000 & -3.000 \end{bmatrix}, \quad L = 0,$$

to be used in (7.2), (7.3), (7.5).

Different responses to the pulse references described above are reported in Figures 7.2 and 7.3, where the two outputs and the two inputs of the plant are represented, respectively. In particular, the dashed curves represent the unconstrained response, which exhibits desirable behavior. The dotted curves, which exhibit an unstable behavior, correspond to the response of the saturated system. The dashed-dotted curves represent the response of the IMC anti-windup and the solid curves represent the response of the Lyapunov-based anti-windup.

While the saturated system exhibits instability, IMC anti-windup is at least able to recover closed-loop stability, but the associated performance is very sluggish: it takes an awfully long time for the dashed-dotted IMC response in Figure 7.2 to converge to the unconstrained dashed response. A much more desirable behavior is exhibited by the Lyapunov-based anti-windup (solid curves in Figure 7.2), which, by way of a nonzero selection of the compensation signal $v$, is able to force the

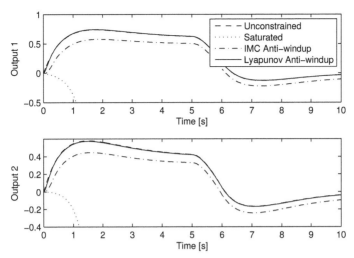

Figure 7.2 Plant output responses of various closed loops for the MIMO academic example in Example 7.2.1.

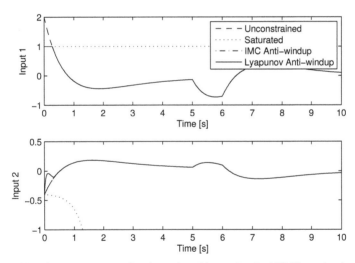

Figure 7.3 Plant input responses of various closed loops for the MIMO academic example in Example 7.2.1.

output response back onto the unconstrained one. The corresponding transients happen in the first 0.5 seconds and are almost invisible in Figure 7.2. They can, however, be appreciated in the lower plot of Figure 7.3, where it appears that the anti-windup system is making use of the second input to quickly "recover" the unconstrained performance. □

**Example 7.2.2** (A turbofan engine) Consider the following turbofan engine model which has three states, three inputs, and three performance outputs (all of them available for measurement):

$$\begin{aligned}
\dot{x} &= A_p x + B_{p,u}\,\mathrm{sat}(u) \\
y &= C_{p,y} x + D_{p,yu}\,\mathrm{sat}(u),
\end{aligned} \tag{7.11}$$

where

$$A_p = \begin{bmatrix} -4.1476 & 1.4108 & 0.0633 \\ 0.2975 & -3.1244 & 0.0623 \\ -0.0429 & -0.1729 & -0.1325 \end{bmatrix},$$

$$B_{p,u} = \begin{bmatrix} 0.2491 & 0.0969 & -0.0112 \\ 0.2336 & 0.0335 & 0.0047 \\ 0.0624 & 0 & 0 \end{bmatrix},$$

$$C_{p,y} = \begin{bmatrix} 8.7379 & 0 & 0 \\ -3.3033 & 3.8052 & 0.0542 \\ 2.1940 & -2.5749 & -0.0295 \end{bmatrix},$$

$$D_{p,yu} = \begin{bmatrix} 0 & 0 & 0 \\ 0.2383 & -0.2748 & 0.0224 \\ -0.1455 & 0.058 & -0.2293 \end{bmatrix}.$$

The eigenvalues of $A_p$ are $-4.4615$, $-2.8055$, and $-0.1375$; hence, the plant is open-loop stable. The three outputs $y \in \mathbb{R}^3$ to be controlled are the *fan speed* (PCN2R), the *core engine pressure ratio* (CEPR), and the *liner engine pressure ratio* (LEPR), respectively. The three inputs $u \in \mathbb{R}^3$ are the *fuel flow rate* (WF), the *nozzle area* (A8), and the *bypass duct area* (A16), respectively.

With the primary goal of guaranteeing decoupled tracking of the set points for PCN2R and CEPR with good regulation of LEPR, the following unconstrained controller has been derived for the unconstrained system:

$$\begin{aligned}
\dot{x}_c &= A_c x_c + B_c (y_{sp} - u_c) \\
y_c &= C_c x_c + D_c (y_{sp} - u_c),
\end{aligned} \tag{7.12}$$

where $y_{sp} \in \mathbb{R}^3$ is the setpoint value for the outputs and the matrix entries are:

$$A_c = \begin{bmatrix} -0.002 & 0 & 0 & -0.0001 \\ 0 & -0.002 & 0 & 0.0001 \\ 0 & 0 & -0.002 & -0.0015 \\ 0 & 0.0004 & -0.0014 & -0.1476 \end{bmatrix},$$

$$B_c = \begin{bmatrix} -0.5393 & -0.5799 & -2.5289 \\ -1.2106 & 2.1547 & -0.2357 \\ -1.6705 & -0.8808 & 0.5579 \\ -0.4447 & -0.5247 & 0.2411 \end{bmatrix},$$

$$C_c = \begin{bmatrix} -0.1466 & 0.2311 & -1.9578 & -0.7258 \\ 0.055 & -2.4711 & -0.1838 & -0.0303 \\ 2.6453 & .0639 & -0.1039 & 0.0597 \end{bmatrix},$$

$$D_c = \begin{bmatrix} 0.8097 & -0.2307 & -0.0013 \\ 0.9338 & 0.7835 & 0.4708 \\ -0.3303 & 0.8211 & 1.8152 \end{bmatrix}.$$

Three of the eigenvalues of $A_c$ are $-0.002$, while the fourth one is $-0.1476$. This suggests that the unconstrained closed-loop system may be subject to windup. Indeed, slow modes in the unconstrained controller often lead to poor performance in the presence of saturation.

The two variants of Algorithm 10 can be applied to the turbofan engine:

1. For the IMC approach, nothing needs to be computed: since $A_p$ is exponentially stable (all the eigenvalues have a negative real part), the anti-windup compensation is feasible and it corresponds to the selection $K = 0$ and $L = 0$ in (7.2), (7.3), (7.5).

2. For the Lyapunov-based synthesis, the matrix $P$ is selected as the solution of the Lyapunov equation $A_p^T P + P A_p = -C_{p,y}^T C_{p,y} < 0$, where the inequality holds because $C_{p,y}$ is a full-rank matrix. Picking $U = I$, the solution then becomes:

$$K = \begin{bmatrix} -271.51 & -71.83 & -8.19 \\ -106.63 & -8.75 & -1.73 \\ 12.51 & -1.78 & 0.04 \end{bmatrix}, \quad L = 0,$$

to be used in (7.2), (7.3), (7.5).

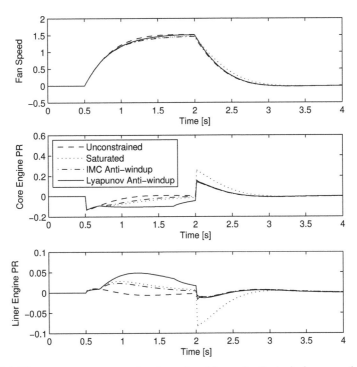

Figure 7.4  Plant output responses of various closed loops for the turbofan example in Example 7.2.2.

For the turbofan engine, a simulation setup is studied where the output set point (or the reference input) is a pulse of amplitude 1.5 and duration 2 seconds at the

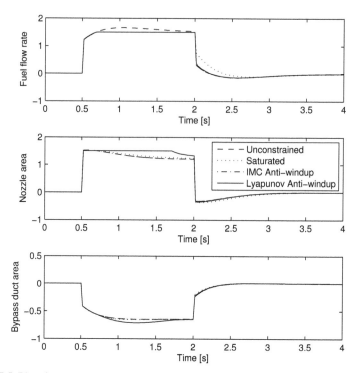

Figure 7.5 Plant input responses of various closed loops for the turbofan example in Example 7.2.2.

PCN2R reference input. All the inputs are assumed to be constrained between $\pm 1.5$. Figures 7.4 and 7.5 represent the closed-loop responses of the unconstrained system (dashed), the saturated system (dotted), the anti-windup closed-loop system with IMC compensation (dashed-dotted), and the anti-windup closed-loop system with Lyapunov-based compensation (solid).

From the simulations it appears that IMC anti-windup already does better than the saturated response especially in the second portion of the simulation. Lyapunov-based compensation is capable of improving even more the output response at the first output (top plot of Figure 7.4) by better exploiting the input authority available on all three inputs (see the middle and lower input plots in Figure 7.5. This ability of the anti-windup compensation to exploit as much as possible the input authority to recover as soon as possible the unconstrained output performance is the key feature that makes it desirable in many practical situations. □

### 7.2.2 Global asymptotic stability

As shown in the previous example, designing the compensation signal $v$ in (7.5) only based on the matrix $K$ (namely, constraining $L = 0$) and selecting this matrix based on a simple Lyapunov argument can already lead to high-performance solutions to the anti-windup problem. A useful generalization of Algorithm 10 is represented by the construction described in this section. The main advantage of this generalization as compared to the previous algorithm is that it also applies to plants which contain marginally stable modes (in addition to possible exponentially stable ones). Marginally stable modes correspond to single poles on the imaginary axis. This class of systems is quite common in real applications because it corresponds to the case of linear (or possibly linearized) mechanical systems without friction forces, where typically the state is formed by positions and velocities of geometrical coordinates, and each degree of freedom characterizes a single pole at the origin.

The dynamics of stable plants with possible marginally stable modes correspond to a state-space representation where the Jordan canonical form of the state transition matrix doesn't have Jordan blocks of size greater than one for any of the eigenvalues on the imaginary axis. To suitably describe the construction algorithm, it will be useful to transform the state-space representation of the plant model reproduced in the MRAW filter into the following form:

$$\dot{x}_{aw} = \begin{bmatrix} \dot{x}_s \\ \dot{x}_0 \end{bmatrix} = \begin{bmatrix} A_s & 0 \\ 0 & A_0 \end{bmatrix} \begin{bmatrix} x_s \\ x_0 \end{bmatrix} + B_{p,u}(\text{sat}(y_c + v_1) - y_c), \quad (7.13)$$

where $A_0$ is a block diagonal matrix characterizing all the marginally stable modes of the plant and, consequently, $A_s$ is exponentially stable. In particular, if a certain marginally stable mode of the plant is real, then the corresponding entry on the diagonal of $A_0$ will be a "0." On the other hand, if the marginally stable mode is a pair of complex conjugate poles, then the entry will be a submatrix of the form $\begin{bmatrix} 0 & \omega \\ -\omega & 0 \end{bmatrix}$. This can be done according to the following procedure.

**Procedure 5** (State-space transformation for marginally stable plants)

> **Step a)** Define two matrices $A_p$ and $B_p$ such that all the eigenvalues of $A_p$ have a nonpositive real part and all the eigenvalues on the imaginary axis have algebraic multiplicity equal to the geometric multiplicity (namely, they are single poles).

> **Step b)** Compute the Jordan form $J_p$ of $A_p$ and the coordinate transformation matrix $T_J$ such that $T_J^{-1} A_p T_J = J_p$ (this can be done in a simple automated way using, e.g., the MATLAB command [Tj,Jp] = jordan(Ap)). The matrix $J_p$ will have all the eigenvalues of $A_p$ on its main diagonal. Denote by $n_0$ the number of (single) eigenvalues on the imaginary axis and by $n_s$ the number of the remaining eigenvalues.

> **Step c)** Construct the matrix $T_0$ column by column as follows:

> > 1. for each column of $T_J$ corresponding to a real eigenvalue in $J_p$, leave that column unchanged; for each pair of complex conjugate columns in $T_J$ corresponding to a pair of complex conjugate eigenvalues in $J_p$, replace that pair with two real columns corresponding to the real and the imaginary part of any of the two complex conjugate columns of $T_J$;

2. stack in $T_0$ the columns corresponding to eigenvalues in $J_p$ with strictly negative real part first; then complete $T_0$ with all the columns related to eigenvalues in $J_p$ with zero real part.

**Step d)** Define the transformed state-space representation for the MRAW compensator matrices as $\bar{A}_p = T_0^{-1} A_p T_0$ and $\bar{B}_{p,u} = T_0^{-1} B_{p,u}$.

$\star$

The following example illustrates, for a nontrivial case, the use of Procedure 5 to transform an anti-windup compensator with neutrally stable modes into the form (7.13).

**Example 7.2.3**   Consider the MRAW compensator (7.2) with the following selections:

$$
A_p = \begin{bmatrix}
8 & 8 & -12 & -12 & -4 \\
18 & -2 & -2 & -22 & 6 \\
13 & 13 & -7 & -27 & 1 \\
-2 & 18 & -2 & -22 & 6 \\
15 & 15 & -5 & -25 & -5
\end{bmatrix}, \quad
B_{p,u} = \begin{bmatrix}
1 \\ 0 \\ 0 \\ 0 \\ 0
\end{bmatrix},
$$

(note that the other matrices in (7.2) are not relevant to the purpose of this example). Step b of Procedure 5 consists in computing the Jordan form $J_p$ of $A_p$ and the coordinate transformation matrix $T_J$ such that $T_J^{-1} A_p T_J = J_p$. The resulting matrices are the following:

$$
J_p = \begin{bmatrix}
0 & 0 & 0 & 0 & 0 \\
0 & -8 & 0 & 0 & 0 \\
0 & 0 & -20 & 0 & 0 \\
0 & 0 & 0 & 20i & 0 \\
0 & 0 & 0 & 0 & -20i
\end{bmatrix}, \quad
T_J = \frac{1}{8} \begin{bmatrix}
2 & 2 & 2 & 1-3i & 1+3i \\
2 & 2 & 0 & -2-4i & -2+4i \\
0 & 2 & 2 & -2-4i & -2+4i \\
2 & 2 & 2 & -3-i & -3+i \\
2 & 0 & 2 & -2-4i & -2+4i
\end{bmatrix},
$$

from which it is evident that $A_p$ has a neutrally stable pole at the origin and a pair of complex conjugate neutrally stable poles at $\pm 20i$. To obtain from $T_J$ a coordinate transformation that generates the block matrix in (7.13), according to step c, it suffices to first replace all the pairs of complex conjugate eigenvectors with the corresponding real and imaginary parts and then to reorder the columns of $T_J$ so that the eigenvectors corresponding to the neutrally stable modes of $A_p$ appear in the last columns of $T_0$. The resulting transformation matrix $T_0$, the resulting block diagonal matrix $\bar{A}_p = T_0^{-1} A_p T_0$, and the corresponding input matrix $\bar{B}_{p,u} = T_0^{-1} B_{p,u}$ are the following:

$$
T_0 = \frac{1}{8} \begin{bmatrix}
2 & 2 & 2 & 1 & -3 \\
2 & 0 & 2 & -2 & -4 \\
2 & 2 & 0 & -2 & -4 \\
2 & 2 & 2 & -3 & -1 \\
0 & 2 & 2 & -2 & -4
\end{bmatrix}, \quad
\bar{A}_p = \begin{bmatrix}
-8 & 0 & 0 & 0 & 0 \\
0 & -20 & 0 & 0 & 0 \\
0 & 0 & 0 & 0 & 0 \\
0 & 0 & 0 & 0 & 20 \\
0 & 0 & 0 & -20 & 0
\end{bmatrix},
$$

$$
B_{p,u} = \begin{bmatrix} 1 & 1 & 1 & 2 & 0 \end{bmatrix}^T,
$$

which complies with the representation required. Note that since in equation (7.13) the matrix $A_s$ is a generic matrix, the first columns in the transformation matrix

$T_0$ can be permuted and/or transformed by linear combinations (as long as they preserve the linear independence of the resulting columns). This will result in a non-diagonal matrix $A_s$ which may correspond to a more desirable representation of the anti-windup compensator dynamics.                                                    □

After having clarified that any anti-windup compensator with simple, neutrally stable poles can be represented in the form (7.13), it is possible to proceed with the description of the anti-windup construction algorithm. In the next algorithm, there are three free design parameters. Two of them, consisting in a positive definite matrix $Q_s$ having the same size as $A_s$ and a positive scalar $\rho$, represent the generalization of the design parameters in step 1 of Algorithm 10. A third one, consisting in another positive scalar $\rho_0$, allows weighing the control authority devoted to the stabilization of the neutrally stable modes, as compared to that devoted to the overall anti-windup action.

**Algorithm 11** (Lyapunov-based MRAW for marginally stable plants)

| Applicability | | | | Architecture | | | Guarantee |
|---|---|---|---|---|---|---|---|
| Exp Stab | Marg Stab | Marg Unst | Exp Unst | Lin/ NonL | Dyn/ Static | Ext/ FullAu | Global/ Regional |
| √ | √ | | | L | D | E | G |

| Comments: Does not create an algebraic loop. Global asymptotic stability is guaranteed but no performance measure is optimized. |
|---|

**Step 1.** Verify that the anti-windup matrices are in the form (7.13), otherwise perform a change of coordinates, following Procedure 5.

**Step 2.** Select a positive definite matrix $Q_s$ of the same size as $A_s$, and positive scalars $\rho, \rho_0$.

**Step 3.** Solve the following Lyapunov equation:

$$A_s^T P_s + P_s A_s = -Q_s$$

in the free variable $P_s > 0$ (for example, use the MATLAB command `Ps = lyap(As',Qs)`).

**Step 4.** Construct the block diagonal matrix

$$P = \begin{bmatrix} P_s & 0 \\ 0 & \rho_0 I_0 \end{bmatrix},$$

where $I_0$ is the identity matrix of the same size as $A_0$.

**Step 5.** Select the compensation gains as $L = 0$ and $K = -\rho B^T P$.

**Step 6.** Construct the anti-windup compensation as (7.2), (7.3), (7.5) with the selection at the previous step. Equivalently, use the filter (7.8) in an external linear anti-windup compensation scheme.

★

The proof of the effectiveness of the construction in Algorithm 11 actually allows for any selection for $v$ of the form given at step 5, where $P$ is a positive definite matrix that solves the Lyapunov inequality

$$A_p^T P + P A_p \le 0. \tag{7.14}$$

However, in general, this solution cannot be determined in a straightforward way, and Algorithm 11 provides a constructive procedure to select such a matrix $P$.

Similar to the case of Algorithm 10, when the plant has more than one input, the constant $\rho$ can be replaced by an arbitrary positive definite diagonal matrix to allow for extra degrees of freedom in affecting the plant inputs. Moreover, the constant $\rho_0$ can be selected as a diagonal positive definite matrix, to allow for different weighting on the different neutrally stable modes. However, special care has to be taken to ensure that for each pair of complex conjugate neutrally stable eigenvalues, the corresponding entries on the diagonal matrix coincide, otherwise the resulting matrix $P$ constructed at step 4 won't satisfy the necessary condition (7.14).

**Example 7.2.4**   (SISO academic example) Consider the the SISO academic example qualitatively discussed in Section 1.2.1 on page 4. The plant consists of a simple integrator with an additive input disturbance. The corresponding matrices in (7.1) are

$$\left[ \begin{array}{c|cc} A_p & B_{p,u} & B_{p,w} \\ \hline C_{p,y} & D_{p,yu} & D_{p,yw} \end{array} \right] = \left[ \begin{array}{c|cc} 0 & 1 & 0 \; 1 \\ \hline 1 & 0 & 0 \; 0 \end{array} \right],$$

where $w = \left[ \begin{smallmatrix} r \\ d \end{smallmatrix} \right]$ comprises the reference and disturbance inputs. For this example, a PID controller is used with all the gains equal to one to induce a desirable closed-loop response and suitable disturbance rejection. The controller can be any type of approximate PID controller as its implementation makes little difference on the resulting closed loop. In addition, the anti-windup compensation acts before and after the controller and is controller independent. The control input in the unconstrained case is selected as the difference $r - y$, which is also taken as the performance output, namely,

$$\left[ \begin{array}{c|cc} C_{p,z} & D_{p,zu} & D_{p,zw} \end{array} \right] = \left[ \begin{array}{c|cc} -1 & 0 & 0 \; 1 \end{array} \right].$$

Following Algorithm 11, the plant dynamics is already in the form (7.13) because it contains only one pole at the origin. Therefore, it is not necessary to apply Procedure 5. At step 2, $A_s$ and $Q_s$ are empty matrices and the selection $\rho = \rho_0 = 1$ is chosen. Then the algorithm provides the selection $L = 0$, $K = -1$. Anti-windup compensation is designed by augmenting the closed loop with the MRAW filter (7.2), (7.3), (7.5) and with this selection.

The simulation of Figure 7.6 is performed by limiting the control input between $\pm 0.1$ and selecting the following reference and disturbance inputs:

$$r(t) = \begin{cases} 0, & \text{if } t < 0, \\ 1, & \text{if } t \ge 0, \end{cases} \qquad d(t) = \begin{cases} 0, & \text{if } t \notin [80, 80.1], \\ 10, & \text{if } t \in [80, 80.1], \end{cases}$$

namely, unit step reference applied at time $t = 0$ and an impulsive disturbance occurring at time $t = 80$.

From the simulation it appears that the anti-windup solution (solid) allows recovering as quickly as possible the unconstrained response (dashed) and eliminating

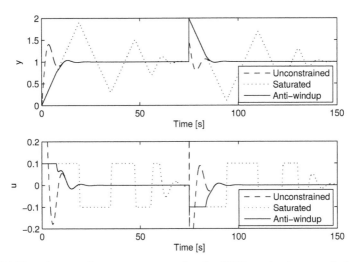

Figure 7.6 Plant input and output responses for the SISO academic example in Example 7.2.4.

the undesirable windup-induced oscillations exhibited by the saturated closed loop (dotted).                                                                                     □

**Example 7.2.5**    Consider the servo-positioning system introduced in Section 1.2.4 on page 10 and represented in Figure 1.10. The design qualitatively illustrated in that section is described in greater detail here.

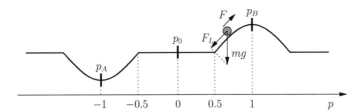

Figure 7.7 Coordinates of the servo-positioning system.

A coordinate system can be defined on the system, as represented in Figure 7.7, where the point $p_0$ corresponds to $p = 0$, the point $p_A$ corresponds to $p = -1$, and the point $p_B$ corresponds to $p = 1$. Assuming that the mass is actuated by an electrical motor, if the arc length of the path represented in Figure 7.7 (namely, distance traveled along the path) is approximated with the horizontal coordinate, then the linear transfer function of the plant can be written as

$$p(s) = \mathcal{P}(s)(F(s) + F_L(s)) = \frac{0.5}{s(1+2s)}(F(s) + F_L(s)),$$

where $\tau_M = 2\ s$ is the mechanical time constant of the motor and of the cart, and $F_L(s)$ is the load torque arising from the gravity action.

To guarantee that the unconstrained closed loop has a 3 dB bandwidth of 6 rad/s and zero steady-state error with constant disturbances, the controller needs to contain a pole in the origin and, based on a frequency domain design, it can be augmented with two coincident lead networks as follows:

$$y_c(s) = \frac{18}{s} \frac{(1+0.67s)^2}{(1+0.042s)^2}(p(s) - F(s)r(s)), \qquad F(s) = \frac{1}{1+\frac{1}{3}s},$$

where $r(s)$ is the reference representing the desired mass position and $F(s)$ is a feedforward filter that improves the transient response. According to the path profile represented in Figure 7.7, by selecting the maximum load force to be equal to 0.5 N, the disturbance input $F_L$ can be chosen as a function of the mass position $p$, as follows:

$$F_L := \begin{cases} -0.5\cos(p/\pi) & \text{if } 0.5 \le p \le 1.5 \\ -0.5\cos(p/\pi) & \text{if } -1.5 \le p \le -0.5 \\ 0 & \text{otherwise.} \end{cases} \qquad (7.15)$$

Consider the presence of saturation at the plant input, which constrains the force exerted on the mass to be between $\pm 1$. The unconstrained closed-loop response of this system when the reference signal corresponds to a step between zero and $\delta = 0.05$ is represented in Figure 1.11, and corresponds to plant inputs that remain smaller than the saturation limits for all times. Hence, the saturated system can reproduce this response. However, the closed-loop system exhibits very different behaviors when the step reference corresponds either to the top of the bounce on the right side (namely, $r = 1$) or to the bottom of the dip on the left side (namely, $r = -1$).

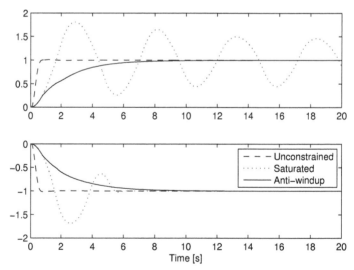

Figure 7.8 Responses of the positioning system of Example 7.2.5 when the step reference signal is 1 (upper plot) and $-1$ (lower plot).

When disregarding input saturation and considering the unconstrained closed-loop responses, these two trajectories are represented by the dashed curves in the two plots of Figure 7.8 (which resemble the responses previously reported in Figures 1.11, 1.12, and 1.14). Note that the effect of the disturbance on the unconstrained closed-loop system is almost negligible. However, the windup effect for this system is very significant, and the saturated response, corresponding to the dotted curves in the same figure, is unacceptable.

To compensate for the undesirable saturated response, an anti-windup compensator can be constructed for this control system according to Algorithm 11 (note that the other algorithms proposed in this chapter and in the previous one are not applicable to this case because the plant is not exponentially stable). To this end, consider the following state-space representation of the plant:

$$
\left[\begin{array}{c|c} A_p & B_{p,u} \\ \hline C_{p,y} & D_{p,yu} \end{array}\right] = \left[\begin{array}{cc|c} -0.5 & 0 & 1 \\ 0 & 0 & 1 \\ \hline -2 & 2 & 0 \end{array}\right].
$$

Then, by selecting $Q_s = 1$, $\rho = 50$, and $\rho_0 = 0.01$, the selection in step 5 corresponds to choosing $v = -\left[\begin{array}{cc} 0.5 & 50 \end{array}\right] x_{aw}$. The response of the resulting anti-windup closed-loop system is represented by the solid curves in Figure 7.8. Note that the anti-windup action eliminates the windup effects experienced on the dotted responses and guarantees a smooth convergence to the desired steady state (in addition to preservation of the local unconstrained performance). □

### 7.2.3 Global exponential stability with local LQ performance

A natural performance measure that one can consider for the design of the stabilizing signal $v$ is a generic quadratic cost of the form:

$$
J = \int_0^\infty (x_{aw}^T Q_P x_{aw} + v^T R_P v)\,dt. \tag{7.16}
$$

With the goal of formulating suitable linear matrix inequalities for anti-windup design, instead of minimizing the linear quadratic (LQ) performance index above evaluated along the real solutions of the anti-windup compensator (7.2), it is possible to minimize it when applied to the solutions of the following linear system:

$$
\begin{aligned} \dot{x}_{aw} &= A_p x_{aw} + B_{p,u} v \\ v &= K_v x_{aw}, \end{aligned} \tag{7.17}
$$

which, especially for the medium signal setting, well approximate the trajectories of the original system (7.2). Several case studies have shown that the resulting compensation gains often lead to highly desirable performance.

**Algorithm 12** (Global LQ-based MRAW)

| Applicability | | | | Architecture | | | Guarantee |
|---|---|---|---|---|---|---|---|
| Exp Stab | Marg Stab | Marg Unst | Exp Unst | Lin/ NonL | Dyn/ Static | Ext/ FullAu | Global/ Regional |
| √ | | | | L | D | E | G |
| **Comments:** A linear quadratic performance index related to URR is optimized subject to guaranteeing global exponential stability. | | | | | | | |

**Step 1.** Select two square positive definite matrices $Q_P$ and $R_P$ of the size of the compensator state and of the compensator input, respectively. Select a constant $v \in [0, 1)$ used to enforce a strong well-posedness constraint (see Section 2.3.7).

**Step 2.** Solve the following LMI optimization problem, in the free variables $Q = Q^T > 0$, $U > 0$ diagonal, $\gamma > 0$, $X_1$, and $X_2$, where $X_1$ has the same dimension as $K_v$ and $X_2$ has the same dimension as $L_v$:

$$\min_{Q, U, X_1, X_2} \gamma \ subject \ to:$$

Strong well-posedness (if $v = 0$, then redundant):

$$X_2^T + X_2 - 2(1 - v)U < 0$$

Global stability:

$$\begin{bmatrix} QA_p^T + A_pQ & B_{p,u}U + X_1^T \\ UB_{p,u}^T + X_1 & X_2^T + X_2 - 2U \end{bmatrix} < 0,$$

Local LQ:

$$\begin{bmatrix} \gamma I & I \\ I & Q \end{bmatrix} > 0,$$

$$\begin{bmatrix} QA_p^T + A_pQ + B_{p,u}X_1 + X_1^T B_{p,u}^T & Q & X_1^T \\ Q & -Q_P^{-1} & 0 \\ X_1 & 0 & -R_P^{-1} \end{bmatrix} < 0.$$

**Step 3.** Based on the optimal solution resulting from the previous step, compute the compensation gains as follows:

$$K = X_1 Q^{-1}, \quad L = X_2 U^{-1}.$$

**Step 4.** Construct the anti-windup compensation as (7.2), (7.3), (7.5) with the selection at the previous step. Equivalently, use the filter (7.8) in an external linear anti-windup compensation scheme.

$\star$

**Example 7.2.6** (Damped mass-spring system) The mass-spring system in Figure 7.9 was briefly discussed and introduced in Section 1.2.5 on page 13. It is here described in much more detail. Its equations of motion are given by

$$\begin{aligned} \dot{x}_p &= \begin{bmatrix} 0 & 1 \\ -k/m & -f/m \end{bmatrix} x_p + \begin{bmatrix} 0 \\ 1/m \end{bmatrix} (u_p + d) \\ z = y &= \begin{bmatrix} 1 & 0 \end{bmatrix} x_p, \end{aligned} \quad (7.18)$$

where $x_p := [q \ \dot{q}]^T$ represents position and speed of the body connected to the spring, $m$ is the mass of the body, $k$ is the elastic constant of the spring, $f$ is the damping coefficient, $u$ represents a force exerted on the mass, and $d$ represents a disturbance at the plant's input. The following values for the parameters can be selected:

$$m_0 = 0.1 \ kg, \quad k_0 = 1 \ \frac{kg}{s^2}, \quad f_0 = 0.005 \ \frac{kg}{s}. \quad (7.19)$$

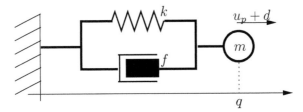

Figure 7.9 The damped mass-spring system.

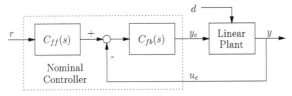

Figure 7.10 The unconstrained (linear) closed-loop system in Example 7.2.6.

Assume that $r \in \mathbb{R}$ is a reference input corresponding to the desired mass position. The linear controller shown in Figure 7.10 corresponds to the equations:

$$y_c = C_{fb}(s)\left(C_{ff}(s)\,r - u_c\right), \tag{7.20a}$$

with

$$C_{fb}(s) = 500\,\frac{(s+15)^2}{s\,(s+80)}, \quad C_{ff}(s) = \frac{5}{2s+5}. \tag{7.20b}$$

This controller has been determined with the aim of guaranteeing a fast response with zero steady-state error to step reference changes, robust to parameter uncertainties. While the performance of the system is very desirable in the absence of saturation (namely when $u_c = y$ and $u_p = y_c$), this control scheme appears to be extremely sensitive to the activation of the saturation nonlinearity.

This is confirmed by the simulations of the unconstrained and saturated closed-loop systems. Consider the case where the disturbance $d$ is chosen as a band-limited white noise of power 0.001 passed through a zero-order holder with sampling time 0.001 seconds. The response of the unconstrained closed-loop system starting from the rest position and with the reference switching between $\pm 0.9$ meters every 5 seconds and going back to zero permanently after 15 seconds is shown by the dashed curve in Figure 7.11. When the force exerted at the plant's input $u$ is limited to between $\pm 1\ \mathrm{kg \cdot m/s^2}$, the corresponding saturated closed-loop response, represented by the dotted curve in Figure 7.11, exhibits persistent oscillations between positive and negative peaks $q_P \approx 370$ meters.

To address and solve this windup problem following the construction proposed in Algorithm 12, the parameters at step 1 are selected as:

$$Q_P = \mathrm{diag}(0.1, 10), \qquad R_P = 10,$$

and $v = 0.1$. Moreover, to improve the numerical conditioning of the LMI solver, the value of the scalar $L$ has been constrained between $\pm 1$ by adding the extra LMI

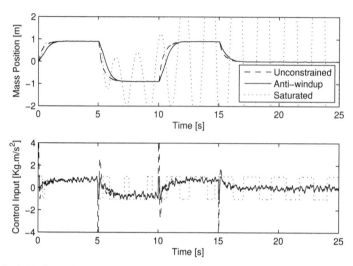

Figure 7.11  Various closed-loop responses for the mass-spring system in Example 7.2.6.

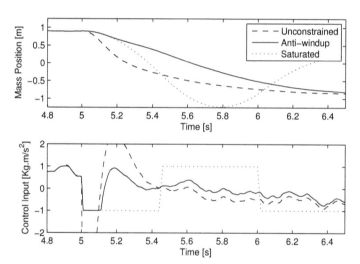

Figure 7.12  An enlarged version of the responses reported in Figure 7.11.

$-2U < X_2 + X_2^T$. This LMI combines with $X_2 + X_2^T < 2U$ (which comes from the lower right term in the global stability LMI at step 2) to imply $|L| < 1$. Note that this trick only works for the case of $L$ being a scalar and otherwise only guarantees a bound on the symmetric part of $L$ suitably scaled by the diagonal matrix $U$. The following solution to the LMI optimization problem at steps 2 and 3 is obtained:

$$K = \begin{bmatrix} -0.0145 & -1.0104 \end{bmatrix}, \qquad L = \begin{bmatrix} -0.35 \end{bmatrix}.$$

The resulting anti-windup response corresponds to the solid curve in Figure 7.11. An enlarged version of the initial transient is shown in Figure 7.12 to better illustrate the anti-windup trajectory and its difference from the unconstrained one. □

### 7.2.4 Global exponential stability with local $\mathcal{H}_2$ performance

An interesting viewpoint for designing the anti-windup compensation gains in (7.5) arises from the observation that, as clarified in Section 6.2, by virtue of the action of the signal $y_{aw}$, the input of the unconstrained controller always coincides with the unconstrained plant response. Therefore, its output $y_c$ also always coincides with the controller output in the unconstrained closed loop. Moreover, in many cases the controller output in the unconstrained closed loop converges to small values after a reasonably short time because it is a signal generated by a linear closed-loop system which is designed to guarantee high performance, and this usually corresponds to imposing fast transients and fast convergence to the steady state. On the other hand, looking at the anti-windup compensator structure (7.2), it appears evident that the controller output $y_c$ acts as a disturbance on the anti-windup dynamics. Hence, in all the cases described above, it can be thought of as a pulse which rapidly drives the state $x_{aw}$ far from the origin and then converges back to zero or to a steady-state value which is smaller than the saturation levels.

It is then of interest to minimize the integral of the performance output $z_{aw} = z - z_\ell$ by guaranteeing that this integral be smaller than the size of the initial state $x_{aw}$ of the anti-windup compensator dynamics, suitably scaled by a variable $\gamma_2$ which measures the anti-windup performance level. Moreover, since the signal $y_c$ affects the dynamics (7.2) through the input matrix $B_{p,u}$, it may be convenient to consider in the optimization problem only the initial conditions that belong to the image of the matrix $B_{p,u}$, because this corresponds to the region where the anti-windup state $x_{aw}$ will be forced into, after a pulse drives the compensator from the input $y_c$.

More precisely, the following algorithm seeks for selections of $K$ and $L$ in (7.5), aimed at minimizing the quantity $\gamma_2$ in the following inequality:

$$\int_T^\infty |z_{aw}(t)|^2 dt \le \gamma_2 |x_{aw}(T)|^2 = \gamma_2 |x_p(T) - x_{p\ell}(T)|^2, \tag{7.21}$$

where $T$ is the smallest time such that $\mathrm{sat}(u_\ell(t)) = u_\ell(t)$ for all $t \ge T$, and where $x_{aw}(T)$ belongs to the image of the matrix $B_{p,u}$.

The abovementioned performance measure corresponds to the $\mathcal{H}_2$ norm of the system

$$\dot{x}_{aw} = A_p x_{aw} + B_{p,u} v + B_{p,u} w$$
$$v = K_v x_{aw} + L_v v$$
$$z_{aw} = C_{p,z} x_{aw} + D_{p,zu} v$$

from the input $w$ to the output $z_{aw}$. This is the motivation for the name given to this particular MRAW construction algorithm.

The algorithm involves the selection of a free design parameter $\rho_v \ge 0$. This scalar penalizes the size of the matrices $K$ and $L$, for implementation purposes. Indeed, especially in the case where $D_{p,zu} = 0$, by selecting $\rho_v = 0$, the convex optimization procedure might result in very high values of $K$ and $L$, while selecting $\rho_v$ to be a small positive number makes the matrix sizes reasonable at the price of a negligible reduction of the performance level.

**Algorithm 13** (Global $\mathcal{H}_2$-based MRAW)

| Applicability | | | | Architecture | | | Guarantee |
|:---:|:---:|:---:|:---:|:---:|:---:|:---:|:---:|
| Exp Stab | Marg Stab | Marg Unst | Exp Unst | Lin/ NonL | Dyn/ Static | Ext/ FullAu | Global/ Regional |
| $\checkmark$ | | | | L | D | E | G |

| |
|---|
| *Comments:* An $\mathcal{H}_2$ performance index related to the unconstrained response recovery is optimized subject to guaranteeing global exponential stability. |

**Step 1.** Select a nonnegative constant $\rho_v \geq 0$, constituting a penalty on the matrices $K$ and $L$. Select a constant $v \in [0,1)$ used to enforce a strong well-posedness constraint (see Section 2.3.7).

**Step 2.** Solve the following LMI optimization problem, in the free variables $Q = Q^T > 0$, $U > 0$ diagonal, $\gamma > 0$, $X_1$, and $X_2$, where $X_1$ has the same dimension as $K$ and $X_2$ has the same dimension as $L$:

$$\min_{Q,U,X_1,X_2} \gamma \ \textit{subject to:}$$

*Strong well-posedness* (if $v = 0$, then redundant):

$$X_2^T + X_2 + 2(1-v)U > 0$$

$$\begin{bmatrix} -X_2 - X_2^T & \star \\ X_2^T & U \end{bmatrix} > 0,$$

*Global stability:*

$$\begin{bmatrix} QA_p^T + A_pQ + B_{p,u}X_1 + X_1^T B_{p,u}^T & \star & \star \\ -UB_{p,u}^T - X_2^T B_{p,u}^T + X_1 & -2U - X_2 - X_2^T & \star \\ \rho_v X_1 & \rho_v X_2 & -I \end{bmatrix} < 0,$$

*Local $\mathcal{H}_2$:*

$$\begin{bmatrix} QA_p^T + A_pQ + B_{p,u}X_1 + X_1^T B_{p,u}^T & \star \\ C_{p,z}Q + D_{p,zu}X_1 & -I \end{bmatrix} < 0$$

$$\begin{bmatrix} \gamma I & B_{p,u}^T \\ B_{p,u} & Q \end{bmatrix} > 0.$$

**Step 3.** Given the optimal solution resulting from the previous step, compute the compensation gains as follows:

$$\begin{aligned} K &= \left(I + X_2 U^{-1}\right)^{-1} X_1 Q^{-1} \\ L &= \left(I + X_2 U^{-1}\right)^{-1} X_2 U^{-1}. \end{aligned} \tag{7.22}$$

**Step 4.** Construct the anti-windup compensation as (7.2), (7.3), (7.5) with the selection at the previous step. Equivalently, use the filter (7.8) in an external linear anti-windup compensation scheme.

$\star$

Note that for the algorithm to be completed, the matrix in the brackets at step 3 needs to be nonsingular. Nonsingularity of this matrix can indeed be guaranteed based on the properties of the LMI constraints at step 2.

The matrix $B_{p,u}$ appearing in the off-diagonal term of the last LMI at step 2 corresponds to the set of initial conditions for which the $\mathcal{H}_2$ performance is minimized. In some cases it might be convenient to replace this matrix with an identity so that the $\mathcal{H}_2$ performance from any initial condition is minimized. Otherwise, alternative subspaces of initial conditions that are of interest for the anti-windup problem under consideration can be incorporated in the optimization constraints by suitably replacing that off-diagonal term.

## 7.3 REGIONAL STABILITY AND PERFORMANCE ALGORITHMS

### 7.3.1 Regional LQ-based design

With the goal in mind of only guaranteeing regional results, a very simple selection for the anti-windup gains in (7.5) consists in selecting $L = 0$, so that no algebraic loops will arise in the scheme, and in selecting $K$ as a stabilizing state feedback gain for the simplified dynamics

$$\dot{x}_{aw} = Ax_{aw} + Bv$$
$$v = Kx_{aw}$$

that corresponds to the actual anti-windup dynamics (7.2) for small enough signals (namely, when saturation is not active) and well approximates the anti-windup dynamics as long as $v$ remains sufficiently small.

Any stabilizing $K$ will guarantee regional stabilization, but since saturation is not taken into account in the design phase, the stability region of the anti-windup closed-loop system may as a result be too small, especially when $K$ is very aggressive. Nevertheless, within this family of solutions, a design paradigm that has proven to work well in practice consists in selecting the gain $K$ as a stabilizing gain via an LQ design, as detailed in the next algorithm.

**Algorithm 14** (Regional LQ-based MRAW)

| Applicability | | | | Architecture | | | Guarantee |
|:---:|:---:|:---:|:---:|:---:|:---:|:---:|:---:|
| Exp Stab | Marg Stab | Marg Unst | Exp Unst | Lin/ NonL | Dyn/ Static | Ext/ FullAu | Global/ Regional |
| $\checkmark$ | $\checkmark$ | $\checkmark$ | $\checkmark$ | L | D | E | R |

*Comments*: Extends the applicability of Algorithm 12 to any loop by requiring only regional exponential stability. Does not create an algebraic loop. A linear quadratic performance index related to small signal unconstrained response recovery is optimized.

**Step 1.** Select two weight matrices $Q = Q^T \geq 0$ and $R = R^T > 0$ for the anti-windup state and input variables.[1]

---

[1] The matrix $Q$ should be selected in such a way that the resulting LQ control law is stabilizing (e.g.,

**Step 2.** Choose the anti-windup gain $K$ as an LQR gain with weights $(Q,R)$ (e.g., use the MATLAB command K=-lqr(A,B,Q,R)). Choose $L = 0$.

**Step 3.** Construct the anti-windup compensation as (7.2), (7.3), (7.5) with the selection at the previous step. Equivalently, use the filter (7.8) in an external linear anti-windup compensation scheme.

★

**Example 7.3.1** (Damped mass-spring system) Consider once again the damped mass-spring system studied in Example 7.2.6 on page 190. In that example, the compensation gains were selected by following Algorithm 12, which can be well understood as the equivalent of Algorithm 14 when global closed-loop stability guarantees are desired.

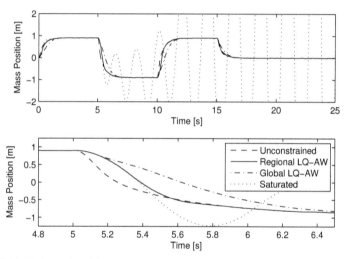

Figure 7.13 Various closed-loop responses for the mass-spring system in Example 7.3.1.

Since no closed-loop stability is ensured by Algorithm 14, it is expected that a more aggressive compensation law will be obtained, possibly inducing a better performance but at the risk of destabilizing the plant in some situations. For this example, by selecting the parameters at step 1 of Algorithm 14 to be $Q = \begin{bmatrix} 500 & 0 \\ 0 & 10 \end{bmatrix}$ and $R = 1$, the gain $K = \begin{bmatrix} -21.38 & -3.78 \end{bmatrix}$ is obtained, which leads to an extremely desirable closed-loop response. The closed-loop response with this anti-windup compensator is compared to the unconstrained (dashed), saturated (dotted) and global LQ anti-windup responses in Figure 7.13, which reports on the equivalents of the two top plots of Figures 7.11 and 7.12 when using Algorithm 14. From the curves in Figure 7.13 it is quite evident that this example greatly benefits from the use of the regional compensation scheme. It should be kept in mind, however, that only local stability guarantees hold when using this scheme. □

---

$Q = C^T C$ with $(A,C)$ detectable.

### 7.3.2 Regional $\mathcal{H}_2$-based design

Based on the global $\mathcal{H}_2$ based MRAW approach of Section 7.2.4, a possible strategy to design the compensation gains $K$ and $L$ for regional results is to follow that same construction and, instead of guaranteeing global exponential stability by constraining the nonlinearity in the global sector, using the regional sector tools in Section 3.5 to obtain extra degrees of freedom in the design phase and only regional guarantees. Based on that regional sector property, it is desirable to solve an optimization problem minimizing $\gamma_2$ in (7.21) by requiring stability guarantees in only a subset of the state space corresponding to an ellipsoid characterized by a sublevel set of a Lyapunov function denoted by $\mathcal{E}(P) = \{x_{aw} : x_{aw}^T P x_{aw} \leq 1\}$. Then, stability would only be guaranteed in a bounded subset of the state space but the anti-windup compensation performance would in general be improved. In addition to this, similar to the previous section, these regional constructions can be applied to any linear windup-prone control system because focusing on a bounded set still leaves enough input authority to make the stability LMIs feasible for any type of plant (rather than only being feasible for exponentially stable plants, as it was the case for the global Algorithm 13).

The next algorithm pursues the goal referred to above. The algorithm requires the selection of three design parameters, one of them being the strong well-posedness parameter $v$ (see Section 2.3.7), the second one being a desired guaranteed stability region, characterized by a matrix $P_s$ so that $\mathcal{E}(P_s)$ is certainly in the domain of attraction, and the third one, $\rho_v$, being a penalty term for the matrices $K$ and $L$ paralleling the one introduced in the global Algorithm 13.

Since the algorithm relies on the use of the regional sector tools of Section 3.5, it involves the knowledge of the saturation levels $\bar{u}_i > 0$, $i = 1,\ldots,n_u$, which may all be different from each other. Due to the properties of those sector tools, the quantities $\bar{u}_i > 0$ only need to be lower bounds on the saturation levels which could be time-varying and/or nonsymmetric. Note, however, that, as extensively commented on in Chapter 6, the problem of designing $v$ in the MRAW compensation scheme is a transformed problem where the saturation nonlinearity is shifted by the presence of the signal $y_c$. Therefore, when dealing with setpoint stabilization problems, one could consider restricting the values $\bar{u}_i$ by an amount corresponding to the worst case steady-state controller output $y_c$, given a bound on the allowable reference signals. In this way, the (regional) stability guarantees of the next algorithm will hold for all references satisfying that bound.

**Algorithm 15** (Regional $\mathcal{H}_2$-based MRAW)

| Applicability | | | | Architecture | | | Guarantee |
|---|---|---|---|---|---|---|---|
| Exp Stab | Marg Stab | Marg Unst | Exp Unst | Lin/ NonL | Dyn/ Static | Ext/ FullAu | Global/ Regional |
| $\checkmark$ | $\checkmark$ | $\checkmark$ | $\checkmark$ | L | D | E | R |

| |
|---|
| *Comments*: Extends the applicability of Algorithm 13 to any loop by requiring only regional exponential stability. An $\mathcal{H}_2$ performance index related to small signal unconstrained response recovery is optimized. |

**Step 1.** Select a symmetric matrix $P_s$ associated with a desired guaranteed stability region $\mathcal{E}(P_s) := \{x_{aw} : x_{aw}^T P_s x_{aw} \leq 1\}$. Select a nonnegative constant $\rho_v \geq 0$, for a penalty on the matrices $K$ and $L$. Select a constant $v \in [0,1)$, used to enforce a strong well-posedness constraint (see Section 2.3.7).

**Step 2.** Solve the following LMI optimization problem, in the free variables $Q = Q^T > 0$, $U > 0$ diagonal, $\gamma > 0$, $Y$, $X_1$, and $X_2$, where $X_1$ and $Y$ have the same dimension as $K$ and $X_2$ has the same dimension as $L$:

$$\min_{Q,U,X_1,X_2} \gamma \; subject \; to:$$

*Strong well-posedness*:
$$X_2^T + X_2 + 2(1-v)U > 0$$
$$\begin{bmatrix} -X_2 - X_2^T & \star \\ X_2^T & U \end{bmatrix} \geq 0,$$

*Regional stability*:
$$He \begin{bmatrix} A_p Q + B_{p,u} X_1 & -B_{p,u}(U+X_2) & 0 \\ X_1 + Y & -U - X_2 & 0 \\ \rho_v X_1 & \rho_v X_2 & -I \end{bmatrix} < 0$$

$$\begin{bmatrix} Q & Y_i^T \\ Y_i & \bar{u}_i^2 \end{bmatrix} > 0, \quad i = 1,\ldots,n_u ,$$

*Local* $\mathcal{H}_2$:
$$\begin{bmatrix} QA_p^T + A_p Q + B_{p,u} X_1 + X_1^T B_{p,u}^T & \star \\ C_{p,z} Q + D_{p,zu} X_1 & -I \end{bmatrix} < 0$$

$$\begin{bmatrix} \gamma I & B_{p,u}^T \\ B_{p,u} & Q \end{bmatrix} > 0,$$

where $Y_i$ denotes the $i$th row of the matrix $Y$ and where $\bar{u}_1,\ldots,\bar{u}_{n_u}$ denotes the symmetric saturation limits.

**Step 3.** Given the optimal solution resulting from the previous step, compute the compensation gains as follows:

$$\begin{aligned} K &= (I + X_2 U^{-1})^{-1} X_1 Q^{-1} \\ L &= (I + X_2 U^{-1})^{-1} X_2 U^{-1}. \end{aligned} \tag{7.23}$$

**Step 4.** Construct the anti-windup compensation as (7.2), (7.3), (7.5) with the selection at the previous step. Equivalently, use the filter (7.8) in an external linear anti-windup compensation scheme.

$\star$

Note that for the procedure to be completed, the matrix in the brackets at step 3 needs to be nonsingular. Similar to the preceding Algorithm 13, this matrix is guaranteed to be invertible by the properties of the LMI constraints at step 2.

## 7.4 NOTES AND REFERENCES

The IMC anti-windup scheme seen as a special case of Algorithm 10 is a reinterpretation (see the notes and references of Chapter 6) of the IMC anti-windup first described in [G21] and then well characterized in [FC7]. The Lyapunov-based synthesis seen as a special case of the same algorithm was suggested in [MR3] for global stability and was there applied to Example 7.2.1 using the gains reported here. The synthesis of Algorithm 11 for global asymptotic stability for marginally stable plants also derives from the ideas in [MR3] and has been illustrated in more detail in the application paper [AP10], where also its discrete-time counterpart is discussed. The global LQ-based synthesis LMIs given in Algorithm 12 come from [MR15], while the regional LQ-based approach is a natural extension of the results in [MR3]. The global and regional $\mathcal{H}_2$-based designs of Algorithms 13 and 15, respectively, come from the ideas in [MR21], even though a different sector characterization was used for the regional design therein proposed. Here, in contrast to [MR21], the LMIs for regional performance follow from the ideas presented in Section 3.5 of Chapter 3, which come from [SAT11] and [MI27].

The plant model used in the MRAW compensation architecture can be an approximated model, and the intrinsic robustness of the scheme guarantees a certain level of tolerance with respect to unmodeled dynamics. See [MR31] for a discrete-time case study with reduced-order models.

The MIMO academic example treated in Example 7.2.1 was introduced in [CS7] and later revisited in [MR3] to illustrate the MRAW construction discussed in this chapter. The turbofan engine example in Example 7.2.2 is taken from [AP1], where details can be found on the design of the unconstrained controller. Moreover, in [AP1] the solution presented here is compared to the use of the conditioning technique from [FC6]. Finally, [AP1] also reports an interesting discussion on the projection of infeasible references onto feasible ones while preserving certain performance requirements. The SISO academic system of Example 7.2.4 was used by Åström in [CS2] while Example 7.2.5 was cooked up by the authors of this book to illustrate how disturbances can lead to windup effects. The damped mass-spring system in Examples 7.2.6 and 7.3.1 was introduced in [MR15] and later used in many anti-windup papers (see, e.g., [MI23, MR23, MR27, MR28]).

# Chapter Eight

## Nonlinear MRAW Synthesis

### 8.1 INTRODUCTION

In this chapter the degrees of freedom available within the model recovery anti-windup (MRAW) structure, which was introduced in Chapter 6, are exploited using nonlinear control ideas that account for actuator constraints. Recall that the MRAW structure turns the unconstrained response recovery objective into the task of choosing a feedback $v$ to drive to zero the state and performance variable of the system

$$
\begin{aligned}
\dot{x}_{aw} &= A_p x_{aw} + B_{p,u} \left[ \text{sat}(v + y_c) - y_c \right] \\
z_{aw} &= C_{p,z} x_{aw} + D_{p,zu} \left[ \text{sat}(v + y_c) - y_c \right],
\end{aligned}
$$

at least for signals $y_c$ that satisfy $||y_c(\cdot) - \text{sat}(y_c(\cdot))||_2 < \infty$ or $||y_c(\cdot) - \text{sat}_\delta(y_c(\cdot))||_2 < \infty$ for some $\delta > 0$. Thus, the MRAW structure makes direct designs for actuator saturation applicable to the anti-windup problem. Here, three particular approaches will be discussed for synthesizing $v$. The common feature of these approaches is that $v$ will be a nonlinear function of $x_{aw}$ and, possibly, of $\text{sat}(v + y_c) - y_c$. This is in contrast to the previous chapter, which focused on generating feedbacks $v$ that were linear in $x_{aw}$ and $\text{sat}(v + y_c) - y_c$. In considering nonlinear synthesis techniques for $v$, the user can expect a tradeoff between anti-windup performance, which has the potential to improve since less structure is imposed on the anti-windup augmentation, and design and implementation complexity. The tradeoff can be extreme in the case of global anti-windup for plants that are marginally unstable. In this case, it is often impossible to solve the anti-windup problem globally using linear $v$'s without algebraic loops, whereas such problems are always solvable using nonlinear $v$'s. For exponentially unstable plants, a design that requires extra state measurements from the plant and then yields improved performance compared to that achieved using linear designs will be illustrated.

Thus, the starting point for this chapter is anti-windup augmentation that has the form

$$
\mathcal{F} \begin{cases}
\dot{x}_{aw} &= A_p x_{aw} + B_{p,u} \left( \text{sat}(y_c + v) - y_c \right) \\
z_{aw} &= C_{p,z} x_{aw} + D_{p,zu} \left( \text{sat}(y_c + v) - y_c \right) \\
y_{aw} &= C_{p,y} x_{aw} + D_{p,yu} \left( \text{sat}(y_c + v) - y_c \right) \\
v &= \text{to be designed.}
\end{cases}
\tag{8.1}
$$

It is connected with the other closed-loop signals in the usual way:

$$
u_c = y - y_{aw}, \qquad u = y_c + v, \qquad \sigma = \text{sat}(u).
\tag{8.2}
$$

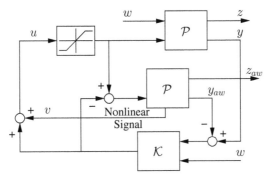

Figure 8.1  The nonlinear MRAW compensation scheme.

The algorithms address picking $v$, typically in the form

$$v = f_v(x_{aw}, \text{sat}(y_c + v) - y_c), \qquad (8.3)$$

where $f_v(\cdot, \cdot)$ is a suitable static map ensuring well-posedness of the resulting algebraic loop and enforcing desirable closed-loop properties on the anti-windup augmented closed loop. Eventually, extensions will be considered where $v$ uses extra
signals coming from the closed loop, such as measurements of the exponentially
unstable part of the plant's state and direct measurement of the unconstrained controller output $y_c$.

## 8.2  SWITCHING AND SCHEDULING LINEAR CONTROLLERS

The most natural extension of the linear anti-windup schemes is derived from observing that in general, because of saturation, the most desirable control action
should be aggressive close to the origin and less and less aggressive as the anti-
windup state $x_{aw}$ goes farther away from the origin. This fact resembles the observation that since the saturation function is bounded, to keep the compensation
signal $v$ within a reasonable range, the gains involved should shrink as the state
becomes large.

In the next two sections this idea is pursued by relying on families of linear
gains being effective in nested ellipsoids centered at the origin. The first approach
involves switching among the various compensators while the second one performs
a smarter scheduling based on a scheduling parameter which is computed online by
integrating a differential equation.

### 8.2.1  Switching nonlinear anti-windup design

In this section the nonlinear function $f_v(\cdot, \cdot)$ is selected as a switched linear function
that can provide global stability guarantees together with extreme performance locally. This construction is based on a generalization of the global $\mathcal{H}_2$-based MRAW
design of Section 7.2.4 and the regional $\mathcal{H}_2$-based MRAW design of Section 7.3.2.

The design paradigm relies on the following switched linear selection for the compensation signal $v$:

$$v = K_{\sigma(t)} x_{aw} + L_{\sigma(t)} (\text{sat}(y_c + v) - y_c), \qquad (8.4)$$

where $\sigma(\cdot)$ is a switching signal taking values in the positive integer set and, depending on the current (and past) values of the state and the switching gains, $K_i, L_i$, $i = 1, \ldots, N$, are a set of $N$ pairs, each of them determined by solving the regional $\mathcal{H}_2$-based algorithm of Section 7.3.2 with a suitable sector bound (which in turn corresponds to a certain operating region).

The advantage of using the switched selection (8.4) versus the linear ones proposed in Sections 7.2.4 and 7.3.2 arises from the fact that typically the linear compensation gains are determined based on a tradeoff between guaranteed performance level and guaranteed stability region of the closed loop. Requiring global stability guarantees, as in Section 7.2.4, typically leads to very conservative gains enforcing slow responses so that stability is guaranteed for arbitrarily large signals. On the other hand, restricting the stability region to a smaller set allows generating more aggressive and performing closed loops, but large enough signals may drive the closed-loop system unstable. The switching strategy in (8.4) allows obtaining both the global stability and the extreme local performance guarantees by selecting more conservative gains if signals grow too large and more aggressive ones when signals become sufficiently small. Special care needs to be taken in the scheduling action to ensure that the switched system guarantees stability property; however, following the approach reported in the next algorithm should be sufficient to obtain an exponentially stable closed loop.

The nonlinear switching design algorithm is based on the definition of a family of $N$ gains $K_i, L_i$, corresponding to a finite number of nested stability regions $\mathcal{E}(P_i) = \{x_{aw} : x_{aw}^T P_i x_{aw} \leq 1\}$, each of them guaranteed to be in the basin of attraction when using the corresponding pair $K_i, L_i$. These $N$ gains and the corresponding matrices $P_i, K_i, L_i$ are determined based on three parameters to be selected at the initialization step of the procedure: a desirable region where the nesting should start (corresponding to $P_{start}$), characterizing the desired size for the largest ellipsoid of the switching design; a shrinking factor $\beta$, characterizing how far each nested ellipsoid should be from the previous one; and the number $N$ of ellipsoids to be constructed. Based on another parameter which is a small constant $h \ll 1$, the switching between the different regions is performed based on hysteresis switching associated with nested ellipsoids in the state space of the anti-windup filter (8.1). The hysteresis switching strategy is better illustrated after the algorithm with the aid of Figure 8.2.

As stated below, the algorithm provides a regionally stabilizing solution with the stability region being $\mathcal{E}(P_N) = \{x_{aw} : x_{aw}^T P_N x_{aw} \leq 1\}$, where $P_N$ is determined at step 2. This region is maximized in the optimization and is always smaller than the region $\mathcal{E}(P_{start}) = \{x_{aw} : x_{aw}^T P_{start} x_{aw} \leq 1\}$. It may, however, be very small when the plant is strongly exponentially unstable and it is guaranteed to be tangent to $\mathcal{E}(P_{start})$ when the plant is not exponentially unstable. When addressing problems with exponentially stable plants, the algorithm can be modified to be globally stabilizing by selecting the matrix $P_N$ and the gains $K_N, L_N$ as globally stabilizing gains.

This is easily achieved by setting $Y = 0$ at that step (namely, replacing $Y$ with zeros wherever it appears). Note, however, that if the plant is not exponentially stable, constraining $Y = 0$ will make the constraints of step 2 impossible to solve.

**Algorithm 16** (Switched MRAW)

| Applicability | | | | Architecture | | | Guarantee |
|---|---|---|---|---|---|---|---|
| Exp Stab | Marg Stab | Marg Unst | Exp Unst | Lin/ NonL | Dyn/ Static | Ext/ FullAu | Global/ Regional |
| $\checkmark$ | $\checkmark$ | $\checkmark$ | $\checkmark$ | NL | D | E | R/G |

*Comments*: Augmentation based on a family of nested ellipsoids and switching among a corresponding family of linear feedbacks. Regional unconstrained response recovery gain is optimized in each ellipsoid. Global exponential stability is guaranteed for any loop with an exponentially stable plant. Otherwise, regional exponential stability is guaranteed.

**Step 1.** Define a matrix characterizing the desired guaranteed region $P_{start}$, the shrinking factor $\beta$, the number $N$ of ellipsoids to be constructed, and a small constant $h \ll 1$ to perform hysteresis switching among the various gains.

**Step 2.** Solve the following LMI eigenvalue problem in the free variables $\{Q, X_1, X_2, U, Y\}$, where $Q = Q^T$ and $U > 0$ diagonal:

$$\max_{Q,U,X_1,X_2,Y} \quad \kappa \quad \text{subject to:}$$

Set inclusion:

$$\kappa I < Q < P_{start}^{-1},$$

Regional stability:

$$\text{He} \begin{bmatrix} A_p Q + B_{p,u} X_1 & B_{p,u}(X_2 - U) \\ X_1 - Y & X_2 - U \end{bmatrix} < 0,$$

$$\begin{bmatrix} \bar{u}_i^2 & Y_i \\ Y_i' & Q \end{bmatrix} \geq 0, \quad i = 1, \ldots, n_u ,$$

and set the matrix characterizing the outermost ellipsoid as $P_N = Q_N^{-1}$ with $Q_N = Q$. Also set $i = N - 1$ and the outer ellipsoid gains as

$$\begin{aligned} K_N &= (I - X_2 U^{-1}) X_1 Q^{-1} \\ L_N &= -(I - X_2 U^{-1}) X_2 U^{-1}. \end{aligned}$$

**Step 3.** Denoting by $\sigma_{i+1}$ the smallest eigenvalue of $Q_{i+1}$, solve the following generalized eigenvalue problem in the free variables $\{Q, X_1, X_2, U, Y, Z\}$, where $Q = Q^T$, $U > 0$ diagonal, and $Z = Z^T$:

$$\max_{Q,U,X_1,X_2,Y,Z} \quad \lambda \quad subject\ to:$$

Nesting condition:

$$\beta \sigma_{i+1} I < Q < Q_{i+1}$$

Regional stability:

$$Z > \lambda Q$$

$$\mathrm{He} \begin{bmatrix} A_p Q + Z + B_{p,u} X_1 & B_{p,u}(X_2 - U) \\ X_1 - Y & X_2 - U \end{bmatrix} < 0,$$

$$\begin{bmatrix} \bar{u}_i^2 & Y_i \\ Y_i' & Q \end{bmatrix} \geq 0, \quad i = 1, \ldots, n_u,$$

and set $P_i = Q^{-1}$ and the associated gains as

$$\begin{aligned} K_i &= (I - X_2 U^{-1}) X_1 Q^{-1} \\ L_i &= -(I - X_2 U^{-1}) X_2 U^{-1}. \end{aligned}$$

Decrease $i = i - 1$. If $i = 0$ then go to the next step, otherwise repeat this step.

**Step 4.** Select the switched value of $v$ as follows:

$$\begin{aligned} v &= K_\sigma x_{aw} + L_\sigma \left( \mathrm{sat}(y_c + v) - y_c \right) \\ s_i &= 1 - x_{aw}^T P_i x_{aw}, \quad i = 1, \ldots, N \\ \rho(s) &= \max\{i : s_i \geq 0\} \\ \sigma(t) &= \phi(\sigma^-(t), s), \quad \sigma(0) = 1 \\ \phi(i, s) &:= \begin{cases} i & \text{if } s_i \geq -h_i \text{ and } i \geq \rho(s) \\ \rho(s) & \text{otherwise,} \end{cases} \end{aligned} \tag{8.5}$$

where $h \ll 1$ has been selected at step 1, $s = [s_1, \ldots, s_N]^T$ and $\sigma^-(t)$ is the limit of $\sigma(\tau)$ from below as $\tau \to t$.

**Step 5.** Construct the anti-windup compensation as (8.1), (8.2) with the selection for $v$ given at the previous step.

$\star$

In equation (8.5), the stabilizing signal $v$ is switched depending on the switching function $\sigma$. Each one of the $N$ indicators $s_i(x_{aw})$ is positive when the anti-windup state $x_{aw}$ is within the set $\mathcal{E}(P_i) = \{x_{aw}^T P_i x_{aw} \leq 1\}$ (by the LMI construction, the sets $\mathcal{E}(P_i)$ turn out to be nested). The function $\rho(s)$ returns the index of the smallest one among the nested sets $\mathcal{E}(P_i)$ containing the anti-windup state $x_{aw}$. Finally, the function $\phi(\cdot, \cdot)$ implements hysteresis switching among the indexes of the nested sets $\mathcal{E}(P_i)$, with hysteresis intervals dependent on the (small) constant $h$. This switching idea is better illustrated in Figure 8.2, where the boundary of a set $\mathcal{E}(P_i)$ and the boundary of its inflated version $\mathcal{E}_h(P_i) = \{x_{aw}^T P_i x_{aw} \leq 1 + h\}$ are represented, together with a possible trajectory. In the solid part of the trajectory, $\sigma(t) = i - 1$, while in its dashed part $\sigma(t) = i$.

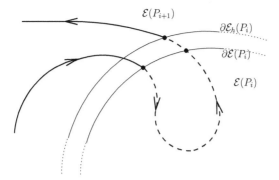

Figure 8.2  Illustration of the hysteresis switching strategy of the compensator (8.5) of the switching anti-windup design in Algorithm 16.

## 8.2.2  Scheduled nonlinear anti-windup design

One of the disadvantages of the switching approach of the last section is that during the switching phase, the control input exhibits discontinuities and might excite unmodeled plant dynamics and induced undesirable transients at the plant output. Moreover, since the anti-windup gains in each ellipsoid are computed by solving a set of linear matrix inequalities, it is quite intuitive that when the anti-windup state lies between the boundaries of two sets $\mathcal{E}(P_i)$ and $\mathcal{E}(P_{i+1})$, the use of a suitable convex combination of the $i$th and $(i+1)$th solutions would lead to improved performance.

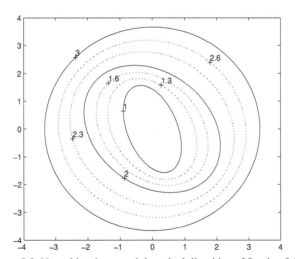

Figure 8.3  Nested level sets and the scheduling idea of Section 8.2.2.

This type of idea is pursued in this section. The main intuition behind this idea is illustrated in Figure 8.3 where the solid ellipsoids represent the boundaries of the sets $\mathcal{E}(P_1)$, $\mathcal{E}(P_2)$ and $\mathcal{E}(P_3)$. The dotted ellipsoids represent intermediate ellipsoids resulting from suitably interpolating the inner and the outer ones via convex combination. The scheduling parameter $\alpha$, used in the next algorithm, is constant on each ellipsoid and is selected to be equal to one inside and on the boundary of the innermost ellipsoid and equal to $N$ outside and on the boundary of the outermost ellipsoid. In the annulus region between these two sets, the scheduling parameter $\alpha$ continuously rises from 1 to $N$.

From Figure 8.3 it appears that $\alpha$ is a static function of the state $x_{aw}$. However, determining its value directly from $x_{aw}$ is hard in general and involves solving an optimization problem corresponding to answering to this question: "What is the convex combination $x_{aw}^T P_\alpha x_{aw}$ of the inner and outer functions for which the state $x_{aw}$ lies on the level set $x_{aw}^T P_\alpha x_{aw} = 1$?" Due to this reason and to simplify the computational burden, in the next algorithm $\alpha$ is not computed but rather estimated by the use of a dynamic equation that can be thought of as a state observer.

The algorithm is a generalization of the switched Algorithm 16. As such, its applicability is exactly the same and the parameters and available degrees of freedom are the same too, except for the hysteresis switching constant $h$, which is not necessary here. A new parameter appears in this algorithm instead, which is the observer gain $\ell$ used to ensure convergence of the state variable $\alpha$ to the static function illustrated in Figure 8.3. Once again, the parameter $P_{start}$ characterizes the desired size for the largest ellipsoid of the scheduled design, the shrinking factor $\beta$ characterizes how far each nested ellipsoid should be from the previous one, and $N$ is the number of ellipsoids to be constructed. In contrast to the switching Algorithm 16, good performance is obtained here even with very few ellipsoids because the solution is interpolated between them.

Similar to the switching case, the stability region is maximized in the optimization and is always smaller than the region $\mathcal{E}(P_{start}) = \{x_{aw} : x_{aw}^T P_{start} x_{aw} \leq 1\}$, however, it may become very small with strongly exponentially unstable plants. Globally stabilizing solutions can be also obtained for exponentially stable plants by replacing $Y$ with zeros wherever it appears at step 2.

**Algorithm 17** (Scheduled MRAW)

| Applicability | | | | Architecture | | | Guarantee |
|:---:|:---:|:---:|:---:|:---:|:---:|:---:|:---:|
| Exp Stab | Marg Stab | Marg Unst | Exp Unst | Lin/ NonL | Dyn/ Static | Ext/ FullAu | Global/ Regional |
| √ | √ | √ | √ | NL | D | E | R/G |

| |
|---|
| *Comments*: Augmentation based on a family of nested ellipsoids and continuous scheduling among a corresponding family of linear feedbacks. Regional unconstrained response recovery gain is optimized in each ellipsoid. Global exponential stability is guaranteed for any loop with an exponentially stable plant. Otherwise, regional exponential stability is guaranteed. |

**Step 1.** Define a matrix characterizing the desired guaranteed region $P_{start}$, the shrinking factor $\beta$, the number $N$ of ellipsoids to be constructed, and a small constant $h \ll 1$ to perform hysteresis switching among the various gains.

**Step 2.** Solve the following LMI eigenvalue problem in the free variables $\{Q, X_1, X_2, U, Y\}$, where $Q = Q^T$ and $U > 0$ diagonal:

$$\max_{Q, U, X_1, X_2, Y} \quad \kappa \quad \text{subject to:}$$

*Set inclusion*:

$$\kappa I < Q < P_{start}^{-1},$$

*Regional stability*:

$$\text{He} \begin{bmatrix} A_p Q + B_{p,u} X_1 & B_{p,u}(X_2 - U) \\ X_1 - Y & X_2 - U \end{bmatrix} < 0,$$

$$\begin{bmatrix} \bar{u}_i^2 & Y_i \\ Y_i' & Q \end{bmatrix} \geq 0, \quad i = 1, \dots, n_u,$$

and set the matrix characterizing the outermost ellipsoid as $P_N = Q_N^{-1}$ with $Q_N = Q$. Also set $i = N - 1$ and the outermost ellipsoid gains as

$$\begin{aligned} K_N &= (I - X_2 U^{-1}) X_1 Q^{-1} \\ L_N &= -(I - X_2 U^{-1}) X_2 U^{-1}. \end{aligned}$$

**Step 3.** Denoting by $\sigma_{i+1}$ the smallest eigenvalue of $Q_{i+1}$, solve the following generalized eigenvalue problem in the free variables $\{Q, X_1, X_2, U, Y, Z\}$, where $Q = Q^T$, $U > 0$ diagonal, and $Z = Z^T$:

$$\max_{Q, U, X_1, X_2, Y, Z} \quad \lambda \quad \text{subject to:}$$

*Nesting condition*:

$$\beta \sigma_{i+1} I < Q < Q_{i+1},$$

*Regional stability*:

$$Z > \lambda Q$$

$$\text{He} \begin{bmatrix} A_p Q + Z + B_{p,u} X_1 & B_{p,u}(X_2 - U) \\ X_1 - Y & X_2 - U \end{bmatrix} < 0,$$

$$\begin{bmatrix} \bar{u}_i^2 & Y_i \\ Y_i' & Q \end{bmatrix} \geq 0, \quad i = 1, \dots, n_u,$$

and set $Q_i = Q$, $X_{1i} = X_1$, $X_{2i} = X_2$, and $U_i = U$. Decrease $i = i - 1$. If $i = 0$ then go to the next step, otherwise repeat this step.

**Step 4.** Define the following functions

$$\begin{aligned} \lfloor \alpha \rfloor &:= \max\{1, \text{floor}(\alpha)\} \\ \lceil \alpha \rceil &:= \min\{N, \text{ceil}(\alpha)\}, \end{aligned}$$

so that for $\alpha \in [1, N]$, $\lfloor \alpha \rfloor$ denotes the greatest integer smaller than or equal to $\alpha$, whereas $\lceil \alpha \rceil$ denotes the least integer greater than or equal to $\alpha$. Moreover, define

$$\begin{aligned} Q(\alpha) &= (\lceil \alpha \rceil - \alpha) Q_{\lceil \alpha \rceil} + (\alpha - \lfloor \alpha \rfloor) Q_{\lfloor \alpha \rfloor} \\ U(\alpha) &= (\lceil \alpha \rceil - \alpha) U_{\lceil \alpha \rceil} + (\alpha - \lfloor \alpha \rfloor) U_{\lfloor \alpha \rfloor} \\ X_1(\alpha) &= (\lceil \alpha \rceil - \alpha) X_{1\lceil \alpha \rceil} + (\alpha - \lfloor \alpha \rfloor) X_{1\lfloor \alpha \rfloor} \\ X_2(\alpha) &= (\lceil \alpha \rceil - \alpha) X_{2\lceil \alpha \rceil} + (\alpha - \lfloor \alpha \rfloor) X_{2\lfloor \alpha \rfloor} \end{aligned}$$

and $P(\alpha) := Q(\alpha)^{-1}$. Also define the following functions

$$M(\alpha) = P(\alpha)(Q_{\lceil\alpha\rceil} - Q_{\lfloor\alpha\rfloor})P(\alpha)$$

$$g(x_{aw}, \alpha) = \ell(x_{aw}^T P(\alpha) x_{aw} - 1).$$

**Step 5.** Select the compensation signal $v$ as:

$$v = K(\alpha) x_{aw} + L(\alpha)(\text{sat}(y_c + v) - y_c)$$

where

$$
\begin{aligned}
K(\alpha) &= (I - X_2(\alpha)U(\alpha)^{-1})X_1(\alpha)Q(\alpha)^{-1} \\
L(\alpha) &= -(I - X_2(\alpha)U(\alpha)^{-1})X_2(\alpha)U(\alpha)^{-1},
\end{aligned}
$$

and $\alpha$ is the state variable obtained from the following dynamical equation, which should be embedded in the anti-windup compensation system:

$$
\dot{\alpha} = \begin{cases}
\max\{0, g(x_{aw}, \alpha)\}, & \text{if } \alpha \le 1 \\[2mm]
\dfrac{2x_{aw}'P(\alpha)x_{aw,d}}{x_{aw}'M(\alpha)x_{aw}} + g(x_{aw}, \alpha), & \text{if } \alpha \in (1, N) \quad \alpha(0) = 1, \\[2mm]
\min\{0, g(x_{aw}, \alpha)\}, & \text{if } \alpha \ge N,
\end{cases}
$$

where $x_{aw,d} = A_p x_{aw} + B_{p,u}(\text{sat}(y_c + v) - y_c)$.

**Step 6.** Construct the anti-windup compensation as (8.1), (8.2), with the selection for $v$ given at the previous step.

$\star$

## 8.3 MODEL PREDICTIVE CONTROL FOR ANTI-WINDUP DESIGN

### 8.3.1 Basics of MPC design

One tool that is tailor-made for control problems with input constraints is model predictive control (MPC). MPC has a long and colorful history, details of which may be found in the references mentioned in the notes to this chapter. While it has it roots in classical optimal control theory, its practical developments incubated within the process control community, where plant time constants were sufficiently slow to allow ample time for online computation. More recently, fueled by significant increases in computer speed and the development of certain efficient optimization algorithms, interest in MPC has started to spread to other application areas. While MPC deals directly with input (as well as state) constraints, it does not address directly the anti-windup problem as described in Chapter 2. Nevertheless, using the MRAW structure, MPC can be applied to the anti-windup problem as long as the control hardware used can support the computations required for its implementation.

In MPC, a constrained quadratic optimization problem is solved, either online or off-line in a parametrized form, to determine an optimal feedback for steering the state of a system to zero. When applied within the MRAW structure, this system will be the anti-windup compensator which, before MPC feedback, corresponds to a copy of the plant model. The MPC computations are tractable when the optimization problem is based on a discrete-time plant model and the time horizon over which the future model state must be predicted is not too long.

The basic idea of discrete-time MPC, as will be needed here, is the following: Given are a discrete-time linear control system

$$\xi^+ = A\xi + Bv \qquad \xi(0) = \zeta \tag{8.6}$$

and decentralized constraints on the input $v$ that are parametrized by a parameter $\beta$. The constraints are given as

$$u_m \le v + \beta \le u_M . \tag{8.7}$$

For vectors, this inequality should be taken component-wise. The parameter $\beta$ should satisfy

$$u_m - \beta \le 0 \le u_M - \beta \tag{8.8}$$

so that the value zero satisfies the constraints. In order for problems involving marginally stable or unstable systems to be feasible, the inequalities in (8.8) should be strict so that zero is in the interior of the constraint set.

A feedback function $\kappa$, depending on $\zeta$ and $\beta$, is synthesized, either in explicit form off-line or in implicit form online, by first finding a minimizing value, denoted $\mathbf{v}^*$, for the function

$$\mathbf{v} \mapsto |\xi(N,\zeta,\mathbf{v})|_P^2 + \sum_{i=0}^{N-1} |C\xi(i,\zeta,\mathbf{v}) + Dv(i,\mathbf{v})|_Q^2 + |v(i,\mathbf{v})|_R^2 ,$$

where $P$, $Q$, and $R$ are symmetric, positive semidefinite matrices, the notation $|w|_P^2$ means $w^T P w$ (similarly for $Q$ and $R$), and

$$\begin{aligned}
\xi(0,\zeta,\mathbf{v}) &= \zeta \\
\xi(i+1,\zeta,\mathbf{v}) &= A\xi(i,\zeta,\mathbf{v}) + Bv(i,\mathbf{v}) \qquad i=0,\dots,N-1 \\
\begin{bmatrix} v(0,\mathbf{v}) \\ \vdots \\ v(N-1,\mathbf{v}) \end{bmatrix} &= \mathbf{v}
\end{aligned} \tag{8.9}$$

subject to the constraints

$$u_m \le v(i,\mathbf{v}) + \beta \le u_M \qquad i=1,\dots,N-1 , \tag{8.10}$$

and then setting

$$\kappa(\zeta,\beta) = v_0(\mathbf{v}^*) .$$

Suppose that the pair $(Q^{1/2}C,A)$ is detectable. One sufficient condition for this is that the eigenvalues of $A$ have magnitude less than one. Another sufficient condition is that $C^T QC$ is positive definite. Then, if either the matrix $P$ has appropriate properties or the horizon length $N$ is sufficiently long, the feedback law $v = \kappa(\xi,\beta)$ locally exponentially stabilizes the origin of the system (8.6) while satisfying the constraints (8.7). More precisely,

1. If all of the eigenvalues of the matrix $A$ have magnitude less than one then:

   (a) No matter what $N \geq 1$ is used, if $P$ satisfies

   $$A^T PA - P < -C^T QC,$$

   then the feedback law makes the origin globally exponentially stable.

   (b) No matter what $P = P^T \geq 0$ is used, if $N$ is sufficiently large, then the feedback law makes the origin globally exponentially stable.

2. Otherwise, as long as zero is in the interior of the constraint set, then:

   (a) No matter what $N \geq 1$ is used, if $P$ satisfies $P = X^{-1}$ where $X = X^T > 0$ and for some $Z$,

   $$\begin{bmatrix} X & A+BZ & 0 \\ (A+BZ)^T & X & Q^{1/2}CX \\ 0 & XC^T Q^{1/2} & I \end{bmatrix} > 0,$$

   then the feedback law $\kappa$ makes the origin locally exponentially stable. (This condition implies the existence of $K$, in fact, $K = ZX^{-1}$ can be chosen such that $(A+BK)^T P(A+BK) - P < -C^T QC$.)

   (b) No matter what $P = P^T \geq 0$ is used, if $N$ is sufficiently large and the pair $(Q^{1/2}C, A)$ is detectable, then the feedback law makes the origin locally exponentially stable.

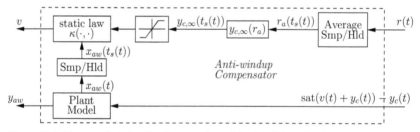

Figure 8.4  The sampled-data implementation of the MPC-based MRAW compensator.

### 8.3.2 Sampled-data implementation

Discrete-time MPC can be used to synthesize the remaining degrees of freedom within the MRAW structure by taking the following preliminary steps, which are also illustrated in Figure 8.4:

1. Sample the state of the continuous-time anti-windup compensator, which corresponds to a copy of the plant model, and is driven by $\text{sat}(v+y_c) - y_c$.

2. Sample a signal related to the input to the anti-windup filter. In the anti-windup algorithm to be given, an averaged version of the reference signal $r$ will be sampled. The role of this signal is to give information to the MPC algorithm about the constraints faced by $v$, which are dependent upon $y_c$ since the argument of the saturation function is $v + y_c$ in the MRAW structure.

The sampled reference signal will be fed to a function that maps constant references to steady-state unconstrained plant input values in the absence of additional exogenous inputs. This map will be denoted $r_a \mapsto y_{c,\infty}(r_a)$. It is a linear map that can be determined from the matrices of the unconstrained closed-loop system. Then the values of the sampled state and $y_{c,\infty}(r_a)$ (actually, a suitably saturated version of it, so that it certainly remains within the allowable limits) will be fed as the values $(\zeta, \beta)$ to the static nonlinear function that results from solving the MPC problem described in the previous subsection for a discrete-time model $(A, B, C, D)$ given by

$$A := e^{A_p T}, \qquad B := e^{A_p T} \left( \int_0^T e^{-A_p T} dt \right) B_{p,u}, \qquad C = C_{p,z}, \qquad D = D_{p,zu},$$

where $(A_p, B_{p,u})$ come from the unconstrained plant and $(C_{p,z}, C_{p,z})$ come from the performance variable of the unconstrained plant in the anti-windup problem.

When choosing the sampling period, which is a design parameter, the user should keep in mind that very small sampling periods accentuate the effect of rounding errors and typically require a larger value for the horizon length $N$ in the MPC problem in order to be effective for high performance. In turn, this increases the computational burdens of determining the static, nonlinear MPC feedback function. On the other hand, very long sampling periods likely will not produce satisfactory anti-windup performance, as the behavior between samples can be unpredictable.

Due to the MRAW structure, the application of MPC to the anti-windup problem in the way outlined here guarantees the small signal preservation property required of anti-windup augmentation. Moreover, even though the MPC algorithm is designed for a discrete-time system and ignores the time-varying behavior of $y_c$, the exponential stability that it produces for its discrete-time model leads to input-output $\mathcal{L}_2$ stability and a certain amount of unconstrained response recovery for the anti-windup augmented closed loop. In particular, it can be established that when $d(\cdot) \in \mathcal{L}_2$ and $r(\cdot) - r_\infty \in \mathcal{L}_2$, where $r_\infty$ is such that $v = 0$ satisfies the parametrized constraints with $\beta = y_{c,\infty}(r_\infty)$ then $z(\cdot) - z_\ell(\cdot) \in \mathcal{L}_2$. This statement applies locally when MPC induces local exponential stability for its discrete-time model. It applies globally when MPC induces global exponential stability for its discrete-time model.

### 8.3.3 Constrained quadratic programming for MPC

In this section it will be assumed for simplicity that $C^T Q C$ and $R$ are positive definite. This is reasonable to assume since $C$, $Q$, and $R$ will correspond to design parameters of the anti-windup compensation.

#### 8.3.3.1 The optimization problem

For MPC's constrained, quadratic optimization problem, $\zeta$ and $\beta$ are given and a solution is required of

$$\text{minimize} \quad s \tag{8.11a}$$

$$\text{subject to:} \quad \begin{cases} \begin{bmatrix} \zeta \\ \mathbf{v} \end{bmatrix}^T X \begin{bmatrix} \zeta \\ \mathbf{v} \end{bmatrix} \leq s^2 \\ \mathbf{u}_m \leq \mathbf{v} + \bar{\beta} \leq \mathbf{u}_M, \end{cases} \tag{8.11b}$$

where

$$\mathbf{u}_m := \begin{bmatrix} u_m \\ \vdots \\ u_m \end{bmatrix}, \quad \mathbf{u}_M := \begin{bmatrix} u_M \\ \vdots \\ u_M \end{bmatrix}, \quad \bar{\beta} := \begin{bmatrix} \beta \\ \vdots \\ \beta \end{bmatrix} \tag{8.12}$$

and the symmetric, positive definite matrix $X$ is determined by the matrices $(A, B, C, D)$ and $(P, Q, R)$ according to the equations

$$\bar{A} := \begin{bmatrix} I \\ A \\ \vdots \\ A^N \end{bmatrix}$$

$$\bar{B} := \begin{bmatrix} I & 0 & \cdots & 0 \\ A & \ddots & \ddots & \vdots \\ \vdots & \ddots & \ddots & 0 \\ A^{N-1} & \cdots & A & I \end{bmatrix} \text{diag}(B, \ldots, B)$$

$$\tag{8.13}$$

$$\bar{X}_x := \text{diag}(C^T Q C, \ldots, C^T Q C, P)$$

$$\bar{X}_v := \text{diag}(R + D^T Q D, \ldots, R + D^T Q D)$$

$$\bar{X}_{xv} := \begin{bmatrix} \text{diag}(C^T Q D, \ldots, C^T Q D) \\ 0 \end{bmatrix}$$

$$\bar{X} := \begin{bmatrix} \bar{X}_x & \bar{X}_{xv} \\ \bar{X}_{xv}^T & \bar{X}_v \end{bmatrix}$$

$$X := \begin{bmatrix} \bar{A}^T & 0 \\ \bar{B}^T & I \end{bmatrix} \bar{X} \begin{bmatrix} \bar{A} & \bar{B} \\ 0 & I \end{bmatrix}.$$

Because $X$ is positive definite, this quadratic problem has a unique solution.

### 8.3.3.2 Online solution techniques

1. *Quadprog*: When running simulations using MATLAB, it is possible to use the MATLAB function `quadprog()` to solve this quadratic program online, i.e., in the course of the simulation.

2. *Fast model predictive control.* Recent work by Boyd and co-workers has led to a collection of methods for improving the speed of MPC, using online optimization. The corresponding MATLAB function is called `fast_mpc()` and can be easily downloaded from the web and integrated into an existing MATLAB installation.

3. *LMI solver:* Another possibility is to cast the optimization problem as a non-strict LMI and call on MATLAB's LMI solver. This is done as follows: first, factor $X$ as $X = LL^T$, perhaps by means of its Cholesky factorization (e.g., using the MATLAB `chol()` function), and then use the fact that

$$w^T w \le s^2 \quad \Longleftrightarrow \quad \begin{bmatrix} sI & w \\ w^T & s \end{bmatrix} \ge 0. \tag{8.14}$$

Then the quadratic program is equivalent to the LMI

$$\text{minimize} \quad s$$

$$\text{subject to:} \quad \begin{cases} \begin{bmatrix} sI & L^T \begin{bmatrix} \zeta \\ \mathbf{v} \end{bmatrix} \\ \begin{bmatrix} \zeta \\ \mathbf{v} \end{bmatrix}^T L & s \end{bmatrix} \ge 0 \\ \mathbf{u}_m \le \mathbf{v} + \overline{\beta} \le \mathbf{u}_M , \end{cases} \tag{8.15}$$

which has a unique solution because $X$ is positive definite.

4. An alternative approach to solving the quadratic program (8.11) is to solve a corresponding algebraic loop using the methods described earlier in the book. The algebraic loop has the form

$$\tilde{\mathbf{v}} = (I - X_{22}^{-1})\overline{\beta} - X_{22}^{-1} X_{12}^T \zeta + \left(I - X_{22}^{-1}\right) \left[\tilde{\mathbf{v}} - \mathrm{sat}(\tilde{\mathbf{v}})\right] , \tag{8.16}$$

where $X_{12}$ and $X_{22}$ are generated from $X$ according to

$$X = \begin{bmatrix} X_{11} & X_{12} \\ X_{12}^T & X_{22} \end{bmatrix}, \tag{8.17}$$

where the partitioning is conformal to the partitioning of $\zeta$ and $\mathbf{v}$ in the vector

$$\begin{bmatrix} \zeta \\ \mathbf{v} \end{bmatrix}.$$

The optimal solution $\mathbf{v}^*$ to the quadratic program (8.11) is generated from the solution $\tilde{\mathbf{v}}^*$ to the algebraic loop (8.16) according to the relation

$$\mathbf{v}^* = \mathrm{sat}(\tilde{\mathbf{v}}^*) - \overline{\beta} . \tag{8.18}$$

The algebraic loop matches the form of the algebraic loops discussed in Sections 2.3.7 and 3.4.2 of this book by defining $\Lambda := I - X_{22}^{-1}$. It is well-posed because

$$2I - \Lambda^T - \Lambda = 2X_{22}^{-1} > 0 . \tag{8.19}$$

The reason why equation (8.18) generates the solution to the quadratic program is the following:

The gradient of the cost function with respect to the free variable $\mathbf{v}$ is given by

$2\left(X_{22}\mathbf{v}+X_{12}^T\zeta\right)$. A necessary and sufficient condition for optimality for this problem is that, at the point $\mathbf{v}^*$ which minimizes the quadratic cost, the gradient of the cost should be a nonpositive linear combination of the gradients of the functions that correspond to active constraints, written in the form

$$\mathbf{v}+\bar{\beta}-\mathbf{u}_M \leq 0 \qquad (8.20)$$

and

$$-\mathbf{v}-\bar{\beta}+\mathbf{u}_m \leq 0 . \qquad (8.21)$$

These gradient conditions are usually referred to as the Karush-Kuhn-Tucker (KKT) conditions for optimality. The condition on the gradient vectors and the condition that the constraints are not violated are satisfied by the relations

$$\begin{aligned} X_{22}\mathbf{v}^* + X_{12}^T\zeta &= \mathbf{v}^* + \bar{\beta} - \tilde{\mathbf{v}}^* \\ \mathbf{v}^* &= \mathbf{sat}\,(\tilde{\mathbf{v}}^*) - \bar{\beta} . \end{aligned} \qquad (8.22)$$

Indeed, when the $i$th constraint is not active, this condition insists that the $i$th component of the gradient vector be zero. When the $i$th component of the constraint (8.20) is active, it insists that the $i$th component of the gradient be negative. When the $i$th component of the constraint (8.21) is active, it insists that the $i$th component of the gradient vector be positive.

The two conditions in (8.22) can be combined to get

$$\mathbf{v}^* = \mathbf{sat}\left(-X_{12}^T\zeta + (I - X_{22})\mathbf{v}^* + \bar{\beta}\right) - \bar{\beta} . \qquad (8.23)$$

With the definition of $\tilde{\mathbf{v}}^*$ in terms of $\mathbf{v}^*$, this is equivalent to the solution to the algebraic loop given in (8.16).

There can be numerical advantages, especially for unstable plants and moderately large $N$, to solving the optimization problem by first invoking a stabilizing feedback transformation. In particular, one may consider first picking $K_\circ$ so that the eigenvalues of $A_\circ := A + BK_\circ$ have magnitude less than one, and then solving the optimization problem

minimize   $s$

$$\text{subject to:} \quad \begin{cases} \begin{bmatrix} \zeta \\ \hat{\mathbf{v}} \end{bmatrix}^T X_\circ \begin{bmatrix} \zeta \\ \hat{\mathbf{v}} \end{bmatrix} \leq s^2 \\ \mathbf{u}_m \leq \hat{\mathbf{v}} - \mathrm{diag}(K_\circ, \dots, K_\circ)\left[\bar{A}_\circ\zeta + \bar{B}_\circ\hat{\mathbf{v}}\right] + \bar{\beta} \leq \mathbf{u}_M, \end{cases} \qquad (8.24)$$

where the triple $(X_\circ, \bar{A}_\circ, \bar{B}_\circ)$ is defined like the triple $(X, \bar{A}, \bar{B})$ was previously but with $A_\circ$ in place of $A$ at all locations, and with the definitions

$$\begin{aligned} C_\circ &:= C + DK_\circ \\ \bar{X}_{x,\circ} &:= \mathrm{diag}(C_\circ^T Q C_\circ + K_\circ^T R K_\circ, \dots, C_\circ^T Q C_\circ + K_\circ^T R K_\circ, P) \qquad (8.25) \\ \bar{X}_{xv,\circ} &:= \begin{bmatrix} \mathrm{diag}(C_\circ^T Q D + K_\circ^T R, \dots, C_\circ^T Q D + K_\circ^T R) \\ 0 \end{bmatrix} . \end{aligned}$$

The solution $\mathbf{v}^*$ to the original optimization problem is related to the solution $\hat{\mathbf{v}}^*$ to the feedback-transformed optimization problem through

$$v_\circ(\mathbf{v}^*) = v_\circ(\hat{\mathbf{v}}^*) + K_\circ\zeta . \qquad (8.26)$$

### 8.3.4 The MPC-based MRAW algorithm

The MPC-based MRAW algorithms are now made explicit. The first algorithm does not require the horizon length of the MPC optimization problem to be long because it imposes special properties on the terminal cost matrix $P$. On the other hand, it should be noted that larger horizon lengths can lead to larger basins of attraction. An exception is the case where $A^T PA - P < -C^T C$, which requires that the eigenvalues of $A$ have magnitude less than one. Under this condition, the basin of attraction is global regardless of the horizon length.

**Algorithm 18** (MPC-based MRAW with special terminal cost)

| Applicability | | | | Architecture | | | Guarantee |
|---|---|---|---|---|---|---|---|
| Exp Stab | Marg Stab | Marg Unst | Exp Unst | Lin/ NonL | Dyn/ Static | Ext/ FullAu | Global/ Regional |
| $\checkmark$ | $\checkmark$ | $\checkmark$ | $\checkmark$ | NL | D | E | R/G |

| |
|---|
| *Comments*: Sampled-data augmentation based on model predictive control for constrained, discrete-time linear systems. An appropriate terminal cost function is used to guarantee global exponential stability for any loop with an exponentially stable plant. Regional exponential stability is guaranteed for any loop. |

**Step 1.** Select any positive sampling period $T$ and any positive horizon length $N$ for the MPC optimization problem.

**Step 2.** Determine the function $\kappa(\cdot,\cdot)$ either in explicit form off-line, or online as a function of the inputs to $\kappa$ shown in the block diagram in Figure 8.4 as follows:

1. Define $A := e^{A_p T}$, $B_v := \int_0^T e^{A_p(\tau)} B_{p,u} d\tau$, $C := C_{p,z}$, $D_v := D_{p,zu}$. (For the sake of using the LMI algorithm as described above, assume that $C^T C$ is positive definite.)
2. Select symmetric, positive definite matrices $R$, $Q$, and $P$ of appropriate dimension such that $P = X^{-1}$ where $X > 0$ and, for some $Z$,

$$\begin{bmatrix} X & A + BZ & 0 \\ (A+BZ)^T & X & Q^{1/2}CX \\ 0 & XC^T Q^{1/2} & I \end{bmatrix} > 0 .$$

(This condition, which is feasible whenever stabilization is possible, is an LMI in $X$ and $Z$. It implies the existence of $K$, in fact $K = ZX^{-1}$ can be chosen such that $(A + BK)^T P(A + BK) - P < -C^T C$.)
3. Define the map $\kappa$ through $\kappa(\zeta,\beta) := v_\circ(\mathbf{v}^*)$ where $\mathbf{v}^*$ solves the MPC optimization problem for $(A,B,C,D)$, $(P,Q,R)$, $u_m$, $u_M$, $N$ as parametrized by $(\zeta,\beta)$.

**Step 3.** Given the sampling interval $T$, an initial time $t_0$, and the sampling times $t_0 = 0, t_1 = T, \ldots, t_i = t_{i-1} + T, \ldots$, denote by $t_s(t)$ the closest sample time smaller or equal to $t$. According to Figure 8.4, define the sampled and averaged reference as

$$r_a(t_s(t)) := \frac{1}{T} \int_{t_{k-1}}^{t_k} r(\tau) d\tau, \quad \forall t \in [t_k, t_{k+1}), \; \forall k \geq 0.$$

**Step 4.** Determine the map $\bar{r} \mapsto y_{c,\infty}(\bar{r})$ providing the steady-state output $y_{c,\infty}$ generated by the unconstrained closed loop under the action of a constant reference $\bar{r}$. For example, if the unconstrained controller is linear and the (exponentially stable) linear closed loop from $r$ to $y_c$ is characterized by the equations

$$\begin{bmatrix} \dot{x}_p \\ \dot{x}_c \end{bmatrix} = A_{c\ell} \begin{bmatrix} x_p \\ x_c \end{bmatrix} + B_{c\ell} r$$

$$y_c = C_{c\ell} \begin{bmatrix} x_p \\ x_c \end{bmatrix} + D_{c\ell} r,$$

then this map is given by

$$y_{c,\infty}(\bar{r}) = \left( D_{c\ell} - C_{c\ell} A_{c\ell}^{-1} B_{c\ell} \right) \bar{r},$$

where $A_{c\ell}$ is always invertible because the unconstrained closed loop is exponentially stable. Also given the largest possible reference $r_{\max}$ allowed in the control system, define the vector $M$ as the excess of saturation available when that reference is applied, namely, if $\bar{u}$ is the vector of the (symmetric) saturation limits, then $M = \bar{u} - |y_{c,\infty}(r_{\max})|$, with the absolute value to be applied component-wise.

**Step 5.** According to Figure 8.4, select the stabilizing signal $v$ of the MRAW compensator as

$$v(t) = \kappa(x_{aw}(t_s(t)), \mathrm{sat}_M(y_{c,\infty}(r_a(t_s(t))))),$$

where $\mathrm{sat}_M(\cdot)$ denotes the saturation function with the saturation limits defined at the previous step.

**Step 6.** Construct the anti-windup compensation as (8.1), (8.2) with the selection for $v$ given at the previous step.

$\star$

The next algorithm requires that a sufficiently long horizon length be used in the MPC problem because no extra properties are assumed for $P$.

**Algorithm 19** (MPC-based MRAW with sufficiently long horizons)

| Applicability | | | | Architecture | | | Guarantee |
|---|---|---|---|---|---|---|---|
| Exp Stab | Marg Stab | Marg Unst | Exp Unst | Lin/ NonL | Dyn/ Static | Ext/ FullAu | Global/ Regional |
| $\checkmark$ | $\checkmark$ | $\checkmark$ | $\checkmark$ | NL | D | E | R/G |

> *Comments*: Sampled-data augmentation based on model predictive control for constrained, discrete-time linear systems. A sufficiently long optimization horizon is used to guarantee global exponential stability for any loop with an exponentially stable plant. Regional exponential stability is guaranteed for any loop.

**Step 1.** Select any positive sampling period $T$ (smaller is better, up to a point) and a sufficiently large horizon length $N$ for the MPC optimization problem. (It is best to determine what is sufficiently large by trial and error.)

**Step 2.** Determine the function $\kappa(\cdot, \cdot)$ either in explicit form off-line, or online as a function of the inputs to $\kappa$ shown in the block diagram in Figure 8.4 as follows:

1. Define $A := e^{A_p T}$, $B_v := \int_0^T e^{A_p(\tau)} B_{p,u} \, d\tau$, $C := C_{p,z}$, $D_v := D_{p,zu}$. (For the sake of using the LMI algorithm as described above, assume that $C^T C$ is positive definite.)
2. Select symmetric, positive definite matrices $R$, $Q$ and a symmetric, positive semidefinite matrix $P$ of appropriate dimensions.
3. Define the map $\kappa$ through $\kappa(\zeta, \beta) := v_\circ(\mathbf{v}^*)$ where $\mathbf{v}^*$ solves the MPC optimization problem for $(A, B, C, D)$, $(P, Q, R)$, $u_m$, $u_M$, $N$ as parametrized by $(\zeta, \beta)$.

**Step 3.** Define the sampled averaged reference $r_a(t_s(t))$ as in step 3 of Algorithm 18.

**Step 4.** Determine the map $\bar{r} \mapsto y_{c,\infty}(\bar{r})$ and the limits vector $M$ as in step 4 of Algorithm 18.

**Step 5.** According to Figure 8.4, select the stabilizing signal $v$ of the MRAW compensator as

$$v(t) = \kappa(x_{aw}(t_s(t)), \mathrm{sat}_M(y_{c,\infty}(r_a(t_s(t))))),$$

where $\mathrm{sat}_M(\cdot)$ denotes the saturation function with the saturation limits defined at the previous step.

**Step 6.** Construct the anti-windup compensation as (8.1), (8.2) with the selection for $v$ given at the previous step.

$\star$

## 8.4 GLOBAL DESIGNS FOR NON-EXPONENTIALLY UNSTABLE PLANTS

So far no algorithms have been given that guarantee global unconstrained response recovery (URR) when the plant is marginally unstable, i.e., no modes of the plant are exponentially unstable. It is important to note that such algorithms exist. On the other hand, finite gain URR is impossible.

Recall from Section 2.2.4 that nonlinear, URR with respect to some $\delta > 0$ corresponds to the relation

$$\|u_\ell - \mathrm{sat}_\delta(u_\ell)\|_2 < \infty \qquad \Longrightarrow \qquad \|z - z_\ell\|_2 < \infty.$$

Using the MRAW structure, all that is required to achieve this relation is an algorithm for synthesizing a state feedback control $v = \kappa(x)$ for the linear system

$$
\begin{aligned}
\dot{x}_{aw} &= A_p x_{aw} + B_{p,u}(v+w) \\
z_{aw} &= C_{p,z} x_{aw} + D_{p,zu}(v+w)
\end{aligned}
\tag{8.27}
$$

with disturbance $w$ so that, for all $x_{aw}$,

$$-\delta \cdot \mathbf{1} \leq \kappa(x_{aw}) \leq \delta \cdot \mathbf{1} \tag{8.28}$$

and the system

$$
\begin{aligned}
\dot{x}_{aw} &= A_p x_{aw} + B_{p,u}(\kappa(x_{aw}) + w) \\
z_{aw} &= C_{p,u} x_{aw} + D_{p,zu}(\kappa(x_{aw}) + w)
\end{aligned}
\tag{8.29}
$$

is $\mathcal{L}_2$ input-output stable. It turns out that such algorithms exist whenever $(A_p, B_{p,u})$ is stabilizable, the eigenvalues of $A_p$ have a nonpositive real part and $\delta > 0$, but that it is impossible to achieve a finite $\mathcal{L}_2$ gain unless the eigenvalues of $A_p$ all have a negative real part. Also, the achievable $\mathcal{L}_2$ performance improves as $\delta$ increases. Of course, the only values of $\delta$ for which the problem specification make sense are those satisfying

$$u_m + \delta \cdot \mathbf{1} \leq \mathbf{0} \leq u_M - \delta \cdot \mathbf{1}. \tag{8.30}$$

Several algorithms for achieving the task mentioned above have appeared in the literature. The major ones include "nested saturation," "Lyapunov forwarding," and "scheduled Riccati" solutions. The structures of the nested saturation and scheduled Riccati solutions are easy to describe and are given below. Lyapunov forwarding is a little more involved and will not be discussed here, but it is certainly relevant for use in the MRAW structure. The nested saturation algorithm is somewhat more explicit than the scheduled Riccati solution. Nevertheless, the scheduled Riccati solution tends to perform better than the nested saturation algorithm.

### 8.4.1 Nested saturation

The algorithm described here is the simplest nested saturation algorithm available. Other more complicated variants that can be found in the literature are able to provide higher performance typically.

**Algorithm 20** (Global MRAW using nested saturation)

| Applicability | | | | Architecture | | | Guarantee |
|---|---|---|---|---|---|---|---|
| Exp Stab | Marg Stab | Marg Unst | Exp Unst | Lin/ NonL | Dyn/ Static | Ext/ FullAu | Global/ Regional |
| ✓ | ✓ | ✓ | | NL | D | E | G |

| |
|---|
| *Comments*: Augmentation using feedback consisting of multiple nested saturation functions. Does not create an algebraic loop. Global exponential stability is guaranteed for any loop with an exponentially stable plant. For a loop with a marginally unstable plant, global asymptotic stability is guaranteed. No performance measure is optimized. |

**Step 1.** Find a transformation matrix $T$ such that $A := T A_p T^{-1}$ is block upper triangular and define $B := T B_{p,u}$ as follows (the symbols $\star$ denote "don't care" entries):

$$A = \begin{bmatrix} \tilde{A}_k & \star & \cdots & \star & \cdots & \cdots & \star \\ 0 & \tilde{A}_{k-1} & \ddots & \vdots & \ddots & \ddots & \vdots \\ \vdots & \ddots & \ddots & \star & \star & \cdots & \star \\ 0 & \cdots & 0 & \tilde{A}_i & \star & \cdots & \star \\ 0 & \ddots & 0 & 0 & \tilde{A}_{i-1} & \ddots & \vdots \\ \vdots & \ddots & \vdots & \vdots & \ddots & \ddots & \star \\ 0 & \cdots & 0 & 0 & \cdots & 0 & \tilde{A}_0 \end{bmatrix}, \quad B = \begin{bmatrix} \tilde{B}_k \\ \tilde{B}_{k-1} \\ \vdots \\ \tilde{B}_i \\ \tilde{B}_{i-1} \\ \vdots \\ \tilde{B}_0 \end{bmatrix}.$$

where each block on the diagonal $\tilde{A}_i$ corresponds to a marginally stable matrix. (For the best performance, all exponentially stable modes should appear in the upper left block.) Define $A_i$ to be the matrix obtained by keeping only that part of $A$ that corresponds to starting at the $i$th block and moving to the right and moving down. Partition $B$ conformally to $A$, and define $B_i$ conformally to $A_i$, as follows:

$$A_i = \overbrace{\begin{bmatrix} \tilde{A}_i & \star & \cdots & \star \\ 0 & \tilde{A}_{i-1} & \ddots & \vdots \\ \vdots & \ddots & \ddots & \star \\ 0 & \cdots & 0 & \tilde{A}_0 \end{bmatrix}}^{n_i}, \quad B_i = \overbrace{\begin{bmatrix} \tilde{B}_i \\ \tilde{B}_{i-1} \\ \vdots \\ \tilde{B}_0 \end{bmatrix}}^{n_u}.$$

**Step 2.** As in the above equation, use $n_i$ to denote the dimension of $A_i$, $i = 1, \ldots, k$.

For $i = 1, \ldots, k$,

Define $\begin{cases} \overline{A}_1 := A_1, & \text{if } i = 1, \\ \overline{A}_i := A_i + B_i \begin{bmatrix} 0_{n_i - n_{i-1}} & \overline{F}_{i-1} \end{bmatrix}, & \text{otherwise.} \end{cases}$

Find $P_i = P_i^T > 0$ such that $\overline{A}_i^T P_i + P_i \overline{A}_i \leq 0$.
Define $\overline{F}_i := -B_i^T P_i$.
Define $F_i := \begin{bmatrix} 0_{n_k - n_i} & \overline{F}_i \end{bmatrix}$

end.

**Step 3.** Set

$$v = \mathrm{sat}_1 \left( F_1 T^{-1} x_{aw} + \mathrm{sat}_2 \left( F_2 T^{-1} x_{aw} + \cdots + \mathrm{sat}_k \left( F_k T^{-1} x_{aw} \right) \cdots \right) \right),$$

where the saturation levels for $\mathrm{sat}_i$ are to be tuned for stability and performance. The saturation level for $\mathrm{sat}_1$ should be $\delta > 0$, the parameter used to characterize the level of URR. (Formal stability proofs require that each inner saturation level be some suitably small fraction of its outer saturation level, but simulations suggest that this isn't necessary.)

**Step 4.** Construct the anti-windup compensation as (8.1), (8.2) with the selection for $v$ given at the previous step.

★

### 8.4.2 Scheduled Riccati

It is a nontrivial fact that if $(A_p, B_{p,u})$ is stabilizable and the eigenvalues of $A_p$ have a nonpositive real part then for each $\gamma > 1$ and $\rho > 0$ there exist $X = X^T$ and $\mu > 0$ satisfying

$$\begin{array}{rcl} \rho I & < & X \\ \left[ \begin{array}{cc} XA_p^T + A_p X - \left(1 - \frac{1}{\gamma^2}\right) B_{p,u} B_{p,u}^T & -X \\ -X & -\mu I \end{array} \right] & < & 0. \end{array} \tag{8.31}$$

This is an LMI in $X$ and $\mu$. It is equivalent to the statement that for each $\gamma > 1$ and $\varepsilon > 0$ there exist $P = P^T > 0$ and $\delta > 0$ such that

$$\begin{array}{rcl} P & < & \varepsilon I \\ A_p^T P + P A_p - \left(1 - \frac{1}{\gamma^2}\right) P B_{p,u} B_{p,u}^T P + \delta I & < & 0. \end{array} \tag{8.32}$$

Indeed, the two conditions are related through $P = X^{-1}$, $\varepsilon = \rho^{-1}$, and $\delta = \mu^{-1}$. With $\gamma > 1$ fixed in (8.31), as $\rho$ increases the minimum $\mu$ such that (8.31) holds is nondecreasing. Similarly, with $\gamma > 1$ fixed in (8.32), as $\varepsilon$ decreases the maximum $\delta$ such that (8.32) holds is nonincreasing. It is clear from (8.32) that this maximum $\delta$ converges to zero as $\varepsilon$ converges to zero. Similarly, the minimum $\mu$ in (8.31) must grow unbounded as $\rho$ grows unbounded.

For the system

$$\dot{x}_{aw} = A_p x_{aw} + B_{p,u}(v + w) \tag{8.33}$$

with the linear control $v = -k B_{p,u}^T P(\varepsilon) x_{aw} = -k B_{p,u}^T X^{-1}(\rho) x_{aw}$, where $k \geq 1$, it can be shown that the closed-loop system is $\mathcal{L}_2$ stable from exogenous input $w$ to the output $x_{aw}$ with gain $\gamma \sqrt{\mu(\rho)}$. Thus, the guaranteed gain becomes arbitrarily large as $\rho$ becomes arbitrarily large in (8.32). On the other hand, the $\mathcal{L}_2$ to $\mathcal{L}_\infty$ gain from $w$ to $B_p^T P(\varepsilon) x_{aw}$ is $\gamma \cdot \sqrt{\text{trace}(B_{p,u}^T P(\varepsilon) B_{p,u})} = \gamma \cdot \sqrt{\text{trace}(B_{p,u}^T X^{-1}(\rho) B_{p,u})}$. Thus, this guaranteed gain becomes arbitrarily small as $\rho$ becomes arbitrarily large.

If the control $v = \text{sat}(-k B_{p,u}^T X^{-1}(\rho) x_{aw})$ is applied to the system (8.33), then the statements about $\mathcal{L}_2$ stability with gain $\gamma \sqrt{\mu(\rho)}$ and $\mathcal{L}_2$ to $\mathcal{L}_\infty$ stability with gain $\gamma \cdot \sqrt{\text{trace}(B_{p,u}^T X^{-1}(\rho) B_{p,u})}$ hold locally. In fact, they hold for all initial conditions and exogenous inputs such that $\text{sat}(B_{p,u}^T X^{-1}(\rho) x_{aw}(t)) = B_{p,u}^T X^{-1}(\rho) x_{aw}(t)$ for all $t \geq 0$. (The coefficient $k$ is omitted intentionally from this equation.) For example, if the initial conditions are zero and the saturation levels are unity, then this condition holds for all exogenous inputs satisfying

$$\|w\|_2 \leq \frac{1}{\gamma \sqrt{\text{trace}(B_{p,u}^T X^{-1}(\rho) B_{p,u})}}. \tag{8.34}$$

Note also that by the first equation in (8.31), the condition

$$||w||_2 \leq \frac{\sqrt{\rho}}{\gamma\sqrt{\text{trace}(B_{p,u}^T B_{p,u})}} \tag{8.35}$$

guarantees that condition (8.34) is satisfied.

This leads to the following algorithm for semiglobal unconstrained response recovery using linear feedback.

**Algorithm 21** (Semiglobal MRAW using Riccati inequalities)

| Applicability | | | | Architecture | | | Guarantee |
|---|---|---|---|---|---|---|---|
| Exp Stab | Marg Stab | Marg Unst | Exp Unst | Lin/ NonL | Dyn/ Static | Ext/ FullAu | Global/ Regional |
| √ | √ | √ | | L | D | E | R |

Comments: Linear augmentation providing an arbitrarily large stability region for any loop containing a marginally unstable plant. Does not create an algebraic loop. Provides a quantified characterization of the regional unconstrained response recovery gain.

**Step 1.** Pick $\gamma > 1$, $\delta > 0$, $v > 0$.

**Step 2.** Solve the LMI

$$\begin{bmatrix} XA_p^T + A_p X - \left(1 - \frac{1}{\gamma^2}\right) B_{p,u} B_{p,u}^T & -X \\ -X & -\mu I \end{bmatrix} \begin{matrix} \frac{4\gamma^2 v^2 \text{trace}(B_{p,u}^T B_{p,u})}{\delta} I & < & X = X^T \\ \\ < & 0, \end{matrix} \tag{8.36}$$

for $X$ and $\mu$.

**Step 3.** Pick $v = -kB_{p,u}^T X^{-1} x_{aw}$ for any $k \geq 1$.

**Step 4.** Construct the anti-windup compensation as (8.1), (8.2), with the selection for $v$ given at the previous step. The construction guarantees $||z - z_\ell||_2 \leq \gamma\sqrt{\mu}||u_\ell - \text{sat}_\delta(u_\ell)||_2$ if $||u_\ell - \text{sat}_\delta(u_\ell)||_2 \leq v$ and $x_{aw}(0) = 0$. The value $\mu$ increases to infinity with $v$.

★

The previous discussion and Algorithm 21 suggest the possibility of scheduling the feedback gains based on the size of the state of the anti-windup filter in order to get good URR performance locally and some URR performance globally.

**Algorithm 22** (Global MRAW using scheduled Riccati inequalities)

| Applicability | | | | Architecture | | | Guarantee |
|---|---|---|---|---|---|---|---|
| Exp Stab | Marg Stab | Marg Unst | Exp Unst | Lin/ NonL | Dyn/ Static | Ext/ FullAu | Global/ Regional |
| √ | √ | √ | | NL | D | E | G |

> *Comments*: Augmentation based on a family of Riccati inequalities and continuous scheduling among a corresponding family of linear feedbacks. Global exponential stability is guaranteed for any loop with an exponentially stable plant. Global asymptotic stability is guaranteed for any loop with a marginally unstable plant.

**Step 1.** Pick $\gamma > 1$ and $\delta > 0$.

**Step 2.** Let $\rho_i$ be a strictly increasing, unbounded sequence and let the sequences $\mu_i$ and $X_i$ satisfy the LMIs

$$\rho_i I < X_i = X_i^T$$

$$\begin{bmatrix} X_i A_p^T + A_p X_i - \left(1 - \frac{1}{\gamma^2}\right) B_{p,u} B_{p,u}^T & -X_i \\ -X_i & -\mu_i I \end{bmatrix} < 0. \qquad (8.37)$$

**Step 3.** For each $\rho \in [\rho_i, \rho_{i+1}]$, define

$$X(\rho) = \frac{\rho_{i+1} - \rho}{\rho_{i+1} - \rho_i} X_i + \frac{\rho - \rho_i}{\rho_{i+1} - \rho_i} X_{i+1}.$$

**Step 4.** Define

$$\rho(z) := \min\left\{\rho : \ \text{trace}(B_{p,u}^T X^{-1}(\rho) B_{p,u}) \cdot z^T X^{-1}(\rho) z \leq \delta\right\}$$

or

$$\rho(z) := \min\left\{\rho : \ \rho^{-1} \cdot \text{trace}(B_{p,u}^T B_{p,u}) \cdot z^T X^{-1}(\rho) z \leq \delta\right\}.$$

**Step 5.** Pick $v = -(\tau(x_{aw}) + 1) B_{p,u}^T X^{-1}(\rho(x_{aw})) x_{aw}$ for any continuous, nonnegative function $\tau$.
(It also works to sample $x_{aw}$ and implement $\rho(\cdot)$ as a function of the sampled $x_{aw}$, as long as the sampling period is small enough.)

**Step 6.** Construct the anti-windup compensation as (8.1), (8.2), with the selection for $v$ given at the previous step.

$$\star$$

## 8.5 DESIGNS FOR EXPONENTIALLY UNSTABLE PLANTS THAT MAXIMIZE THE BASIN OF ATTRACTION

When addressing bounded stabilization of exponentially unstable plants, the bounded input authority only allows to achieve stability regions that are bounded in the directions of the exponentially unstable modes. The size of these regions depends on how far from the origin the bounded input is able to dominate the exponential instability term which grows linearly with the size of the unstable state. Since anti-windup can be seen as a bounded stabilization problem with constraints on the small signal behavior, it is prone to the same type of limitation.

So far, all the anti-windup constructions that have been given and that apply to exponentially unstable plants don't directly address the issue of maximizing the stability region of the compensated closed loop. This is done here by way of an extra-strong requirement which corresponds to having available a measurement of all the plant states corresponding to eigenvalues with a nonnegative real part. Based on this measurement and based on the knowledge of the unconstrained controller output $y_c$, the algorithm proposed here generates signals affecting the closed loop in the following two cases:

1. When the controller output exceeds the saturation limits: this is the condition where all the algorithms seen so far start injecting signals in the closed loop.
2. When the plant state approaches the boundary of the stability region: this is the novelty of this algorithm which allows it to anticipate for a future situation when the unconstrained controller won't be able to bring back the plant state to a safe region.

To effectively perform the second action above, the anti-windup compensator requires knowledge of the plant state in the exponentially stable directions. This is why those measurements are required here. However, an observer could also be used to estimate those quantities and the scheme would still work locally, where locally means that it would work for small initial observation errors. Clearly, the closer the initial plant state is to the dangerous null controllability boundaries, the smaller the initial observation error should be for the scheme to ensure stability.

The algorithm only focuses on a subset of the plant dynamics containing all the modes that are not exponentially stable but possibly also some slow modes that need to be driven to zero faster than their free response. For this reduced plant, a stabilizing feedback is computed, which guarantees forward invariance[1] of a suitable ellipsoid centered at the origin, that can be seen as a safety region for the plant states. Based on that feedback gain, the scheme then results in a signal $v$ that, whenever the plant state gets too close to the boundary of that ellipsoid, is allowed to fully disregard the unconstrained controller output and keep the plant state within the safe region. Due to this architecture, the compensated responses have the peculiarity to follow the unconstrained ones as much as possible and then slide along the boundary of that safety region until the unconstrained response comes back in it and the compensated one is able to track it again.

The algorithm focuses on ellipsoidal sets and linear stabilizers, but much more general, perhaps less constructive, solutions could be used too.

**Algorithm 23** (MRAW with guaranteed region of attraction)

| **Applicability** | | | | **Architecture** | | | **Guarantee** |
|---|---|---|---|---|---|---|---|
| Exp Stab | Marg Stab | Marg Unst | Exp Unst | Lin/ NonL | Dyn/ Static | Ext/ FullAu | Global/ Regional |
| $\checkmark$ | $\checkmark$ | $\checkmark$ | $\checkmark$ | NL | D | E | R |

*Comments*: Augmentation that, in order to achieve a large operating region for a loop containing an exponentially unstable plant, uses measurements of the plant's exponentially unstable modes.

---

[1]Forward invariance of a set means that each initial condition in that set generates a trajectory completely contained in the set.

**Step 1.** Transform the anti-windup state space representation so that $A_p = \begin{bmatrix} A_s & A_{12} \\ 0 & A_u \end{bmatrix}$ and $B_{p,u} = \begin{bmatrix} B_s \\ B_u \end{bmatrix}$, where all the eigenvalues of $A_s$ have a negative real part. Ideally, one would choose to store in $A_s$ all the sufficiently fast plant modes.

**Step 2.** Solve the following LMI optimization problem in the variables $G$, $Q = Q^T > 0$ and $\gamma$ to obtain a state feedback stabilizer $F_i x_u$ that makes the set $\mathcal{E}(P) = \{x_u : x_u^T P x_u \leq 1\}$ forward invariant:

$$\min \gamma \text{ s.t.}$$
$$\begin{bmatrix} 1 & G_i \\ G_i^T & Q \end{bmatrix} \geq 0, \quad i = 1, \dots, n_u,$$
$$\begin{bmatrix} \gamma I & I \\ I & Q \end{bmatrix} \geq 0,$$
$$QA_u^T + A_u Q + G^T B_u^T + B_u G < 0,$$

where $G_i$ denotes the $i$th row of $G$. Select $P = Q^{-1}$, $F_u := GP$.

**Step 3.** Fix a small $\varepsilon \ll 1$ (e.g., $\varepsilon = 0.05$) and define $P_\varepsilon := P/(1-\varepsilon)^2$ and the following functions:

$$\beta(x_u) = \min \left\{ 1, \max \left\{ 0, \frac{1 - x_u^T P x_u}{1 - (1-\varepsilon)^2} \right\} \right\}$$

$$\Psi(x_u, x_u^*) = \frac{(x_u^*)^T P v + \sqrt{((x_u^*)^T P v)^2 + (v^T P v) w}}{w},$$

$$\Psi_\varepsilon(x_u, x_u^*) = \frac{(x_u^*)^T P_\varepsilon v + \sqrt{((x_u^*)^T P_\varepsilon v)^2 + (v^T P_\varepsilon v) w_\varepsilon}}{w_\varepsilon},$$

where $v = x_u - x_u^*$, $w = 1 - (x_u^*)^T P x_u^*$, and $w_\varepsilon = 1 - (x_u^*)^T P_\varepsilon x_u^*$.

**Step 4.** Define the following additional functions:

$$P(x_u) = \frac{1}{\max\{1, \Psi_\varepsilon(x_u, 0)\}} x_u,$$
$$P_u(x_u) = -(B_u^T B_u)^{-1} B_u^T A_u x_u$$
$$\gamma(x_u, x_u^*) = \Psi(x_u, x_u^*) F_u \tilde{x}_u + (1 - \Psi(x_u, x_u^*)) P_u(x_u^*),$$
$$\alpha(x_u, \xi_u, y_c) = \gamma(x_u, P(\xi_u)) + \beta(x_u)(y_c - P_u(\xi_u)).$$

**Step 5.** Select the compensation signal $v$ as

$$v = \alpha(x_u, x_u - x_{aw,u}, y_c) - y_c, \tag{8.38}$$

where, according to the block separation at step 1, $x_u$ and $x_{aw,u}$ are the second block components of $x$ and $x_{aw}$, respectively.

**Step 6.** Construct the anti-windup compensation as (8.1), (8.2), with the selection for $v$ given at the previous step.

$\star$

## 8.6 NOTES AND REFERENCES

The switching anti-windup approach of Section 8.2.1 arises from using the switching strategy proposed in [MR21] with the nested regions and linear gains selections proposed in [MR29]. The linear gains originally proposed in [MR21] relied on a less effective regional characterization of the saturation nonlinearity whereas the gains selection given here results from the ideas of [SAT11, MI27]. In Algorithm 16, the hybrid notation for the switching signal is taken from [G3]. Section 8.2.2 is based on the ideas of [MR29] to transform a switching solution into a scheduling one. The recent work in [MR28], [MR27] and [MR29] contains further developments on alternative selections of the family of linear gains and of the nested Lyapunov level sets. Moreover, it proposes solutions for achieving global asymptotic stability with finite or infinite regions and switched and scheduled designs. Those algorithms are not reported here because some of them are not fully constructive. This is still a very active research field.

The model predictive control literature is extensive but some key references can be found in [G13]. The general constrained MPC results discussed in Section 8.3.1 derive from the results of [SAT10]. The fast model predictive control techniques mentioned on page 213 appeared in the recent paper [G16]. An extended journal version of it will soon appear in the *IEEE Transactions on Control Systems Technology*.

Several papers have drawn connections to "anti-windup" by noting that MPC gives corrections away from the operating point for linear LQR designed for the operating point. Papers along these lines include [OS9, SAT12, SAT13]. The extensive literature on the reference governor solutions (see [RG8, RG7, RG10] and references therein) is actually based on the use of a type of MPC scheme to rescale the reference so that the actuators' limits are not reached by the controller output. To the best of our knowledge, the first application of MPC to the anti-windup problem for general (even nonlinear) unconstrained controllers was in [MR17]. The connections between the quadratic programs that appear in input-constrained MPC and algebraic loops with input saturation were pointed out in [SAT12, SAT13], which also provided simple iterative algorithms for their solution.

The Lyapunov forwarding methods discussed at the beginning of Section 8.4, which can be used in the MRAW architecture, may be extracted from the work in [G10, G9]. The nested saturation Algorithm 20 for anti-windup synthesis was suggested in [MR3], based on its use for direct design in [SAT2], [SAT3], and [SAT8]. More recent nested saturation algorithms with improved performance can be found in the literature. The scheduled Riccati solution to nonlinear anti-windup synthesis is based on the direct design suggested by Megretski in [SAT7]. The fact that the LMI (8.31) is feasible for any stabilizable plant with no eigenvalues with positive real part is proven in [SAT5, Lemma 3.1]. Algorithm 23 is taken from [MR30] and is written here in a simplified form, whereas in its most general form several extra degrees of freedom are available for the shape of the stability region and the type of stabilizers used. Extensions and clarifications can be found in [MR30].

Discrete-time counterparts of these nonlinear anti-windup solutions can be found in [MR32], where the discrete-time versions of the scheduled Riccati approach of Algorithms 21 and 22 are suggested based on the results in [SAT9]. With discrete-time MRAW, the MPC approach can be used in a straightforward way to select $v$ because no sampling is needed. Based on this fact, an MPC solution to the discrete-time anti-windup problem is also described in [MR32].

# Chapter Nine

## The MRAW Structure Applied to Other Problems

### 9.1 RATE- AND MAGNITUDE-SATURATED PLANTS

When dealing with plants with both input rate and magnitude saturation, the model recovery anti-windup (MRAW) scheme can be employed to recover the performance of a controller synthesized for the system without input magnitude and rate saturation. From a mathematical viewpoint, input rate saturation can be modeled by augmenting the plant equations, reported here for completeness,

$$\mathcal{P} \begin{cases} \dot{x}_p &= A_p x_p + B_{p,u}\sigma + B_{p,w} w \\ y &= C_{p,y} x_p + D_{p,yu}\sigma + D_{p,yw} w \\ z &= C_{p,z} x_p + D_{p,zu}\sigma + D_{p,zw} w, \end{cases} \tag{9.1}$$

with an additional state equation having the same size as the control input and associated with the following discontinuous dynamics for the new state variable $\sigma$:

$$\dot{\sigma} = R\,\mathrm{sign}(\mathrm{sat}(u) - \sigma)), \tag{9.2}$$

where $u$ is the input before saturation, $\mathrm{sat}(\cdot)$ denotes the input saturation, and $\sigma$ is the rate-saturated input that becomes part of the augmented state of the plant. Finally, $R$ is a diagonal matrix having as diagonal entries the maximal rates $\bar{u}_{r,1}, \ldots, \bar{u}_{r,n_u}$ of each input.

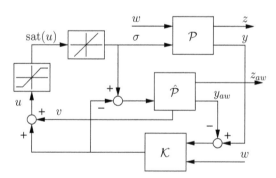

Figure 9.1 The MRAW scheme for rate- and magnitude-saturated systems.

When dealing with magnitude- and rate-saturated inputs, the MRAW scheme can be easily generalized by using the same anti-windup interconnection adopted in the magnitude-saturated case, namely,

$$u_c = y - y_{aw}, \qquad u = y_c + v, \qquad \sigma = \mathrm{sat}_{MR}(u), \tag{9.3}$$

where $\text{sat}_M R$ is a shortcut to denote a complicated nonlinear and dynamic effect comprising both magnitude and rate saturation. Moreover, the structural interconnection of the anti-windup compensator is preserved, namely, its input is the difference between the plant input and the unconstrained controller output:

$$\mathcal{F} \begin{cases} \dot{x}_{aw} &= A_p x_{aw} + B_{p,u}\left(\sigma - y_c\right) \\ z_{aw} &= C_{p,z} x_{aw} + D_{p,zu}\left(\sigma - y_c\right) \\ y_{aw} &= C_{p,y} x_{aw} + D_{p,yu}\left(\sigma - y_c\right) \\ v &= \text{to be designed.} \end{cases} \tag{9.4}$$

This compensation scheme is represented in Figure 9.1. The only difference compared to the MRAW scheme for magnitude saturation discussed in Chapter 6 and in subsequent chapters lies in the more complicated nonlinearity affecting the plant input. The bottom line is that while the model recovery properties of the MRAW scheme in Figure 9.1 remain the same as those discussed in Section 6.3.2 for the magnitude-saturated case, the mismatch equation (6.14) that is to be stabilized via the feedback signal $v$ becomes now:

$$\dot{x}_{aw} = A_p x_{aw} + B_{p,u}\left[\text{sat}_{MR}(v + y_c) - y_c\right], \tag{9.5}$$

so that extra care has to be taken in designing $v$.

In light of this, most of the algorithms for the design of $v$ discussed in Chapters 7 and 8 cannot be used with rate and magnitude saturation because they don't account for the rate-saturated input. In general, designs for $v$ which are linear functions of the anti-windup state $x_{aw}$ and input $(\sigma - y_c)$ are not sufficient to guarantee global stability or performance results. Nevertheless, the regional design techniques given in Section 7.3 and corresponding to Algorithms 14 and 15 can be used also in the rate-saturated case, where they lead, once again, to regional results. Another solution that works well when the plant is exponentially stable with fast enough modes is the IMC special case of Algorithm 10, which induces global exponential stability but typically poor performance. In the following two examples, the regional linear quadratic (LQ) technique of Algorithm 14 is used to address the windup phenomenon in two relevant flight control systems.

**Example 9.1.1**    (The Caltech ducted fan) The Caltech ducted fan consists of a short wing and a ducted fan engine with a high-efficiency electric motor and a 6-inch diameter blade, capable of generating up to 15 Newtons of thrust. Paddles on the fan allow the thrust to be vectored from side to side. Additional inputs are the motor speed and the angle of a flap attached to the wing, which acts as an elevator. The engine and wing are mounted on a stand with three degrees of freedom, which allows horizontal and vertical translations as well as an unrestricted pitch angle.

Taking the linearization of the system's dynamics around a trim condition associated with a forward velocity of 8 meters per second, the following model is obtained:

$$\overbrace{x - x_{trim}}^{\cdot} = A_p\left(x - x_{trim}\right) + B_{p,u}\left(u - u_{trim}\right),$$

where the state $x \in \mathbb{R}^6$ and the input $u \in \mathbb{R}^3$ represent the following physical quantities:

$$
x := \begin{bmatrix} \text{horizontal position [m]} \\ \text{vertical position [m]} \\ \text{pitch angle [rad]} \\ \text{horizontal velocity [m/s]} \\ \text{vertical velocity [m/s]} \\ \text{pitch angle rate [rad/s]} \end{bmatrix}, \quad u := \begin{bmatrix} \text{voltage to motor [V]} \\ \text{paddle deflection [rad]} \\ \text{elevator deflection [rad]} \end{bmatrix}.
$$

The nominal trim condition associated with a forward velocity of 8 meters per second is given by

$$
x_{trim} \approx \begin{bmatrix} 8t \\ 0 \\ 0.11 \\ 8 \\ 0 \\ 0 \end{bmatrix}, \quad u_{trim} \approx \begin{bmatrix} 1.1 \\ 0 \\ -0.021 \end{bmatrix}, \tag{9.6}
$$

and the associated system matrices are given by

$$
A_p = \begin{bmatrix} 0 & 0 & 0 & 1 & 0 & 0 \\ 0 & 0 & 0 & 0 & 1 & 0 \\ 0 & 0 & 0 & 0 & 0 & 1 \\ 0 & 0 & -2.7301 & -0.0991 & -0.1983 & 0 \\ 0 & 0 & -6.9299 & -0.1239 & -0.8694 & 0 \\ 0 & 0 & -32.8784 & 0.1853 & -4.1098 & 0 \end{bmatrix}
$$

$$
B_{p,u} = \begin{bmatrix} 0 & 0 & 0 \\ 0 & 0 & 0 \\ 0 & 0 & 0 \\ 1.0859 & -0.1369 & 0.8535 \\ -.0482 & -0.3783 & 2.1523 \\ 0.6915 & -12.1145 & 49.1147 \end{bmatrix}.
$$

A challenging problem associated with the Caltech ducted fan is that the second and the third control inputs are subject to rate saturation, imposed by the maximal velocity of the paddle and of the elevator. Indeed, these devices are also subject to magnitude saturation but the problems induced by the rate saturation are typically more harmful in flight control systems. To address the problem, a composite controller based on two LQR designs is synthesized by exploiting the MRAW approach for magnitude- and rate-saturated plants and the construction in Algorithm 14.

For the trim condition (9.6), two linear quadratic regulators have been designed to minimize two performance indexes depending both on the input $u - u_{trim}$ and on the state $x - x_{trim}$. The first one leads to an aggressive control action and is chosen to be the unconstrained controller (note that no dynamics are associated with this design, hence the controller state $x_c$ is empty). The second design corresponds to an index where the control input is significantly more penalized than the state, thus leading to a less aggressive controller, suitable to be used as the stabilizing action to be used in Algorithm 14 for $v$. Note that by Algorithm 14, only regional stability is guaranteed by this anti-windup action, however the basin of attraction of the resulting closed-loop system is sufficiently large to guarantee exponential stability under reasonable operating conditions. Based on the discussion above, the unconstrained controller is selected as

$$y_c - u_{trim} = K_{hi}(x - x_{trim} - r),$$

where $K_{hi}$ is selected as an LQR gain with weights $Q_{hi} = \text{diag}\{0.1, 20, 10, 0.1, 1, 10\}$ and $R_{hi} = \text{diag}\{1, 3, 3\}$ and corresponds to

$$K_{hi} = \begin{bmatrix} -.3049 & 1.1834 & -1.3409 & -0.7877 & 0.9323 & -0.1153 \\ -0.0174 & -1.0212 & 1.9178 & 0.01 & -0.7356 & 0.4881 \\ 0.0451 & 2.2709 & -5.5627 & -0.0338 & 1.8239 & -1.9056 \end{bmatrix},$$

and the stabilizing signal $v$ is selected as

$$v = K_{lo}x_{aw},$$

where $K_{lo}$ is selected as an LQR gain with weights $Q_{lo} = Q_{hi}/100$ and $R_{lo} = R_{hi}$ and corresponds to

$$K_{lo} = \begin{bmatrix} -0.0313 & 0.0604 & -0.0312 & -0.2022 & 0.1067 & -0.0121 \\ -0.0008 & -0.0675 & 0.0811 & 0.0032 & -0.0738 & 0.0542 \\ 0.0023 & 0.2468 & -0.3025 & -0.0149 & 0.2755 & -0.2154 \end{bmatrix}.$$

Figure 9.2 reports the simulation results when using a step reference of amplitude 0.5 meters in the vertical displacement (namely, $r = [0\, 0.5\, 0\, 0\, 0\, 0]^T$). In the simulation and according to the specifications of the experimental system, the first input is not subject to saturation, and both the second and the third inputs are rate saturated, with a maximal rate of 0.6 rad/s (approximately corresponding to 34.4 deg/s). The top plot of Figure 9.2 shows the vertical displacement output while the middle and lower plots show the two rate-saturated inputs. Notice how the anti-windup response almost recovers the unconstrained performance and eliminates the unpleasant oscillations exhibited by the saturated closed loop.                     □

**Example 9.1.2**     (The F-16's daisy chain control allocator)     The short-period longitudinal dynamics of the VISTA/MATV F-16 at Mach 0.2 and altitude 10,000 feet (corresponding to a dynamic pressure value of 40.8 psf) at a trim angle of attack of 28 degrees is described locally by the linear model

$$\dot{x}_p := \begin{bmatrix} \dot{\alpha} \\ \dot{q} \end{bmatrix} = A_p x_p + B_{p,u}\,\delta, \tag{9.7}$$

where $x_p := [\alpha\, q]^T$ is the state representing deviations of the angle of attack and pitch rate from the trim condition, $\delta = [\delta_e\, \delta_{ptv}]^T$ is the input representing respectively the deviations of the elevator deflection and of the pitch thrust vectoring from the trim condition, and the entries of the matrices $A_p$ and $B_{p,u}$ are:

$$A_p = \begin{bmatrix} -0.23 & 1 \\ -0.2 & -0.5 \end{bmatrix}, \quad B_{p,u} = \begin{bmatrix} b_{11} & b_{12} \\ b_{21} & b_{22} \end{bmatrix} = \begin{bmatrix} -0.04 & -0.05 \\ -1.95 & -2.74 \end{bmatrix}. \tag{9.8}$$

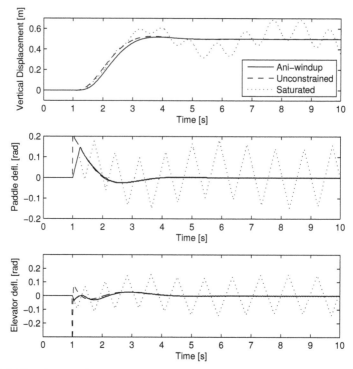

Figure 9.2 Plant input and output responses of various closed loops for the Caltech ducted fan in Example 9.1.1.

Both inputs are subject to rate and magnitude saturation. By using the model (9.2) for the rate saturation, if $u = [u_e \, u_{ptv}]^T$ is the input to the saturation block, the following dynamic equations complete the saturated model of the F-16:

$$\begin{aligned}
\dot{\sigma}_e &= -R_e \operatorname{sign}\left(\sigma_e - M_e \operatorname{sat}\left(\tfrac{u_e}{M_e}\right)\right) \\
\dot{\sigma}_{ptv} &= -R_{ptv} \operatorname{sign}\left(\sigma_{ptv} - M_{ptv} \operatorname{sat}\left(\tfrac{u_{ptv}}{M_{ptv}}\right)\right),
\end{aligned} \tag{9.9}$$

where the values $M_e = 21$ deg, $M_{ptv} = 17$ deg, $R_e = 50$ deg/s, $R_{ptv} = 45$ deg/s have been adopted.

According to the results in [SAT4], a daisy-chained allocation of the inputs can be adopted; in particular, if $r \in \mathbb{R}$ is the desired angle of attack, the desired value for $[b_{21} \, b_{22}] \sigma$ is given by

$$\kappa(x_p, r) = K_{inner} x_p + K_{outer}(s)(r - \alpha),$$

where $K_{inner} = [-8.8, \; -4.3]$ and

$$K_{outer}(s) = 11.296 \, \frac{(s+10043)(s^2+4.803 s+9.078)}{(s+0.0593)(s+5.102)(s+99.09)(s+125.64)}.$$

According to the matrices in (9.8), the daisy chain allocation is then given by:

$$y_c = \begin{bmatrix} y_{ce} \\ y_{cptv} \end{bmatrix} = \begin{bmatrix} b_{21}^{-1} \kappa(x, r) \\ b_{22}^{-1} b_{21} \left( b_{21}^{-1} \kappa(x, r) - \sigma_e \right) \end{bmatrix}. \tag{9.10}$$

Note that, with the above allocation, in the direct connection between the controller and the plant, $u = y_c$, if $\sigma_{ptv} \equiv y_{cptv}$, then $[b_{21}\ b_{22}]\,\sigma \equiv \kappa(x, r)$; moreover, if $\sigma_e \equiv y_{ce}$, then $y_{cptv} \equiv 0$. In other words, only the elevator deflection input is used to enforce the desired angle of attack.

Following Algorithm 14, the anti-windup compensation signal $v$ is selected by computing a linear gain $K$ optimizing an LQ cost function with $Q = 5I_2$ and $R = I_2$. The resulting gain is

$$K = \begin{bmatrix} 0.9930 & 1.3202 \\ 1.3687 & 1.8524 \end{bmatrix},$$

to be used, according to the algorithm, in $v = Kx_{aw}$.

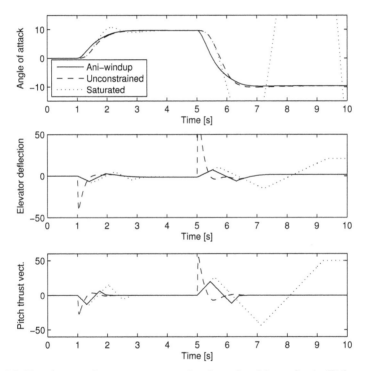

Figure 9.3 Plant input and output responses of various closed loops for the F16 case study in Example 9.1.2.

In Figure 9.3, the responses of the unconstrained, saturated, and anti-windup closed loops are compared to each other. The top plot shows the angle of attack and the middle and lower plots represent the two saturated inputs. The reference signal is chosen to be

$$r(t) = \begin{cases} 0 & deg, & 0 \ge t < 1 \\ 10 & deg, & 1 \ge t < 5 \\ -10 & deg, & t \ge 5. \end{cases}$$

The simulations show that rate limitation induces instability and oscillations. Magnitude limitation is actually reached by the saturated inputs at time $t = 9$ after the angle of attack has reached a value of 80 degrees, which is definitely already an intolerable flight condition (and is significantly out of scale in the first plot). Anti-windup successfully eliminates the closed-loop instability and induces a desirable closed-loop response slightly deteriorated as compared to the unconstrained one.

□

## 9.2 ANTI-WINDUP FOR DEAD-TIME PLANTS

When a linear plant is subject to the so-called "dead-time" effect—namely it is subjected to a pure delay at its input or, equivalently at its output—the MRAW scheme can be nicely adapted to the situation without much extra effort as compared to the delay-free case. In the most general situation, suppose that there's an input delay $\tau_I$ at the plant control input and an output delay $\tau_O$ at the plant measurement output. The plant equations can be then written as

$$\mathcal{P}\begin{cases} \dot{x}_p &=& A_p x_p + B_{p,u}\,\sigma(t-\tau_I) + B_{p,w}\,w \\ y(t+\tau_O) &=& C_{p,y} x_p + D_{p,yu}\,\sigma(t-\tau_I) + D_{p,yw}\,w \\ z &=& C_{p,z} x_p + D_{p,zu}\,\sigma(t-\tau_I) + D_{p,zw}\,w, \end{cases}$$

where the dependence on time has been omitted wherever the functions depend on the current time $t$.

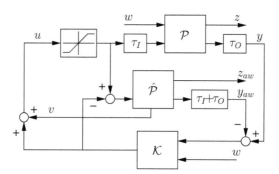

Figure 9.4 The MRAW scheme with dead-time plants.

In this situation, the natural MRAW solution to the windup problem in the presence of input magnitude saturation consists in inserting an identical anti-windup filter to the one used in the delay-free case (see equation (7.4) in Chapter 7.4):

$$\mathcal{F}\begin{cases} \dot{x}_{aw} &=& A_p x_{aw} + B_{p,u}\left(\sigma - y_c\right) \\ z_{aw} &=& C_{p,z} x_{aw} + D_{pz}\left(\sigma - y_c\right) \\ y_{aw} &=& C_{p,y} x_{aw} + D_{py}\left(\sigma - y_c\right) \\ v &=& \text{to be designed}, \end{cases} \qquad (9.11)$$

and modifying the interconnection equations as follows

$$u_c = y - y_{aw}(t-\tau_I-\tau_O), \qquad u = y_c + v, \qquad \sigma = \text{sat}(u).$$

The block diagram of this compensation scheme is represented in Figure 9.4. The great advantage of this solution is that the delay has been virtually extracted from within the saturation loop and is only left in the loop with delay and without saturation. In other words, the saturation problem has been fully decoupled from the dead-time problem. One way to appreciate this fact is to notice that the model recovery properties of the scheme correspond to the ones discussed in Section 6.2.2 for the delay-free case with the peculiarity that the top system in the cascade representation of Figure 6.2 on page 159 will involve the input and output plant delays, whereas the bottom system, which is affected by saturation, remains the same.

It should be emphasized that due to this fact, the anti-windup problem for dead-time system addresses the problem of recovering the performance of an unconstrained control system which has been designed disregarding actuator saturation but fully taking into account the delay effects. Therefore, the unconstrained closed loop will correspond to the closed loop between the unconstrained controller and the plant without saturation and *with* delay. No constraints are imposed on the structure of this unconstrained controller indeed, so that, for example, a linear controller equipped with a Smith predictor would satisfy the requirement of being stabilizing for the delayed (but not saturated) plant.

The nice feature of the solution illustrated in Figure 9.4 is that, as in the dead-time free case, the anti-windup problem is transformed into a bounded stabilization problem with disturbances, namely, that of designing $v$ as a suitable linear or non-linear state feedback signal to drive to zero the state of the dynamics

$$\dot{x}_{aw} = A_p x_{aw} + B_{p,u}\left[\text{sat}(v + y_c) - y_c\right]$$

which is dead-time free. Due to this reason, almost all the model recovery anti-windup algorithms given in Chapters 7 and 8 can be directly applied to this case, allowing one to address many different situations. An exception to this is Algorithm 23, where extra measurements are required from the plant (namely, the unstable states). That algorithm cannot be generalized to the dead-time case because it would require the availability at time $t$ of the unstable plant states values at time $t + t_I$. This leads to a noncausal solution which requires extra work (e.g., the use of predictors) to be directly applicable.

**Example 9.2.1**    (An integrating dead-time system)  Consider a simple scalar simulation example whose unconstrained closed-loop system is represented in Figure 9.5. The plant is an integrator with a large delay of 5 seconds at its input and the unconstrained controller has a modified Smith predictor structure (the original Smith predictor scheme only applies to exponentially stable dead-time plants).

Since the plant is marginally stable, the construction in Algorithm 11 can be used to design the signal $v$. In particular, the following state-space representation for the plant can be used:

$$\left[\begin{array}{c|c|c} A_p & B_{p,u} & B_{p,w} \\ \hline C_{p,y} & D_{p,yu} & D_{p,yw} \end{array}\right] = \left[\begin{array}{c|c|c} 0 & 1 & 1 \\ \hline 1 & 0 & 0 \end{array}\right],$$

and to induce a desirable performance, the parameters in Algorithm 11 are selected as $\rho_0 = 1$, $\rho = 10$. Therefore the compensation signal is $v = -10x_{aw}$, where $x_{aw}$ is

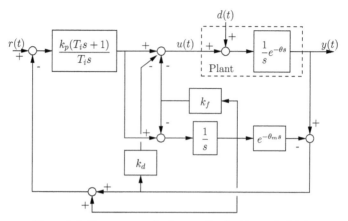

Figure 9.5  The unconstrained controller for Example 9.2.1.

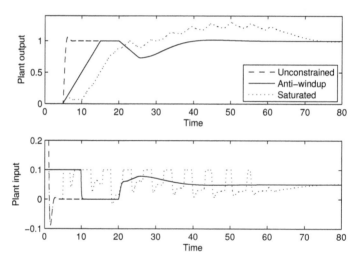

Figure 9.6  Plant input and output responses of various closed loops for the dead-time system in Example 9.1.1 (nominal case).

the only state of the anti-windup compensator. Note that if $\rho$ is increased, a faster performance recovery and more aggressive anti-windup action are obtained.

Simulation results for the unconstrained, saturated, and anti-windup closed-loop systems are reported in Figures 9.6 and 9.7, where the following selections for the disturbance and reference inputs have been used:

$$r(t) = \begin{cases} 0, & t < 0 \\ 1, & t \geq 0, \end{cases} \qquad d(t) = \begin{cases} 0, & t < 15 \\ -0.05, & t \geq 15, \end{cases}$$

and a saturation level of $\pm 0.1$ is enforced at the plant input.

The first simulation, reported in Figure 9.6 corresponds to the following selection of the parameters of the plant and unconstrained controller of the scheme in

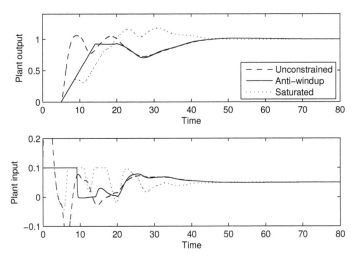

Figure 9.7  Plant input and output responses of various closed loops for the dead-time system in Example 9.1.1 (perturbed case).

Figure 9.5:

$$\theta = 5, \ \theta_m = 5, \ k_p = 0.1, \ T_i = 0.01, \ k_f = 4.131, \ k_d = 0.105,$$

which describes the situation where all the plant parameters are perfectly known and can be incorporated in the unconstrained controller and anti-windup design. In the figure it appears that the anti-windup response (solid) successfully recovers in an efficient way the unconstrained response (dashed). The saturated response (dotted) instead exhibits unpleasant transients. Note also that since the disturbance does not cause input saturation for the unconstrained trajectory, the anti-windup response perfectly reproduces the unconstrained response in the second part of the simulation.

A second simulation is reported in Figure 9.7 where the plant parameters are perturbed by a 10% uncertainty both affecting the unconstrained controller and the anti-windup design. The parameter selection in the scheme in Figure 9.5 then becomes

$$\theta = 5, \ \theta_m = 5.5, \ k_p = 0.1, \ T_i = 0.1, \ k_f = 1.247, \ k_d = 0.095.$$

Also in this case the anti-windup action is capable of significantly improving the closed-loop response as compared to the saturated case. □

## 9.3 BUMPLESS TRANSFER IN MULTICONTROLLER SCHEMES

Bumpless authority transfer among different controllers is a control goal that was formulated qualitatively in the very early stages of research in applied automatic control. The goal of "bumpless transfer" arose from experimental needs faced by industrialists when the application of classic control techniques led to undesired transients and even instability after power-on and after switching control authority among controllers that induce different closed-loop performance properties.

As an example, due to input saturation, plants are often activated by a preliminary phase where precomputed inputs drive the plant close to the desired operating point (while the automatic controller is still disconnected). Subsequently the controller is switched on and connected to the plant to start the automatic control phase. However, if special care is not taken when interconnecting the controller, the closed loop may experience transients where the plant output reaches unacceptable values and/or the controller output exceeds the saturation limits thus possibly causing stability loss already at power-on. Plant startup is indeed a well-known challenge within the industrial community, especially in the chemical process industry.

Bumpy switches typically correspond to controller-induced transients, after the switch, that unnecessarily affect the plant output and that could be avoided if the controller was appropriately initialized. For this reason, the bumpless transfer problem can be seen as an initial condition problem (for the controller). Indeed, all that matters after switching is that the plant and controller states "match up" and thus lead to a graceful output response where authority is transferred to the controller. Although controller re-initialization is one possible bumpless transfer strategy, alternative approaches, which are typically more robust, correspond to interconnecting the controller to some closed-loop system signals before the switching time (namely, before the time when the controller output is connected to the plant input) using a pre-conditioning scheme that suitably drives the controller dynamics into a safe configuration, ready for the switch. This is the nature of many solutions proposed in the literature, and it is the nature of the approach proposed here.

In the literature, the bumpless transfer problem is often addressed in conjunction with the anti-windup problem. The association between the two problems is mainly motivated by the two following facts. 1) The bumpless transfer problem can be seen as a problem of input substitution, whereas the anti-windup problem deals with input saturation. In both cases, for whatever reason, the type of phenomenon to be addressed is that of dealing with a mismatch between the actual and the commanded plant input. 2) Very often, solutions to the anti-windup problem turn out to be useful as a starting point to provide solutions to the bumpless transfer problem, so that a unifying theory could be formulated.

For MRAW it is indeed true that an accurate implementation of the switching results in successful bumpless transfer, in addition to anti-windup compensation, without any extra complication of the compensation scheme. Denote by $t_s$ the switching time, characterizing the time when the unconstrained controller is hooked up to the plant in the closed-loop system, then successful bumpless transfer is achieved by performing the switch as in Figure 9.8. This switching is described by the following equations:

$$
\begin{aligned}
u(t) &= \begin{cases} \hat{u}(t), & t < t_s, \\ y_c(t) + v(t), & t \geq t_s, \end{cases} \\
u_c(t) &= y(t) - y_{aw}(t), \quad t \in (-\infty, +\infty),
\end{aligned}
\tag{9.12}
$$

where $\hat{u}(t)$ is the plant input before the switch. This input may come from a manual

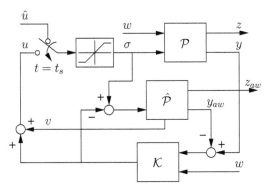

Figure 9.8  The bumpless transfer extension of the MRAW scheme.

control input or from the output of a different controller having authority over the plant input before the switch.

The $(-\infty, +\infty)$ specification in equations (9.12) denotes the fact that the scheme of Figure 9.8 should be implemented by interconnecting the controller to the plant output long before the switching time $t_s$, so that the closed-loop transient induced by this interconnection will only happen virtually between the controller state $x_c$ and the anti-windup state $x_{aw}$. In this way, the plant won't be affected by it. After the controller transient has expired, the switch can occur and bumpless transfer will be enforced by the stabilizing signal $v$, which had remained disconnected until that time.

Clearly, the length of this transient (therefore an estimate of how long the controller should be connected to the plant before the switching time $t_s$) depends on the closed-loop system dynamics and on the nature of the disturbances affecting the closed loop. It might be in general a strong limitation to wait for the transient to expire before turning on the controller. To address this situation, denote by $t_0$ the power-on time for the controller-filter pair represented by the two lower blocks in Figure 9.8. If the switching time $t_s$ is relatively close to the power-on time $t_0$, the following two strategies can be useful to speed-up the transient controller response commented above:

1. The controller state at the initial time $t_0$ may be initialized with a value which matches (in an averaged sense) the current plant state. Note that no requirement is imposed on this initial state, although a good guess may reduce the transient response.
2. In cases where the plant is operating around the steady state and the disturbances and the references have a sufficiently uniform frequency content (or are constant), the controller and filter dynamics may be run at a faster rate during the interval before the switching time $t_s$. This strategy will accordingly speed up the controller transient and prepare the controller for the switch in a faster way. The correct timing of both controller and filter could then be reestablished gracefully just before the switching time $t_s$.

Once the controller-anti-windup filter pair in Figure 9.8 have completed their

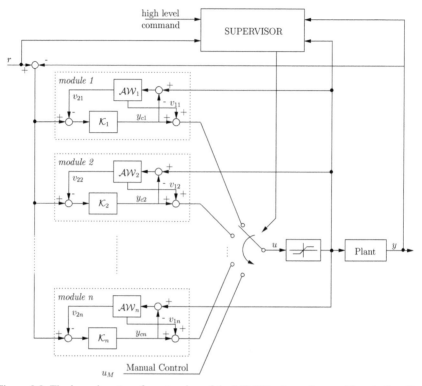

Figure 9.9 The bumpless transfer extension of the MRAW scheme in a multicontroller situation.

transient, the control scheme may be switched on and off multiple times. After each switch on time, bumpless transfer will occur, based on the information stored into the state $x_{aw}$ of the AWBT filter. This feature may be useful in a multicontroller scheme where certain task oriented controllers are required to operate during some control phases. The bumpless transfer scheme would guarantee a smooth authority transfer among the different controllers. Notice, however, that the AWBT filter should be replicated for each controller involved in the multicontroller scheme. A block diagram illustrating this fact is represented in Figure 9.9. Another situation where this scheme might be useful is the case when certain controllers need to be tested in the feedback loop. In that case, a security transfer to a previously tested controller could be commanded at any time, guaranteeing the bumpless transfer feature.

Finally, for the case where there is some flexibility in the switching time $t_s$, an advantageous property of the model recovery AWBT solution is that the amount of transient to be experienced just after the switching phase is proportional to the size of the anti-windup filter state $x_{aw}$. Indeed, because of the model recovery properties discussed in Section 6.3.2 on page 162 (see equation (6.14)), this state keeps track of the mismatch between the actual plant response and the target plant response that the controller would like to enforce on the plant. It is then reasonable to perform an "intelligent switch" by monitoring the size (e.g., the Euclidean norm) of $x_{aw}(\cdot)$ to select the switching instant as an instant when this size is small, subject also to the constraint that $t_0 \ll t_s$. The corresponding performance output will almost instantaneously follow the target output response. Ideally, if $t_0 = -\infty$ and one could find a time $T$ where $x_e(T) = 0$, then the switching transient would be completely removed by the bumpless transfer scheme. For an example, see the bold solid curve in the simulation results of Figure 9.10.

**Example 9.3.1**   (Bumpless plant startup)   Consider a simple marginally stable SISO system characterized by an input saturation level of 0.25 and by the following state-space matrices:

$$A_p = \begin{bmatrix} -500 & 1 \\ 0 & 0 \end{bmatrix}, \ B_{p,u} = \begin{bmatrix} 0 \\ 1 \end{bmatrix}, \ C_{p,y} = \begin{bmatrix} 1 & 1 \end{bmatrix}, \ D_{p,yu} = \begin{bmatrix} 0 \end{bmatrix},$$

and design two controllers. The first controller is an LQG controller which enforces a slow closed-loop behavior and is used at startup. The second controller is more aggressive and is designed with two internal models aimed at guaranteeing zero steady-state error for step references and at rejecting sinusoidal disturbances with unitary frequency (namely, the controller has three poles, at $0, +j, -j$). To suitably stabilize the closed-loop system the second controller is completed by designing an LQG stabilizer for the plant augmented with the internal models.

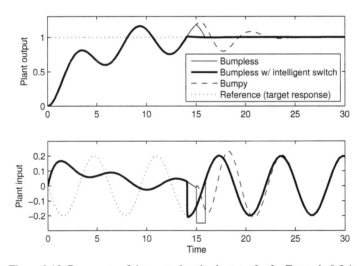

Figure 9.10 Responses of the control authority transfer for Example 9.3.1.

Since the plant is marginally stable, MRAW can be designed for the system following Algorithm 11 for the synthesis of $v$. The resulting selection is $v = -Kx_{aw}$ with $K = \begin{bmatrix} 0.002 & -100 \end{bmatrix}$. Note that this scheme will work for any (even nonlinear) controller designed for this plant.

The simulation scenario is that of a plant startup where the first controller is used first and then authority is switched to the second controller to guarantee asymptotic reference tracking and disturbance rejection. The reference is a unit step and the disturbance is a sine wave of amplitude 0.2 affecting the plant input (namely, $B_{p,w} = B_{p,u}$ and $D_{p,yw} = D_{p,yu}$). Figure 9.10 represents three responses. The dotted line in the upper plot represents the reference signal. The dotted line in the lower plot corresponds to the input response that the second controller would have generated if it had been connected to the plant since $t_0 = -\infty$. This can be seen as the target response that the controller output should reach after the switching transient.

For the thin solid and dashed curves the switching time is fixed at $t_s = 15$ and it is evident from the upper plot that the bumpless scheme rapidly drives the plant response to the reference value (note that the input authority is fully exploited by the scheme because in the lower plot, the plant input is kept into saturation until the transfer error becomes very small). The bold solid curve represents the results arising from applying the "intelligent switching" idea outlined above. In this case, the switching time is commanded upon detection of a very small Euclidean norm of the state $x_{aw}$. In particular, slightly after time $t = 14$, this state is almost zero and the bumpless switch happens without any transient.                                          □

## 9.4  RELIABLE CONTROL VIA HARDWARE REDUNDANCY

Following the discussion on bumpless transfer, a natural idea to consider is the possibility of assigning the $n$ controllers in Figure 9.9 as multiple copies of the same controller. This may be desirable as a means for increasing the reliability of control schemes subject to actuator, sensor or other hardware failures.

When all of the controllers in Figure 9.9 reproduce the same dynamics, all of the controller outputs $y_{ci}$ coincide at all times and switching among the different modules does not affect the system response. However, with the aim of increasing the reliability of the nominal control scheme, the multiple controller structure is a useful tool. In particular, if each controller has a dedicated sensor located on the plant for the measurement of the output $y$, then there are two main reasons for differences among the state responses $x_{ci}$ of the controllers:

1. differences among measurement noise associated to each sensor;
2. controller failures, either corresponding to failures of the sensors or failures of the controlling device (a DSP board, a PLC or any other digital or analog device that implements the controller dynamics).

Due to the stability of the unconstrained closed-loop system and to the presence of an AWBT compensator for each controller, bounded measurement disturbances correspond to bounded differences among the controllers' responses. On the other hand, if a failure occurs in the $i$th controller, typically the control objective is no longer met by the $i$th unconstrained closed-loop system, whose information is stored in the $i$th AWBT compensator. In this case, differences will eventually be observed between the input (or, similarly, the output) of the $i$th controller and the inputs (or, respectively, the outputs) of the other controllers. Based on these differences, the "supervisor" block in Figure 9.9 can be designed as a system detecting controller failures based on the measurement of the input-output pairs $(u_{ci}, y_{ci})$ $i = 1, \ldots, n$. Upon a failure detection, the supervisor can switch off the defective controller and guarantee correct functioning of the overall control system.

Although the supervisor block can be implemented in many different ways, depending on the particular application, a possible general scheme is to define a maximum tolerance relating the differences among inputs and outputs of the controllers and, provided a majority of controllers are operating correctly, failures can be detected by comparing the responses $(u_{ci}, y_{ci})$. The responses corresponding to the majority of controllers that are operating correctly will differ by a small amount, while their difference with the malfunctioning ones will exceed the tolerance, thus allowing the failure to be detected.

A controller failure can occur in two main situations.

1. *The failing controller is not connected to the plant.* In this case, after the failure is detected, the controller can be shut off and repaired without affecting the control performance. As a matter of fact, in Figure 9.9 it is clear that any disconnected controller is affected by the plant but does not affect the plant, because its output is disconnected. Once the controller is repaired, it can be reconnected to the reliable scheme and, after a transient (which, once again, does not affect the plant input), its state will converge to the values of the other controllers' states thus reconstituting its normal operation. Note that the whole maintenance process for the controller is invisible to the plant.

2. *The failing controller is connected to the plant.* In this case, before the failure is detected, the state of the plant could be driven far from the desired trajectory. However, due to the bumpless scheme, once the control input is switched to a functioning controller, the state of the plant is driven back to the unconstrained trajectory (whose information is kept in the AWBT compensator of the new controller). Once the malfunctioning controller is disconnected from the plant input, it can be repaired and reconnected to the reliable scheme as described in the previous item.

In summary, the proposed reliable scheme guarantees the following properties:

- In addition to guaranteeing fault detection, the design also provides bumpless transfer between the malfunctioning controller and a functioning one.
- The proposed method is able to detect any fault occurring in the control loop: even when this corresponds to a slight malfunctioning that slowly drives the plant output far from the desired trajectory; as a matter of fact, the information about the desired trajectory is embedded in the reliable scheme and constantly compared to the plant response.
- Once the reliable design is implemented, the anti-windup property of the resulting controller with respect to possible plant input saturation is automatically guaranteed by the reliable scheme.

In the following example, the controller designed for the damped mass-spring system discussed in Example 7.2.6 on page 190 is replicated in a reliable scheme. Sensor failures are assumed in the system to demonstrate the abilities of the reliable scheme.

**Example 9.4.1** (Reliable control for the damped mass-spring system) Consider the unconstrained control design for the damped mass-spring system of Example 7.2.6 on page 190 and assume that three copies of that unconstrained controller are combined in a reliable scheme. Assume that the reference signal is a sawtoothed wave of period 5 seconds and amplitude 1.6 meters. Assume that the measurement of each sensor is affected by band-limited white noise with noise power $10^{-4}$.

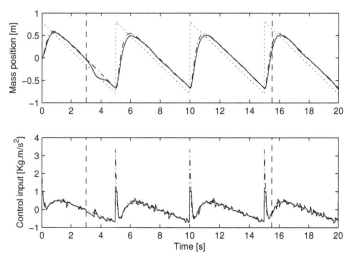

Figure 9.11 Input and output of the plant in the reliable control scheme. The dashed vertical lines correspond to the sensor failures times.

Assume that the sensor of the first controller fails at time $t_{F1} = 3$ seconds giving a zero output from time $t_{F1}$ until its repair time $t_{R1} = 10$ seconds. Assume also that the sensor of the second controller fails at time $t_{F2} = 15.5$ seconds, giving a large constant output value after that time.

Figure 9.11 represents the response of the system when the supervisor block only compares the various controller outputs $y_{ci}$, $i = 1, 2, 3$, with tolerance value 0.4. The dotted line represents the reference signal, the dashed-dotted line represents the unconstrained response (in the absence of measurement noise), and the solid line represents the response of the reliable control scheme. Sensor failures occur at the vertical dashed lines. The three controller outputs in the reliable scheme are shown in Figure 9.12, where solid lines are used when the controller is connected to the plant and dotted lines are used when the controller is disconnected from the plant.

At the system power-on, the first controller is active. In Figure 9.12 it is shown that the sensor failure at time $t_{F1}$ is detected with a significant delay and this causes the plant output to exhibit the bump in the top plot of Figure 9.11. Note that the sensor failure could be detected earlier by comparing the controller inputs $u_{ci}$, $i = 1, 2, 3$ in the supervisor block. However, it is of interest to show how the bumpless scheme is able to steer the plant output back to the desired trajectory even when the malfunctioning detection is delayed.

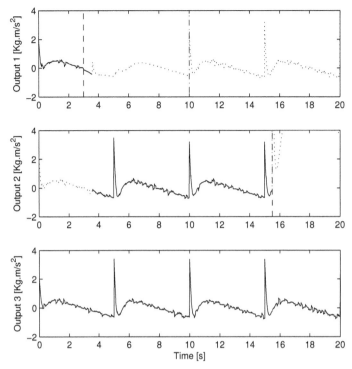

Figure 9.12  Outputs of the three redundant controllers in the reliable control scheme. Solid
           lines mean that the controller is connected to the plant. Controller repairs occur
           at the dashed-dotted vertical lines.

At time $t_{R1}$, the first controller is repaired and reconnected to the plant output.
After a short transient, it is again correctly functioning in the reliable scheme and it
is used, together with the third controller, to detect the failure of the second sensor
at time $t_{F2}$ by majority comparison. After this second failure, it is assumed that
the second sensor gives a large constant response, thus driving the malfunctioning
controller output to large values (see the middle plot in Figure 9.12), hence the
failure is detected almost instantly by the supervisor and no bump is noticeable on
the plant output as seen in Figure 9.11. After the failure is detected, the plant input
is switched to the third controller output until the end of the simulation.          □

## 9.5  NOTES AND REFERENCES

The rate and magnitude saturation compensation scheme discussed here is taken from [AP2] where
the scheme was applied to the example study discussed in Example 9.1.2. The study in [AP2] was
motivated by the fact that previous papers [SAT4, SAT6] illustrated weaknesses of the daisy-chained
scheme for redundant actuators in the presence of rate saturation. As in [AP2], the controller parameters
are taken from [SAT4]. The Caltech ducted fan case study in Example 9.1.1 is taken from [AP3] where
experimental results are also reported. The recent work in [MR33] provides additional algorithms for
MRAW-based solutions of the anti-windup problem with rate- and magnitude-saturated plants. Those

algorithms are not reported here to keep the discussion simple but might be included in a revised version of this book.

The MRAW solution for dead-time systems presented in Section 9.2 is taken from [MR24]. Example 9.2.1 is also taken from [MR24] but was originally presented in [G11, Example 4], where the unconstrained controller of Figure 9.5 was proposed. However, in [G11] the example was discussed in the absence of saturation.

The bumpless transfer extension of the MRAW scheme discussed in Section 9.3 is taken mainly from [MR25], where the corresponding theory is presented both in continuous and discrete time. The reader is also referred to [AP10], where a more sophisticated case study consisting in the dynamics of an open water channel has been studied, both in the discrete- and in the continuous-time setting. The multicontroller scheme of Figure 9.9 is taken from [MR15], which contains preliminary results on model recovery bumpless transfer and also contains the reliable control ideas of Section 9.4.

# Chapter Ten

## Anti-windup for Euler-Lagrange Plants

### 10.1 FULLY ACTUATED EULER-LAGRANGE PLANTS

This chapter focuses on a specific case study where the model recovery anti-windup (MRAW) construction described in Section 6.3 is applied to a class of nonlinear plants consisting of all fully actuated Euler-Lagrange systems. In particular, for all such plants, an anti-windup construction is given here that, provided that some feasibility conditions are satisfied, is able to recover global asymptotic stability of the closed loop with saturation as long as the unconstrained controller guarantees global asymptotic stability and local exponential stability of the unconstrained closed loop.

This case study can be understood as an example of how far the techniques illustrated in this book can reach beyond the linear and arguably simpler situations addressed in the other chapters. Clearly, nonlinear plants require more specific and tailored solutions; however, the model recovery architecture and a carefully designed feedback-stabilizing signal $v$ can guarantee successful anti-windup in most reasonable nonlinear application studies.

Fully actuated Euler-Lagrange systems correspond to mechanical systems characterized by some generalized position variables $q \in \mathbb{R}^n$, a so-called Raleigh dissipation function $(q, \dot{q}) \mapsto \mathcal{R}(q, \dot{q}) = \frac{1}{2}\dot{q}^T R(q)\dot{q}$, and a so-called Lagrangian function $(q, \dot{q}) \mapsto \mathcal{L}(q, \dot{q})$, defined as $\mathcal{L}(q, \dot{q}) = T(q, \dot{q}) - V(q)$, where $T(q, \dot{q})$ is the kinetic energy of the system and $V(q)$ is its potential energy. As usual in classical mechanics, the kinetic energy can be written as $T(q, \dot{q}) = \frac{1}{2}\dot{q}^T B(q)\dot{q}$, where $q \mapsto B(q)$, called *generalized inertia matrix*, is a differentiable function, symmetric positive definite and globally bounded away from zero and from infinity. In other words, there exist positive numbers $\lambda_M$ and $\lambda_m$ such that $\lambda_m I \leq B(q) \leq \lambda_M I$ for all $q \in \mathbb{R}^n$ (where $I$ denotes the identity). For the remaining parameters, it suffices that the *potential energy* $q \mapsto V(q)$ and the *dissipation matrix* $q \mapsto R(q)$ be differentiable, but weaker conditions would also be compatible with the solution given here.

The dynamics of a fully actuated Euler-Lagrange system can be written as

$$\frac{d}{dt}\left(\frac{\partial \mathcal{L}(q, \dot{q})}{\partial \dot{q}}\right) - \frac{\partial \mathcal{L}(q, \dot{q})}{\partial q} + \frac{\partial \mathcal{R}(q, \dot{q})}{\partial \dot{q}} = u, \tag{10.1}$$

where $u$ denotes the generalized external forces applied to the system. Fully actuated means that the number of independent inputs $u$ in the previous equation is equal to the number of generalized position variables $n$.

A special class of the fully actuated Euler-Lagrange systems described above is given by the following equations describing the dynamics of fully actuated robotic manipulators:

$$B(q)\ddot{q} + C(q,\dot{q})\dot{q} + R(q)\dot{q} + h(q) = u, \qquad (10.2)$$

where $C(q,\dot{q})\dot{q}$ represents the generalized centrifugal and Coriolis terms, $h(q)$ is the vector of gravitational forces and the function $R(q)\dot{q}$ represents the friction forces. The relation between equations (10.1) and (10.2) is easily derived as

$$C(q,\dot{q})\dot{q} = \dot{B}(q)\dot{q} - \dot{q}\frac{\partial B(q)}{\partial q}\dot{q}, \quad h(q) = \frac{\partial V(q)}{\partial q}.$$

Since the matrix $B(q)$ is nonsingular for all values of its argument $q$, then equations (10.1) and (10.2) can be rewritten as

$$\dot{x}_p := \begin{bmatrix} \dot{q} \\ \dot{q}_d \end{bmatrix} := \begin{bmatrix} \dot{q} \\ \ddot{q} \end{bmatrix} = F(x_p,u) := \begin{bmatrix} F_1(q,q_d,u) \\ F_2(q,q_d,u) \end{bmatrix}, \qquad (10.3)$$

where $F_1(q,q_d,u) := q_d$ and $F_2(q,q_d,u)$ is uniquely defined by the Lagrangian function and the Raleigh dissipation function. For example, when using the description (10.2),

$$F_2(q,q_d,u) = B(q)^{-1}\left(u - C(q,q_d)q_d + R(q)q_d + h(q)\right).$$

In the rest of the chapter, the two notations (10.1) and (10.3) will be used interchangeably.

Given the Euler-Lagrange system (10.1), assume that the following unconstrained controller

$$\begin{aligned} \dot{x}_c &= f_c(x_c,u_c,r) \\ y_c &= k(x_c,u_c,r) \end{aligned} \qquad (10.4)$$

(where $x_c \in \mathbb{R}^{n_c}$ is the controller state, $u_c \in \mathbb{R}^{2n}$ is the controller input and $r \in \mathbb{R}^n$ is the external reference input) has been designed in such way that for any selection of $r$ within a prescribed set of admissible references, the unconstrained feedback interconnection

$$u_c = (q,\dot{q}) = x, \qquad u = y_c, \qquad (10.5)$$

guarantees global asymptotic stability and local exponential stability of a point $(x^*,x_c^*)$ where $x^* = (q^*,\dot{q}^*) = (r,0)$.

Then the anti-windup problem addressed here is that of recovering the possible stability and performance loss due to input saturation, experienced in the following saturated closed-loop interconnection between the plant (10.1) and the controller (10.4):

$$u_c = (q,\dot{q}) = x, \qquad u = \text{sat}(y_c), \qquad (10.6)$$

where $\text{sat}(\cdot)$ is the magnitude saturation function with symmetric saturation levels $\bar{u}_1, \ldots, \bar{u}_n$.

## 10.2 ANTI-WINDUP CONSTRUCTION AND SELECTION OF THE STABILIZER $V$

Model recovery anti-windup for the saturated closed loop described in the previous section can be performed by generalizing the compensation scheme for linear plants as outlined in Section 6.3 on page 161. In particular, for the problem analyzed here, where $f$ characterizes the plant with input saturation and $F$ characterizes the plant without input saturation, equation (6.8) characterizing the anti-windup filter dynamics becomes

$$\dot{x}_{aw} = F(x_p, \text{sat}(u)) - F(x_p - x_{aw}, y_c), \tag{10.7}$$

where the anti-windup state $x_{aw} = (q_{aw}, \dot{q}_{aw})$ has the same dimension as the state of the plant. The dynamics (10.7) is used in a model recovery anti-windup framework to enforce modifications on the unconstrained controller, as follows:

$$u = \text{sat}(\text{sat}(y_c) + v), \qquad u_c = x_p - x_{aw}, \tag{10.8}$$

where $v$ is the stabilizing signal to be designed in such a way that the anti-windup dynamics (10.7) is driven to zero and model recovery is obtained, as illustrated in Chapter 6.

Similar to the anti-windup solutions for linear plants given in Chapters 7 and 8, also in this case, regardless of the nonlinearity of the plant, the selection of $v$ is a key aspect of successful anti-windup augmentation. The solution proposed next is guaranteed to induce closed-loop global asymptotic stability and performance recovery whenever the following condition holds:

$$h_{Mi} := \sup_{q \in \mathbb{R}^n} \left| \left[ \frac{\partial V(q)}{\partial q_i} \right] \right| < \bar{u}_i, \quad i = 1, \cdots, n. \tag{10.9}$$

The meaning of equation (10.9) is quite intuitive when looking at the special case of robotic manipulators in equation (10.2). Indeed, since in that case, $h(q) = \frac{\partial V(q)}{\partial q}$ denotes the gravitational forces, equation (10.9) requires that there be enough input authority to induce any desired steady-state position $q = r$ on the robot. The construction for $v$ given here can also be carried out when condition (10.9) is not satisfied. However, in that case there is only a local asymptotic stability guarantee on the resulting closed-loop system.

Whenever condition (10.9) is satisfied, the selection of $v$ can be done as follows:

$$\begin{aligned} v &= \gamma(x_p, x_p - x_{aw}), \\ &= \gamma\left( (q, \dot{q}), (q - q_{aw}, \dot{q} - \dot{q}_{aw}) \right) \\ &= \left[ \frac{\partial V}{\partial q}(q) \right]^T - \left[ \frac{\partial V}{\partial q}(q - q_{aw}) \right]^T \\ &\quad - K_g \text{sat}\left( K_g^{-1} K_q q_{aw} \right) - K_{qd}(q_{aw}, \dot{q}_{aw}) \dot{q}_{aw}, \end{aligned} \tag{10.10}$$

where $K_q$ and $K_g$ are diagonal positive definite constant matrices and $K_{qd}(\cdot, \cdot)$ is a diagonal positive definite matrix function which is constant in a sufficiently small neighborhood of the origin and bounded away from zero, [1] and such that the function $(q_{aw}, \dot{q}_{aw}) \mapsto K_{qd}(q_{aw}, \dot{q}_{aw}) \dot{q}_{aw}$ is locally Lipschitz in all its arguments. Moreover, $K_g$ is chosen in such a way that

$$h_{Mi} + \kappa_{gi} \bar{u}_i < \bar{u}_i, \quad i = 1, \cdots, n, \tag{10.11}$$

---

[1] A diagonal positive definite matrix function $K_{qd}(\cdot, \cdot)$ is bounded away from zero if there exists a small enough positive scalar $\kappa_d$ such that $K_{qd}(\cdot, \cdot) \geq \kappa_d I$ for all values of its arguments.

where $\kappa_{gi}$ are the diagonal entries of $K_g$. Note that as long as condition (10.9) holds, there always exists a small enough matrix $K_g$ that satisfies constraints (10.11)). If, however, constraints (10.9) do not hold, then it is possible to select $K_g$ without satisfying (10.11) and the resulting anti-windup compensation will not guarantee global properties.

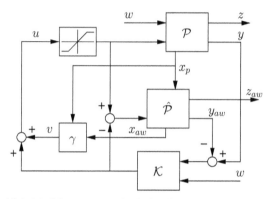

Figure 10.1  Model recovery anti-windup for Euler-Lagrange systems.

The anti-windup construction outlined above is represented in Figure 10.1. Note that as compared to the typical model recovery anti-windup scheme discussed in Chapter 6, full state measurement is required here from the plant because the anti-windup dynamics depends both on $x_{aw}$ and on $x$. The anti-windup construction is summarized step by step in the following algorithms, where two possible solutions are given. The first one corresponds to a constant selection of $K_{qd}(\cdot,\cdot)$ and is therefore a simpler scheme with lower performance. The second one corresponds to to a nonlinear selection of $K_{qd}(\cdot,\cdot)$ and leads to improved performance.

**Algorithm 24** (MRAW for Euler-Lagrange systems with linear injection)

| Applicability | | | | Architecture | | | Guarantee |
|---|---|---|---|---|---|---|---|
| Exp Stab | Marg Stab | Marg Unst | Exp Unst | Lin/ NonL | Dyn/ Static | Ext/ FullAu | Global/ Regional |
| n/a | n/a | n/a | n/a | NL | D | E | G |

*Comments*: Augmentation for a loop containing a plant from a class of nonlinear systems. Global asymptotic stability is guaranteed for any loop containing a plant whose parameters satisfy a certain constraint.

**Step 1.** Choose a diagonal positive definite $n \times n$ matrix $K_g$ whose diagonal entries $\kappa_{gi}$, $i = 1,\ldots,n$ satisfy the constraints (10.11). Typically, select the entries $\kappa_{gi}$ as the maximum allowable within the constraints (10.11). If the compatibility condition (10.9) does not hold, then select any positive definite diagonal $K_g$ and only local asymptotic stability will be guaranteed.

**Step 2.** Augment the saturated closed-loop system with the anti-windup compensator (10.7) interconnected via (10.8) and where $v$ is given in (10.10) with the constant selection $K_{qd}(q_{aw}, \dot{q}_{aw}) = K_0$, where $K_0$ and $K_q$ are still to be defined.

**Step 3.** Select the positive diagonal entries $\kappa_{qi}$ and $\kappa_{0i}$, $i = 1, \ldots, n$ of $K_q$ and $K_0$, respectively, following a standard PD tuning technique at each joint.

⋆

The next algorithm provides a further improved action by selecting $K_{qd}(\cdot, \cdot)$ as a nonlinear function. The idea behind this selection is that it coincides with the linear one proposed in the previous algorithm as long as the saturation function multiplying $K_g$ in equation (10.10) is not active. However, when that saturation becomes active, the nonlinear selection of $K_{qd}(\cdot, \cdot)$ allows to suitably scale down the right term at the last line of (10.10) as much as the left term at that last line is scaled down by the saturation effect. It will be clear from the simulations reported in the example cases studied next that the nonlinear selection leads to highly improved closed-loop performance.

**Algorithm 25** (MRAW for Euler-Lagrange systems with nonlinear injection)

| Applicability | | | | Architecture | | | Guarantee |
|---|---|---|---|---|---|---|---|
| Exp Stab | Marg Stab | Marg Unst | Exp Unst | Lin/ NonL | Dyn/ Static | Ext/ FullAu | Global/ Regional |
| n/a | n/a | n/a | n/a | NL | D | E | G |

| *Comments*: Extends and improves upon the performance in Algorithm 24 by replacing certain linear terms with nonlinear terms. |
|---|

**Step 1.** Choose a diagonal positive definite $n \times n$ matrix $K_g$ whose diagonal entries $\kappa_{gi}$, $i = 1, \ldots, n$ satisfy the constraints (10.11). Typically, select the entries $\kappa_{gi}$ as the maximum allowable within the constraints (10.11). If the compatibility condition (10.9) does not hold, then select any positive definite diagonal $K_g$, and only local asymptotic stability will be guaranteed.

**Step 2.** select $K_{qd}(q_e, \dot{q}_e) = \gamma_E(q_e, \dot{q}_e)K_0$, where the parameter $K_0$ is a diagonal positive definite matrix and the diagonal matrix function $\gamma_E(\cdot, \cdot) = \mathrm{diag}\{\gamma_{E1}(\cdot, \cdot), \ldots, \gamma_{En}(\cdot, \cdot)\}$ has diagonal entries corresponding to

$$\gamma_{Ei} = \begin{cases} 1, & \text{if } q_{ei}\dot{q}_{ei} \geq 0, \\ \dfrac{\kappa_{gi}\mathrm{sat}_{\bar{u}_i}\left(\kappa_{gi}^{-1}\kappa_{qi}q_i\right)}{\kappa_{qi}q_i} & \text{if } q_{ei}\dot{q}_{ei} < 0, \end{cases} \quad i = 1, \ldots, n,$$

where $\mathrm{sat}_{\bar{u}_i}(\cdot)$ is the symmetric scalar saturation function with limits $\bar{u}_i$.

**Step 3.** Augment the saturated closed-loop system with the anti-windup compensator (10.7) interconnected via (10.8) and where $v$ is given in (10.10) with the function $K_{qd}(\cdot, \cdot)$ defined at the previous step, and where $K_0$ and $K_q$ are still to be defined.

**Step 4.** Select the positive diagonal entries $\kappa_{qi}$ and $\kappa_{0i}$, $i = 1, \ldots, n$ of $K_q$ and $K_0$, respectively, following a standard PD tuning technique at each joint.

⋆

## 10.3 SIMULATION EXAMPLES

In this section some simulation examples are reported to demonstrate the proposed anti-windup construction. The first two examples will be useful to appreciate the proposed construction in simple linear case studies. The next example shows the performance of the proposed construction in a simple nonlinear robot arm. Finally, the construction is applied to models of two industrial robots, the PUMA robot and the SCARA robot.

Prior to starting to describe the example studies, the following section clarifies some specific unconstrained controller selections carried out in the examples. Note that no constraint is given on the controller selection as long as it stabilizes the nonlinear Euler-Lagrange system without saturation, therefore the two approaches described next could be replaced by any alternative stabilizing selection.

### 10.3.1 Selection of the unconstrained controller

Since only unconstrained closed-loop stability is required by the unconstrained controller, this could be selected among a large variety of control laws for Euler-Lagrange systems that populate a vast literature on robotics research distributed over the last three decades. It is noteworthy that, due to the model recovery architecture of the anti-windup solution, the anti-windup compensator dynamics and the selection of the parameters in the function $\gamma(\cdot,\cdot)$ illustrated in Algorithms 24 and 25 are only related to the plant dynamics and are independent of the unconstrained controller.

Among the many possible selections of the unconstrained controller, two most typical approaches are reported next and will be used in the following simulation examples. Detailed descriptions of these and other unconstrained control laws can be found in any classical robot control textbook.

1. *Proportional-integral-derivative regulator with a nonlinear gravity compensation* (otherwise called PID+$h(q)$). This controller is typically used for set-point regulation purposes and can be written in the form (10.4) as follows:

$$
\begin{aligned}
\dot{x}_c &= q - r \\
y_c &= -K_i x_c - K_p(q-r) - K_d(\dot{q}-\dot{r}) + \frac{\partial V(q)}{\partial q},
\end{aligned}
\tag{10.12}
$$

where $r$ is the reference input corresponding to the desired joint position, $K_p$, $K_i$ and $K_d$ are suitable positive definite diagonal matrices representing the proportional, integral and derivative gains, and $V(q)$ is the potential energy of the system. When using the notation in (10.2), equation (10.12) can be written as

$$
\begin{aligned}
\dot{x}_c &= q - r \\
y_c &= -K_i x_c - K_p(q-r) - K_d(\dot{q}-\dot{r}) + h(q).
\end{aligned}
$$

2. *Feedback linearization with extra PID action* (otherwise called "computed torque" + PID). This control technique employs feedback linearization to linearize the unconstrained Euler-Lagrange system. Then any arbitrary linear

decoupled behavior is imposed on the resulting unconstrained closed loop by means of an extra decentralized PID compensation action, so that decoupled behavior is obtained, in addition to arbitrary linear performance. This controller is typically used for trajectory tracking purposes and the corresponding controller equations in the form (10.4) are given by

$$\dot{x}_c = q - r$$
$$y_c = \frac{d}{dt}\left(\frac{\partial \mathcal{L}(q,\dot{q})}{\partial \dot{q}}\right) - \frac{\partial \mathcal{L}(q,\dot{q})}{\partial q} + \frac{\partial \mathcal{R}(q,\dot{q})}{\partial \dot{q}} \qquad (10.13)$$
$$-B(q)(\ddot{q} - K_i x_c - K_p(q-r) - K_d(\dot{q} - \dot{r})),$$

where $K_p$, $K_d$, $K_i$ are suitable diagonal matrices, $B(q)$ is the generalized inertia matrix, and $r$ and $\dot{r}$ denote a desired reference profile and its derivative with respect to time, respectively. Note that due to the special structure of the Lagrangian function and of the kinetic energy, the second equation in (10.12) is independent of $\ddot{q}$ because the term $B(q)\ddot{q}$ cancels out with the corresponding term resulting from the first term of the right-hand side. In particular, when using the notation in (10.2), equation (10.12) is equivalent to the following simplified form:

$$\dot{x}_c = q - r$$
$$y_c = C(q,\dot{q})\dot{q} + h(q) + R(q)\dot{q} - B(q)(K_i x_c +$$
$$K_p(q-r) + K_d(\dot{q} - \dot{r})),$$

where $r$ and $\dot{r}$ represent the trajectory to be tracked.

### 10.3.2 Academic examples

The three examples addressed here are simple control systems aimed at illustrating step by step the construction in Algorithm 25.

**Example 10.3.1**   (Double integrator)  Consider a double integrator, which could represent a unit mass moving on a horizontal rail without any friction. Assume that the control input has an input saturation level $M = 0.25$. Select the unconstrained controller as a stabilizing PID controller with gains $(k_p, k_i, k_d) = (8, 4, 4)$. Since the Euler-Lagrange equations for this simple system are linear (they actually are $\ddot{q} = u$), then the resulting anti-windup compensator equation is extremely simplified and, based on (10.7), (10.8), corresponds to

$$\dot{x}_{aw} = A_{aw}x_{aw} + B_{aw}(\text{sat}(\text{sat}(y_c) + v) - y_c),$$

with $A_{aw} = \begin{bmatrix} 0 & 1 \\ 0 & 0 \end{bmatrix}$, $B_{aw} = \begin{bmatrix} 0 \\ 1 \end{bmatrix}$. It should be also noted that in this case, due to the linearity of the plant dynamics, the anti-windup compensator does not require any plant state measurement because the two terms at the right-hand side of equation (10.7) reduce to $F(x_{aw}, u - y_c)$.

Since the potential energy for this example is constant, the selection of $\gamma(\cdot, \cdot)$ in (10.10) only depends on $x_{aw} = (q_{aw}, \dot{q}_{aw})$. Based on the construction in Algorithm 25, the signal $v$ in (10.10) is then selected as

$$v = -K_g \text{sat}\left(K_g^{-1} K_q q_{aw}\right) - \gamma_E(q_{aw}, \dot{q}_{aw}) K_0 \dot{q}_{aw}.$$

Regarding the selection of the scalar parameters $K_g$, $K_q$, and $K_0$, according to step 1 of the algorithm, $K_g$ is selected as $K_g = 0.99$, which is the maximum value satisfying equation (10.11). Then the two other gains are selected similarly to the tuning procedure for PD gains with the goal of obtaining a desirable response recovery. This leads to the selections $K_q = 80$ and $K_0 = 100$.

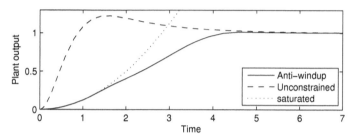

Figure 10.2 The double integrator in Example 10.3.1. Responses to a unit reference.

The closed-loop responses to a unit step reference of the unconstrained, saturated, and anti-windup closed loops are represented in Figure 10.2. The unconstrained response (dashed) rapidly converges to the desired steady-state exhibiting a slight overshoot. The saturated response (dotted) diverges in an oscillatory way and the anti-windup response (solid) shows a fast and graceful recovery of the unconstrained response. □

**Example 10.3.2** (Planar positioning system) Consider a positioning system consisting of a two-link robot, with two linear joints. The model is quite simple: the robot is not subject to the gravitational force, the generalized inertia matrix is linear and decoupled, and so is the matrix containing the friction terms. A schematic diagram of the planar positioning system is reported in Figure 10.3.

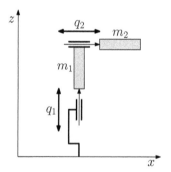

Figure 10.3 The planar positioning system of Example 10.3.2.

The system model is expressed by the following equations:

$$(M_1 + M_2)\ddot{q}_1 + \rho_1 \dot{q}_1 = u_1$$
$$M_2 \ddot{q}_2 + \rho_2 \dot{q}_2 = u_2, \tag{10.14}$$

where $M_i$ is the total mass of each link (including the actuator's mass), $\rho_i$ is the friction coefficient of the $i$th link, and $u_i$ is the $i$th actuator's force. The values of parameters are reported in Table 10.1, where $\bar{u}_i$ is the saturation level of each actuator.

| Link | $M_i$ [kg] | $\bar{u}_i$ [N] | $\rho_i$ [kg/s] |
|------|------------|-----------------|-----------------|
| 1    | 3          | 40              | 2               |
| 2    | 2          | 40              | 1               |

Table 10.1  Parameters of the planar positioning system of Example 10.3.2.

The unconstrained control system is a "computed torque" controller, which is able to induce global exponential stability when saturation does not occur. The performance of the unconstrained controller is obtained choosing suitable values for the diagonal matrices $K_p$, $K_d$, $K_i$. The equations of the unconstrained controller are:

$$
\begin{aligned}
\dot{x}_{c_1} &= \tilde{q}_1 - r_1 \\
\dot{x}_{c_2} &= \tilde{q}_2 - r_2 \\
y_{c_1} &= (M_1 + M_2)\left(-k_{p_1}(\tilde{q}_1 - r_1) - k_{d_1}\dot{\tilde{q}}_1 - k_{i_1}x_{c_1}\right) + \rho_1\dot{\tilde{q}}_1 \\
y_{c_2} &= M_2\left(-k_{p_2}(\tilde{q}_2 - r_2) - k_{d_2}\dot{\tilde{q}}_2 - k_{i_2}x_{c_2}\right) + \rho_2\dot{\tilde{q}}_2,
\end{aligned}
$$

where $(\tilde{q}, \dot{\tilde{q}}) = ([\tilde{q}_1\ \tilde{q}_2]^T, [\dot{\tilde{q}}_1\ \dot{\tilde{q}}_2]^T)$ represents the controller input, so that the unconstrained interconnection corresponds to $(\tilde{q}, \dot{\tilde{q}}) = (q, \dot{q})$, $u = y_c$, the saturated interconnection (without anti-windup) corresponds to $(\tilde{q}, \dot{\tilde{q}}) = (q, \dot{q})$, $u = \text{sat}(y_c)$, and the anti-windup interconnection corresponds to $(\tilde{q}, \dot{\tilde{q}}) = (q - q_{aw}, \dot{q} - \dot{q}_{aw})$, $u = \text{sat}(y_c + v)$, where $(q_{aw}, \dot{q}_{aw})$ is the anti-windup compensator's state. The proportional, integral, and derivative gains of the unconstrained controller have been selected as follows: $K_p = \text{diag}(360, 360)$, $K_d = \text{diag}(30, 30)$, $K_i = \text{diag}(8, 8)$.

The anti-windup construction is carried out following Algorithm 25. In particular, the anti-windup compensator dynamics (10.7) become:

$$
\begin{aligned}
\ddot{q}_{aw_1} &= \tfrac{1}{M_1+M_2}(u_{p_1} - \rho_1\dot{q}_1) + \tfrac{1}{M_1+M_2}(\rho_1(\dot{q}_1 - \dot{q}_{aw_1}) - y_{c_1}) \\
\ddot{q}_{aw_2} &= \tfrac{1}{M_2}(u_{p_2} - \rho_2\dot{q}_2) + \tfrac{1}{M_2}(\rho_2(\dot{q}_2 - \dot{q}_{aw_1}) - y_{c_2}),
\end{aligned}
$$

where $u = [u_1, u_2]^T$ is the force plant input that, according to equation (10.8), is selected as:

$$
\begin{aligned}
u_1 &= \text{sat}\left(\text{sat}(y_{c_1}) - k_{g_1}\text{sat}\left(\tfrac{k_{q_1}q_{aw_1}}{k_{g_1}}\right) - k_{qd_1}(q_{aw}, \dot{q}_{aw})\dot{q}_{aw_1}\right) \\
u_2 &= \text{sat}\left(\text{sat}(y_{c_2}) - k_{g_2}\text{sat}\left(\tfrac{k_{q_2}q_{aw_2}}{k_{g_2}}\right) - k_{qd_2}(q_{aw}, \dot{q}_{aw})\dot{q}_{aw_2}\right),
\end{aligned}
$$

where $\text{sat}(\cdot)$ is the saturation function having symmetric saturation limits $\bar{u}_1 = \bar{u}_2 = 40$ and $k_{qd_1}(\cdot, \cdot)$ and $k_{qd_2}(\cdot, \cdot)$ are the diagonal elements of the matrix function $K_{qd}(\cdot, \cdot)$ selected at step 2 of the algorithm.

As for the selection of the diagonal matrix gains $K_g$, $K_q$, and $K_0$, since there is no gravity effect on this model, it is possible to select $K_g$ arbitrarily close to the identity, e.g., $K_g = \text{diag}(0.99, 0.99)$, which satisfies constraint (10.11). The remaining matrix gains $K_q$ and $K_0$ should be selected with the goal of improving the performance of the anti-windup law during the transient response. For each entry $i = 1,2$ on the diagonal of $K_q$ and $K_0$, the proportional term $K_{qi}$ is selected so as to guarantee a fast enough convergence of the related component of $q_{aw}$ to zero, and the derivative term $K_{0i}$ is selected so as to enforce suitable damping on the terminal part of the trajectory, thus avoiding undesirable oscillations of the anti-windup closed-loop response. Following this approach, the parameters are easily tuned as $K_0 = \mathrm{diag}(650, 250)$, $K_q = \mathrm{diag}(2600, 1600)$.

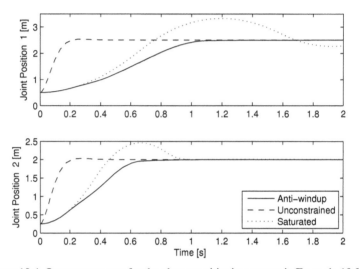

Figure 10.4  Output responses for the planar positioning system in Example 10.3.2.

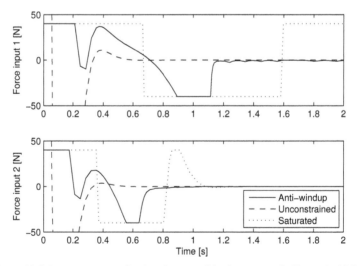

Figure 10.5  Input responses for the planar positioning system in Example 10.3.2.

The anti-windup construction is tested by simulation by selecting the reference signal as the following step input:

$$r = (2.5\ m, 2\ m),\qquad\qquad\qquad (10.15)$$

and initializing both the plant and the controller states at zero. The corresponding simulations are reported in Figures 10.4 and 10.5. In Figure 10.4, the bold curves represent the unconstrained output responses. The output responses of the anti-windup closed-loop system (thin solid) reach the reference positions in less than one second, eliminating the undesired oscillations exhibited by the saturated closed-loop system without anti-windup (dotted). Figure 10.5 represents the plant input responses for the same three closed-loop systems. Note that the anti-windup action exploits the full actuator's power available to allow for the fast output responses of Figures 10.4. This fact becomes evident when noticing that the plant input signals become saturated both during the acceleration and during the deceleration phases.                                                                              □

**Example 10.3.3** (Two-link planar robot with rotational joints) Consider the planar two-link robot arm represented in Figure 10.6, displaced on a vertical plane so that the gravitational vector is oriented as shown in the figure. Differing from the previous example, the robot dynamics is nonlinear and not decoupled, and the gravitational acceleration affects both links. The aim of this example is to show the quasi-decoupled performance of the anti-windup closed-loop system in the presence of input saturation and the easy selection of the anti-windup parameters.

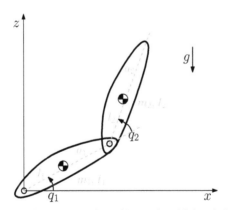

Figure 10.6 The two-link planar robot with rotational joints in Example 10.3.3.

According to the notation used in equation (10.2), the generalized inertia matrix $B(q)$, the matrix $C(q,\dot{q})$ related to the centrifugal and Coriolis terms and the gravitational vector $h(q)$ are given next. For simplicity, the friction forces are set to zero. The generalized inertia matrix $B(q)$ corresponds to:

$$B(q) = \begin{bmatrix} i_{11} & i_{12} \\ i_{12} & i_{22} \end{bmatrix},$$

with

$$i_{11} = I_1 + M_1 l_1^2 + I_2 + M_2(a_1^2 + l_2^2 + 2a_1 l_2 \cos(q_2))$$
$$i_{12} = I_2 + M_2(l_2^2 + a_1 l_2 \cos(q_2))$$
$$i_{22} = I_2 + M_2 l_2^2,$$

where $q = [q_1, q_2]$ contains the two joint variables, $(M_1, M_2)$ are the total masses of the two links (including the actuators' masses), $(a_1, a_2)$ represent the link lengths, $(l_1, l_2)$ represent the distances of the center of mass of each link from the preceding joint, and $(I_1, I_2)$ represent the rotational inertias at the two joints. The matrix $C(q, \dot{q})$ can be written as follows:

$$C(q, \dot{q}) = \begin{bmatrix} -M_2 a_1 l_2 \sin(q_2) \dot{q}_2 & -M_2 a_1 l_2 \sin(q_2)(\dot{q}_1 + \dot{q}_2) \\ M_2 a_1 l_2 \sin(q_2) \dot{q}_1 & 0 \end{bmatrix},$$

and the gravitational forces vector corresponds to

$$h(q) = \begin{bmatrix} g(M_1 l_1 + M_2 a_1) \cos(q_1) + g M_2 l_2 \cos(q_1 + q_2) \\ g M_2 l_2 \cos(q_1 + q_2) \end{bmatrix},$$

where $g$ is the gravitational acceleration value. The physical parameters of the robot have been selected as shown in Table 10.2, where $(\bar{u}_1, \bar{u}_2)$ represent the saturation levels for the torques exerted at the two joints.

| Link | $l_i$ [m] | $M_i$ [kg] | $I_i$ [kg·m²] | $a_i$ [m] | $\bar{u}_i$ [Nm] |
|------|-----------|------------|----------------|-----------|------------------|
| 1    | 0.5       | 6          | 0.2            | 1         | 138 Nm           |
| 2    | 0.25      | 5          | 0.1            | 0.5       | 40 Nm            |

Table 10.2 Parameters of the two-link planar robot in Example 10.3.3.

As in the previous example, the unconstrained controller is selected as a computed torque controller, which induces linear and decoupled closed-loop behavior before saturation is activated. The corresponding equations are:

$$\dot{x}_{c_1} = \tilde{q}_1 - r_1$$
$$\dot{x}_{c_2} = \tilde{q}_2 - r_2$$
$$y_{c_1} = (I_1 + M_1 l_1^2 + I_2 + M_2(a_1^2 + l_2^2 + 2a_1 l_2 \cos(\tilde{q}_2))) \cdot$$
$$(-k_{p_1}(\tilde{q}_1 - r_1) - k_{d_1}\dot{\tilde{q}}_1 - k_{i_1}x_{c_1})$$
$$+ (I_2 + M_2(l_2^2 + a_1 l_2 \cos(\tilde{q}_2)))(-k_{p_2}(\tilde{q}_2 - r_2) - k_{d_2}\dot{\tilde{q}}_2 - k_{i_2}x_{c_2})$$
$$- 2M_2 a_1 l_2 \dot{\tilde{q}}_1 \dot{\tilde{q}}_2 \sin(\tilde{q}_2) - M_2 a_1 l_2 \dot{\tilde{q}}_2^2 \sin(\tilde{q}_2)$$
$$+ g(M_1 l_1 + M_2 a_1) \cos(\tilde{q}_1) + g M_2 l_2 \cos(\tilde{q}_1 + \tilde{q}_2)$$
$$y_{c_2} = (I_2 + M_2(l_2^2 + a_1 l_2 \cos(\tilde{q}_2)))(-k_{p_1}(\tilde{q}_1 - r_1) - k_{d_1}\dot{\tilde{q}}_1 - k_{i_1}x_{c_1})$$
$$+ (I_2 + M_2 l_2^2)(-k_{p_2}(\tilde{q}_2 - r_2) - k_{d_2}\dot{\tilde{q}}_2 - k_{i_2}x_{c_2})$$
$$+ M_2 a_1 l_2 \dot{\tilde{q}}_1^2 \sin(q_2) + g M_2 l_2 \cos(\tilde{q}_1 + \tilde{q}_2),$$

where $(\tilde{q}, \dot{\tilde{q}})$ represents the controller input interconnected to the closed loop in the same way as in the previous example. The proportional, integral, and derivative gains of the unconstrained controller have been selected as $K_p = \text{diag}(240, 255)$, $K_d = \text{diag}(45, 50)$, and $K_i = \text{diag}(4, 4)$.

According to equation (10.7), the anti-windup compensator's dynamics can be written as:

$$\ddot{q}_{aw} = I^{-1}(q)(u_p - C(q,\dot{q})(q,\dot{q}) - h(q))$$
$$+ I^{-1}(q - q_{aw})(C(q - q_{aw}, \dot{q} - \dot{q}_{aw})(q - q_{aw}, \dot{q} - \dot{q}_{aw}) + G(q - q_{aw}) - y_c)$$
$$u = \text{sat}(y_c + v),$$

where $y_c$ is the unconstrained controller output and $v$ is selected according to Algorithm 25 as

$$v_1 = (\sigma_1(y_{c_1}) - y_{c_1}) + g(M_1 l_1 + M_2 a_1)\cos(q_1) + gM_2 l_2 \cos(q_1 + q_2)$$
$$- g(M_1 l_1 + M_2 a_1)\cos(q_1 - q_{aw_1}) - gM_2 l_2 \cos(q_1 - q_{aw_1} + q_2 - q_{aw_2})$$
$$- k_{g_1}\sigma_1\left(\frac{k_{q_1} q_{aw_1}}{k_{g_1}}\right) - k_{qd_1}(q_{aw}, \dot{q}_{aw})\dot{q}_{aw_1}$$
$$v_2 = (\sigma_2(y_{c_2}) - y_{c_2}) + gM_2 l_2 \cos(q_1 + q_2) - gM_2 l_2 \cos(q_1 - q_{aw_1} + q_2 - q_{aw_2})$$
$$- k_{g_2}\sigma_1\left(\frac{k_{q_2} q_{aw_2}}{k_{g_2}}\right) - k_{qd_2}(q_{aw}, \dot{q}_{aw})\dot{q}_{aw_2},$$

where $\sigma_1(\cdot)$ and $\sigma_2(\cdot)$ denote the two input saturation functions and $k_{qd_1}(\cdot,\cdot)$ and $k_{qd_2}(\cdot,\cdot)$ are the diagonal elements of the matrix function $K_{qd}(\cdot,\cdot)$ defined at step 2 of Algorithm 25.

The diagonal elements of the matrix $K_g$ have been chosen to satisfy the constraint (10.11) as $K_g = \text{diag}(0.29, 0.64)$. The diagonal elements of the matrices $K_q$, $K_0$ have been selected once again following the approach at step 4 of Algorithm 25. The resulting matrices are $K_0 = \text{diag}(150, 400)$ and $K_q = \text{diag}(400, 400)$.

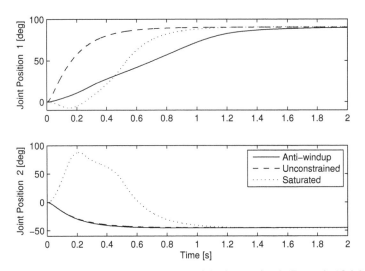

Figure 10.7 Output responses for the two-link planar robot in Example 10.3.3.

For the simulation tests, the reference signal has been selected as a step input, assuming the following values:

$$r = (90\ deg, -45\ deg), \tag{10.16}$$

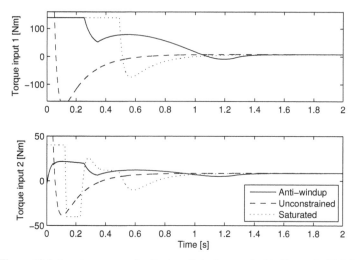

Figure 10.8 Input responses for the two-link planar robot in Example 10.3.3.

and both the plant and the controller states have been initialized at zero.

The corresponding simulations are reported in Figures 10.7 and 10.8. As usual, the dashed curves represent the unconstrained responses, the dotted curves represent the saturated responses (without anti-windup), and the solid curves represent the anti-windup responses. Observe that the undesired undershoot presented by the saturated response is completely eliminated by the anti-windup action. Moreover, the anti-windup compensation is able to almost fully preserve the linear performance at the second joint. This is not the case in the first joint response, which exhibits an inevitable response delay due to the input limitation.

From the input responses reported in Figure 10.8, it appears that the anti-windup compensator makes large use of the available input effort. Nevertheless, the input signal doesn't reach saturation other than for a short time interval during the first 200 milliseconds. This suggests that increasing the anti-windup gains $K_q$ and $K_0$ may lead to a faster response. Additional simulations, which aren't reported here, confirm that increasing the anti-windup gains to the values $K_0 = \mathrm{diag}(600, 400)$ and $K_q = \mathrm{diag}(4000, 400)$ allows to reduce the recovery transient on the first joint from approximately 1.5 seconds to 1 second (the response on the second joint remains the same).                                                                    □

### 10.3.3 SCARA robot

#### 10.3.3.1 Model description and anti-windup design

The SCARA robot (Selective Compliance Assembly Robot Arm) is a four-link robot, with four joints, all of them having vertical axes. The first two links of the robot reproduce the behavior of a planar two-link robot on a horizontal plane (hence, they are not subjected to gravity due to structural constraints); the third link is connected to a linear joint which is used to impose the tilt of the end effector on the working plane; and the last joint is a rotational joint that allows the roll movement of the end effector for manufacturing purposes.

| Link | $l_i$ [m] | $M_i$ [kg] | $I_i$ [kg·m²] | $\bar{u}_i$ |
|------|-----------|------------|---------------|-------------|
| 1    | 0.6       | 12         | 0.36          | 55 Nm       |
| 2    | 0.4       | 6          | 0.08          | 45 Nm       |
| 3    | 1         | 3          | 0.08          | 70 N        |
| 4    | 0         | 1          | 0.08          | 25 Nm       |

Table 10.3  Parameters of the SCARA robot.

Based on the physical parameters of the robot, which are selected as shown in Table 10.3, a quantitative Euler-Lagrange model of the form (10.2) can be derived for the SCARA robot. For simplicity of notation, assume that the center of gravity of each link is located at its middle point. Table 10.3 reports, for each link the length $l_i$ of the link, the total mass $M_i$ of the link (including the actuators' masses), the rotational inertia $I_i$ of the link, and the saturation level $\bar{u}_i$ of each actuator. The viscous friction, corresponding to the matrix $R(q)$ in equation (10.2), is assumed to be zero for the purpose of illustration. In all the following simulations, the performance of the anti-windup closed-loop system has been tested in the case where uncertainties affect the model of the robot. To this aim, the actual masses of the robot have been chosen to be 20% more than the nominal ones listed in Table 10.3. Moreover, to represent a payload held by the end effector, the mass of the last link is 0.5 kg larger than that of the model.

For the SCARA robot, after verifying that this system with the saturation levels in Table 10.3 satisfies equation (10.9), it is possible to apply Algorithm 25 and to design an anti-windup compensator inducing guaranteed global asymptotic stability. In particular, the selection of the parameters of the stabilizing function $\gamma(\cdot,\cdot)$ is carried out as in steps 1 and 4 of the algorithm, thus resulting in

$$K_g = \mathrm{diag}(0.99, 0.99, 0.4, 0.99),$$
$$K_0 = \mathrm{diag}(336, 160, 160, 40),$$
$$K_q = \mathrm{diag}(1120, 560, 560, 140).$$

### 10.3.3.2  Setpoint regulation

The anti-windup closed-loop system is first tested on a setpoint regulation problem. To this aim, the unconstrained controller is chosen to be the $PID + h(q)$ controller in (10.12). The setpoint simulations carried out correspond to the following selection of the reference input:

$$r = [150 \deg, -100 \deg, 1 \text{ m}, 200 \deg]. \tag{10.17}$$

In particular, the PID gains of the unconstrained controller are selected as

$$\begin{aligned}
K_p &= 10^3 \mathrm{diag}(24, 14, 15, 15), \\
K_d &= \mathrm{diag}(820, 370, 470, 330), \\
K_i &= \mathrm{diag}(150, 120, 120, 120).
\end{aligned} \tag{10.18}$$

Moreover, to provide a fair comparison among the different responses, following standard control techniques for industrial robots, the reference signal is smoothened by a cubic interpolation joining the initial zero configuration to the desired vector goal. If the time interval to join the initial configuration to the desired position is large enough, then the corresponding control inputs are regular functions that don't reach the saturation limits, thereby avoiding undesired oscillations and possible stability loss. For the setpoint (10.17), this time interval is chosen as 1.8 seconds, which is the minimum allowable before the occurrence of saturation at the first joint.

On the other hand, the closed-loop system with anti-windup augmentation can be driven directly by the step reference input (10.17) because the anti-windup action ensures closed-loop stability in any operating condition and achieves unconstrained response recovery regardless of the occurrence of saturation. Note that this feature simplifies the regulation task because it doesn't require the computation of the smoothened reference profile, in addition to guaranteeing an improved closed-loop performance, as shown by the following simulations.

The responses of the $PID + h(q)$ controller with smoothened reference and of the $PID + h(q)$ controller with anti-windup action are reported in Figures 10.9 and 10.10. The initial conditions of the plant, controller, and anti-windup blocks are always set to zero. In both figures, the solid curves represent the anti-windup responses while the dotted curves represent the saturated responses with smoothened reference input. As a term of comparison, the dashed curves reproduce the unconstrained responses in the absence of saturation. These responses represent the ideal performance of the system with this unconstrained controller. Indeed, the (virtual) absence of saturation allows driving the closed loop directly with the step input and observing a fast and smooth convergence to the desired steady state. This type of response is, of course, only ideal and not reproducible on the saturated plant. Nevertheless, it represents a good comparison means for evaluating the performance of the two other responses in terms of their mismatch with that unconstrained one.

From the solid and dotted curves in Figures 10.9 and 10.10 it appears that the closed loop with anti-windup compensation, in addition to not requiring the computation of the smoothened reference, guarantees a faster convergence to the desired setpoint. This property is a consequence of the fact that the anti-windup action does not fear the occurrence of saturation but makes use of saturated signals to suitably drive the plant output to the desired values (see Figure 10.10). On the other hand, the reference smoothing strategy deliberately avoids the occurrence of saturation for the closed loop without anti-windup, so that undesired oscillatory behavior may not occur, thereby leading to a slower closed-loop response. Note that the reference smoothing solution is crucial for this particular application because when the saturated closed loop reaches saturation even for a very small amount, persistent and very large oscillations are triggered on the closed loop, thus leading to disastrous effects. This has been preliminarily illustrated in Section 1.2.7 on page 16 and is not reported here due to the choice of using a more industrially relevant smoothed reference solution.

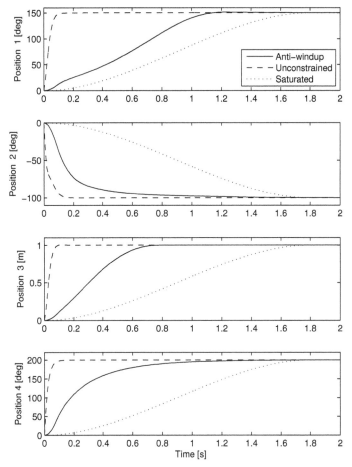

Figure 10.9 Output responses to the reference (10.17) of various closed loops for the SCARA robot.

### 10.3.3.3 Trajectory tracking

In a second set of simulations, the anti-windup scheme is tested on a trajectory tracking task where the control input demanded by the unconstrained controller persistently transits inside and outside the saturation limits. The trajectory that the robot is supposed to track corresponds to a circle on the horizontal plane centered at $(x_0, y_0) = (0.15\,\mathrm{m}, 0.15\,\mathrm{m})$ and having radius 0.1 m. The circle should be tracked by the end effector at an angular speed of $0.9\pi$ rad/s. To guarantee perfect tracking on the unconstrained plant, the unconstrained controller is selected as the computed torque + PID controller (10.13), which is driven by a suitable reference $r(t)$ and its time derivative $\dot{r}(t)$. For simplicity of exposition, the PID gains of the controller have been chosen as in (10.18). Indeed, experience indicates that the performance of the closed loop is almost independent of the selection of the PID gains. What is

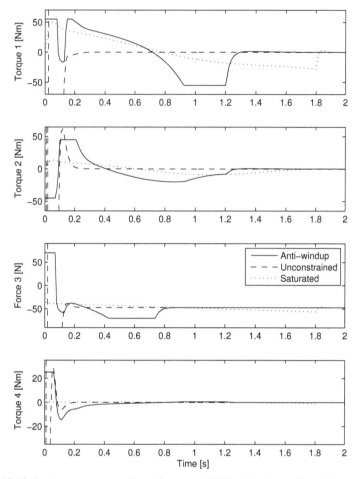

Figure 10.10 Input responses to the reference (10.17) of various closed loops for the
SCARA robot.

mostly responsible for bad saturated responses is the feedback linearizing action of
the computed torque controller.

Figures 10.11 and 10.12 show the three responses of the anti-windup (solid),
unconstrained (dashed), and saturated closed loop. The saturated closed loop is
equipped with a simple "integrator shut-off" strategy, which is a well known heuristic used to combat actuator saturation in classical control systems.

Only the responses of the first two joints are reported because they are the most
interesting ones. The three closed-loop systems are all initialized at the position
$(q_1, q_2, q_3, q_4) = (92 \text{ deg}, -109.23 \text{ deg}, 0 \text{ m}, 0 \text{ deg})$ along the circular trajectory
(with zero initial velocity), so that it is possible to appreciate the steady-state be-
havior, rather than the transient responses, which are well characterized by the
previous set of simulations.

Due to the high coupling between the two links and the aggressive control ac-

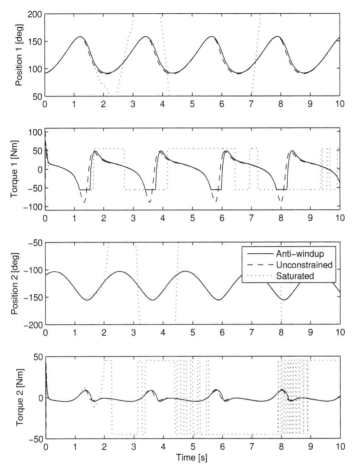

Figure 10.11  Input and output responses of the first two joints in the reference tracking
simulation for the SCARA robot.

tion arising from the feedback linearizing components of the controller, the saturated response is unable to preserve stability and exhibits persistent oscillations of
increasing amplitude. This is apparent from the top and mid-lower plots of Figure 10.11, where the dotted curves depart from the two other ones at time $t = 1.5$.
After this time, the saturated response is most of the time out of range in the plots
of Figure 10.11. Accordingly, in Figure 10.12 the dotted trajectory leaves the circular trajectory in unpredictable directions. (For graphical purposes, the dotted
responses have been limited to the time $t \in [0, 2]$ seconds on this last figure, as this
is sufficient to show its unacceptable behavior that persists in future times.) On the
other hand, the anti-windup closed-loop system performs well, due to the fact that
the anti-windup compensator is able to rapidly drive the plant state back toward
the unconstrained response. This is particularly evident in the two top plots of Figure 10.11, where, from the solid curves, it appears that the control signal is kept in

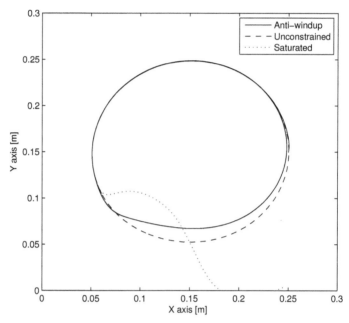

Figure 10.12  End effector trajectory on the horizontal plane in the reference tracking simulation for the SCARA robot.

saturation (lower plot) so as to drive the first joint angle toward the unconstrained
angular response (top plot). Based on this recovery property, the trajectory is cor-
rected by the anti-windup action at each period immediately after saturation and,
as shown in Figure 10.12, the effector position response only slightly deviates from
the ideal circular reference path.

### 10.3.4  PUMA robot

*10.3.4.1 Model description and anti-windup design*

The PUMA robot (Programmable Universal Machine for Assembly) is a six degrees of freedom robot with six rotational joints. By way of the six degrees of
freedom, the end-effector position and orientation can be arbitrarily configured in
the three-dimensional space. In particular, the first joint axis is vertical and the
second and third joint axes are parallel and horizontal. The remaining joint axes in-
tersect at one point: the center of the spherical wrist. The presence of the spherical
wrist allows assigning in a decoupled way the orientation and the position of the
end effector.

The physical parameters of the robot have been selected as shown in Tables 10.4
and 10.5 based on the information available from the literature. Based on these
physical parameters, a dynamical model of the robot can be written in the form
(10.2). For each link, Table 10.4 reports the total mass $M_i$, including the actuators'
masses (the mass of link 1 is irrelevant because the center of gravity of the first link

| Link i | $M_i$ [kg] | $\bar{u}_i$ [Nm] | $r_x$ [cm] | $r_y$ [cm] | $r_z$ [cm] |
|--------|------------|-------------------|------------|------------|------------|
| Link 1 | -          | 97.6              | 6.8        | 6          | -1.6       |
| Link 2 | 17.40      | 186.4             | 0          | -7         | 14         |
| Link 3 | 4.80       | 89.4              | 0          | -14.3      | 1.4        |
| Link 4 | 0.82       | 24.2              | 0          | 0          | -1.9       |
| Link 5 | 0.34       | 20.1              | 0          | 0          | 0          |
| Link 6 | 0.09       | 21.3              | 0          | 0          | 3.2        |

Table 10.4 Parameters of the PUMA robot.

| Link i | $I_{xx}$ [kg·m$^2$] | $I_{yy}$ [kg·m$^2$] | $I_{zz}$ [kg·m$^2$] | $I_{motor}$ [kg·m$^2$] |
|--------|---------------------|---------------------|---------------------|------------------------|
| link 1 | 0                   | 0                   | 0.35                | 1.14                   |
| link 2 | 0.130               | 0.524               | 0.539               | 4.71                   |
| link 3 | 0.066               | 0.0125              | 0.086               | 0.83                   |
| link 4 | 1.80e−3             | 1.80e−3             | 1.30e−3             | 0.200                  |
| link 5 | 0.30e−3             | 0.30e−3             | 0.40e−3             | 0.179                  |
| link 6 | 0.15e−3             | 0.15e−3             | 0.04e−3             | 0.193                  |

Table 10.5 Diagonal terms of the rotational inertias and motor inertias.

never moves), the maximum motor torque $\bar{u}_i$, and the coordinates $r_x$, $r_y$, $r_z$ of the center of gravity of each link, referred to the corresponding link reference frame according to the Denavit-Hartenberg standard notation. For each link, Table 10.5 also reports the rotational inertias $I_{xx}$, $I_{yy}$, $I_{zz}$ and the motor and drive inertia $I_m$, expressed in terms of the Denavit-Hartenberg reference frame, translated onto the link center of gravity. The viscous friction, corresponding to the matrix $R(q)$ in equation (10.2), is assumed to be zero for the purpose of illustration. To better illustrate the anti-windup effectiveness, assume that the actuator saturation level for the PUMA robot is one-third of the actual values reported in Table 10.4. Under these circumstances, controlling the robot becomes a harder task to accomplish, while on the other hand lighter and cheaper actuators may be mounted on the system. To illustrate robustness, in the following simulations the masses of the robot links have been increased by 20% with respect to their nominal values (listed in Table 10.4). Moreover, to represent a payload held by the end effector, the mass of the last link is 0.5 kg larger than that of the model.

This PUMA robot complies with the requirement in equation (10.9), even when the saturation levels correspond to one-third of the values in Table 10.4. Then it is possible to achieve guaranteed global asymptotic stability following the construction in Algorithm 25. According to this construction, the anti-windup parameters are selected as:

$$K_g = \text{diag}(0.99, 0.26, 0.86, 0.99, 0.99, 0.9),$$
$$K_0 = 100 \,\text{diag}(90, 100, 4, 1, 1, 1),$$

$$K_q = 100\,\mathrm{diag}(400, 380, 45, 10, 10, 10).$$

### 10.3.4.2 Setpoint regulation

The anti-windup performance has been tested on the PUMA robot by simulating the response to a setpoint regulation task. The unconstrained controller has been chosen to be a $PID + h(q)$ controller because it has been noted that alternative choices did not lead to aggressive enough responses and would have appeared too slow as compared to the anti-windup ones. On the other hand, the $PID + h(q)$ controller with a suitably smoothened reference signal via cubic interpolation proved to be the best controller selection among the typical control strategies for set-point regulation of industrial robots. To induce a nice unconstrained response on the six robot joints in the small signal case (i.e., in the absence of saturation), the PID gains of the unconstrained controller have been selected as follows:

$$K_p = \mathrm{diag}(16000, 55250, 14400, 200, 200, 200),$$
$$K_d = \mathrm{diag}(1000, 2800, 760, 25, 20, 20),$$
$$K_i = \mathrm{diag}(200, 500, 80, 4, 4, 4).$$

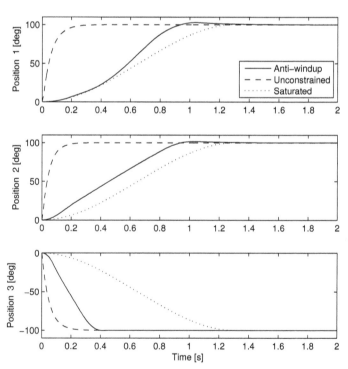

Figure 10.13 Output responses to the reference (10.19) of various closed loops for the PUMA robot.

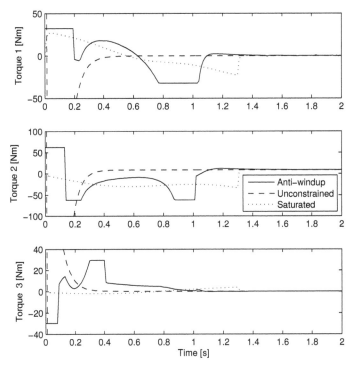

Figure 10.14  Input responses to the reference (10.19) of various closed loops for the PUMA robot.

To suitably appreciate the effects of the anti-windup action on the PUMA dynamics, a very large setpoint value has been selected as:

$$r = [100 \text{ deg}, 100 \text{ deg}, -100 \text{ deg}, 100 \text{ deg}, 100 \text{ deg}], \qquad (10.19)$$

and the following closed loops have been simulated comparatively:

1. The unconstrained closed-loop response obtained by interconnecting the controller directly to the Euler-Lagrange system without any input saturation.
2. The saturated closed-loop system corresponding to the interconnection of the unconstrained controller to the plant with saturation while the reference input is smoothed in a similar way as in Section 10.3.3. To obtain improved performance with the saturated closed loop, the PID gains have also been shrunk to the values

$$K_p = \text{diag}(790, 3980, 1410, 150, 150, 130),$$
$$K_d = \text{diag}(500, 2800, 980, 75, 75, 84),$$
$$K_i = \text{diag}(14, 60, 27, 1.8, 1.8, 1.8),$$

and based on these values the reference has been smoothed via cubic interpolation to join the initial and final joint configurations in 1.3 seconds.
3. The anti-windup closed-loop system corresponding to the interconnection of the unconstrained controller to the plant with saturation and without any ref-

erence smoothing. As in the SCARA example addressed in the previous section, due to the ability of the anti-windup compensator to handle the occurrence of saturation, the anti-windup closed-loop system is driven directly by the bare step reference. The anti-windup action will automatically take care of achieving a closed-loop response, which gracefully converges to the desired steady state regardless of the occurrence of saturation.

The simulation results are reported in Figures 10.13 and 10.14 where the position outputs and the torque inputs of the first three joints are reported. The responses on the remaining three joints are not reported because they are very comparable to the response of the third joint. Indeed, on the last joints the robot dynamics is almost negligible due to the low masses and inertias of the links and it is possible to enforce an almost decoupled behavior on the closed-loop system, which is not of much interest here.

From Figure 10.14 it is evident that the saturated response makes large use of the actuators within the saturated region. This is one of the features that allows achieving the high-performance response of Figure 10.13, where the solid curves converge to the desired set point at a faster speed than the saturated response without anti-windup. It is stressed once again that the high-performance response induced by the anti-windup action corresponds to a control scheme which does not require the extra computational burden arising from determining a cubic interpolant for the reference signal. Also in this case, as in the previous section, the saturated response without reference smoothing is not shown in the figures. That response leads to extremely large and undesirable overshoots and undershoots in the first three joints and is a quite unreasonable solution from an implementation viewpoint.

## 10.4 NOTES AND REFERENCES

The model recovery anti-windup solution illustrated in this chapter was first outlined in [MR4] and then better illustrated and extended in [MR19], which also contains the SCARA simulations reported here. The remaining examples treated here are taken from [MR18]. As in [MR18] and [MR19], the parameters of the SCARA robot have been taken from [G7], while the parameters of the PUMA robot have been taken from [G2] and [G4]. An explanation of the "integrator shut-off" heuristic used in the simulation of Figure 10.12 can be found, e.g., in [SUR5].

# Chapter Eleven

## Annotated Bibliography

### 11.1 OVERVIEW

The history of anti-windup research ranges from the early approaches dating back to the 1950s (where fixes to malfunctioning analog controllers typically corresponded to ad hoc solutions proposed and implemented by industrialists) to the numerous recently reported results (where systematic solutions to suitable formalizations of the problem provide constructive tools with guaranteed performance and stability properties).

In the past fifty years, anti-windup research migrated from the initial ad hoc attempts to more general and systematic constructions and further, to formal definitions of the underlying problem and modern high-performance solutions. Many instances of the anti-windup problem are still unsolved and are the subject of ongoing research activity. Although it is not the aim of this book to provide a survey of the existing literature on anti-windup, the rest of this chapter underlines the main phases characterizing this research strand and lists a number of relevant publications for each one.

### 11.2 PROBLEM DISCOVERY

The anti-windup problem has been known since the very early developments of automatic control. Already in the 1940s, when industries started implementing analog controllers, actuator saturation constituted one of the main pitfalls in control design. The emerging methodology was mainly due to engineering experience within the industrial environment that remained undocumented, at least until the 1950s. However, it should be noted that the control problem for systems with input saturation nonlinearities partially motivated the large research activity in absolute stability theory, which has been an important research topic since the 1940s. The following list of papers corresponds to the first research publications that appeared in this field, mainly reflecting adhoc schemes and dating from the 1950s to the late 1970s.

[PD1] J.C. Lozier. A steady-state approach to the theory of saturable servo systems. *IRE Transactions on Automatic Control*, 1:19–39, May 1956.

[PD2] H.A. Fertik and C.W. Ross. Direct digital control algorithm with antiwindup feature. *ISA Transactions*, 6(4):317–328, 1967.

[PD3] L.C. Kramer and K.W. Jenkins. A new technique for preventing direct digital control windup. In *Proceedings of the Joint Automatic Control Conference*, pages 571–577, St. Louis (MO), USA, August 1971.

[PD4] R.K. Cox and J.P. Shunta. Tracking action improves continuous control. *Chemical Engineering Progress*, 69(9):56–61, 1973.

[PD5] P. Vandenbussche. Digital transposition and extension of classical analogical control algorithms. In *Proceedings of the 6th IFAC World Congress*, pages 5.6/1–4, Boston and Cambridge (MA), USA, August 1975.

[PD6] J. Khanderia and W.L. Luyben. Experimental evaluation of digital algorithms for antireset windup. *Industrial Engineering Chemistry, Process Design and Development*, 15(2):278–285, 1976.

[PD7] N.J. Krikelis. State feedback integral control with "intelligent" integrators. *International Journal of Control*, 32(3):465–473, 1980.

[PD8] R.M. Phelan. *Automatic Control Systems*. Ithaca (NY), USA, Cornell University Press, 1977.

## 11.3 THE FIRST CONSTRUCTIVE TECHNIQUES

From the late 1970s until the late 1980s, a new phase of anti-windup research occurred. Based on the increasing demand for constructive and general anti-windup techniques, several important schemes appeared in the literature. Even though most of these new schemes still were not formally addressing stability and performance, at least they provided techniques that were applicable to a broader class of control systems, with a high probability of achieving successful anti-windup.

[FC1] J. Debelle. A control structure based upon process models. *Journal A*, 20(2):71–81, 1979.

[FC2] R. Hanus. A new technique for preventing control windup. *Journal A*, 21(1):15–20, 1980.

[FC3] A.H. Glattfelder and W. Schaufelberger. Stability analysis of single loop control systems with saturation and antireset-windup circuits. *IEEE Transactions on Automatic Control*, AC-28(12):1074–1081, 1983.

[FC4] I. Horowitz. A synthesis theory for a class of saturating systems. *International Journal of Control*, 38(1):169–187, 1983.

[FC5] P. Kapasouris and M. Athans. Multivariable control systems with saturating actuators antireset windup strategies. In *Proceedings of the American Control Conference*, pages 1579–1584, Boston (MA), USA, June 1985.

[FC6] R. Hanus, M. Kinnaert, and J.L. Henrotte. Conditioning technique, a general anti-windup and bumpless transfer method. *Automatica*, 23(6):729–739, 1987.

[FC7] A. Zheng, M. V. Kothare, and M. Morari. Anti-windup design for internal model control. *International Journal of Control*, 60(5):1015–1024, 1994.

[FC8] K.J. Åström and B. Wittenmark. *Computer Controlled Systems: Theory and Design.* Englewood Cliffs (NJ), USA, Prentice-Hall, 1984.

## 11.4 CALL FOR SYSTEMATIZATION

A landmark paper signaling the final step in the early anti-windup schemes was presented by Doyle et al. at the 1987 American Control Conference. In this paper, the authors pointed to the need for more rigorous and general solutions to the anti-windup problem by means of simple examples where the inadequacy of some of the early schemes was demonstrated. The reaction exploded in the following years, starting in 1988, with the well-known survey paper by Hanus, where most of the existing techniques were reported and commented on. One year later, a large number of papers on anti-windup were presented at the 1989 American Control Conference. The following papers characterize this period of passage:

[CS1] J.C. Doyle, R.S. Smith, and D.F. Enns. Control of plants with input saturation nonlinearities. In *Proceedings of the American Control Conference*, pages 1034–39, Minneapolis (MN), USA, June 1987.

[CS2] K.J. Åström and L. Rundqwist. Integrator windup and how to avoid it. In *Proceedings of the American Control Conference*, volume 2, pages 1693–1698, Pittsburgh (PA), USA, June 1989.

[CS3] R.S. Baheti. Simple anti-windup controllers. In *Proceedings of the American Control Conference*, volume 2, pages 1684–1686, Pittsburgh (PA), USA, June 1989.

[CS4] P.J. Campo, M. Morari, and C.N. Nett. Multivariable anti-windup and bumpless transfer: A general theory. In *Proceedings of the American Control Conference*, volume 2, pages 1706–1711, Pittsburg (PA), USA, June 1989.

[CS5] R. Hanus and M. Kinnaert. Control of constrained multivariable systems using the conditioning technique. In *Proceedings of the American Control Conference*, volume 2, pages 1712–1718, Pittsburgh (PA), USA, June 1989.

[CS6] M.F. Weilenmann, H.P. Geering, L. Guzzella, and A.H. Glattfelder. A comparison of several controllers for plants with saturating power amplifiers. In *Proceedings of the American Control Conference*, volume 2, pages 1719–1724, Pittsburgh (PA), USA, June 1989.

[CS7] P.J. Campo and M. Morari. Robust control of processes subject to saturation nonlinearities. *Computers and Chemical Engineering*, 14(4/5):343–358, 1990.

[CS8] K.S. Walgama, S. Rönnbäck, and J. Sternby. Generalisation of conditioning technique for anti-windup compensators. *IEE Proceedings on Control Theory and Applications*, 139(2):109–118, March 1992.

The following papers represent interesting surveys of the anti-windup techniques dating from the 1970s to the early 1990s.

[SUR1] R. Hanus. Antiwindup and bumpless transfer: A survey. In *Proceedings of the 12th IMACS World Congress*, volume 2, pages 59–65, Paris, France, July 1988.

[SUR2] K.S. Walgama and J. Sternby. Inherent observer property in a class of anti-windup compensators. *International Journal of Control*, 52(3):705–724, 1990.

[SUR3] M. Morari. Some control problems in the process industries. In H.L. Trentelman and J.C. Willems, editors, *Essays on Control: Perspectives in the Theory and Its Applications*, pages 55–77. Boston (MA), USA, Birkhauser, 1993.

[SUR4] M.V. Kothare, P.J. Campo, M. Morari, and N. Nett. A unified framework for the study of anti-windup designs. *Automatica*, 30(12):1869–1883, 1994.

[SUR5] C. Bohn and D.P. Atherton. An analysis package comparing PID anti-windup strategies. *IEEE Control Systems Magazine*, 15(2):34–40, April 1995.

[SUR6] Y. Peng, D. Vrancic, and R. Hanus. Anti-windup, bumpless, and conditioning transfer techniques for PID controllers. *IEEE Control Systems Magazine*, 16(4):48–57, 1996.

[SUR7] C. Edwards and I. Postlethwaite. Anti-windup and bumpless-transfer schemes. *Automatica*, 34(2):199–210, 1998.

[SUR8] A.R. Teel. A nonlinear control viewpoint on anti-windup and related problems. In *Proceedings of the 4th Nonlinear Control Systems Design Symposium (NOLCOS)*, pages 115–120, Enschede, The Netherlands, July 1998.

## 11.5 MODERN ANTI-WINDUP SCHEMES

In the last decade, improvements in technology and the advent of increasingly sophisticated control systems pointed to the need for more accurate and rigorous solutions to windup problems. In addition, the qualitative goals of anti-windup design (already formulated in the 1950s by industrialists) were interpreted and formalized within precise mathematical frameworks, so that constructive designs with guaranteed stability and performance properties could be sought. A comprehensive description and discussion of the many recently proposed modern anti-windup solutions can be found in the following two recent survey papers:

[SM1] S. Tarbouriech and M.C. Turner. Anti-windup design: An overview of some recent advances and open problems. *IET Control Theory and Application*, 3(1):1–19, 2009.

[SM2] S. Galeani, S. Tarbouriech, M.C. Turner, and L. Zaccarian. A tutorial on modern anti-windup design. *European Journal of Control*, 15(3-4):418–440, 2009.

In the remainder of this section, a classification is given of modern anti-windup schemes where the different contributions are grouped based on the underlying compensation architecture.

## Reference and measurement governors

[RG1] E.G. Gilbert and K.T. Tan. Linear systems with state and control constraints: The theory and application of maximal output admissible sets. *IEEE Transactions on Automatic Control*, 36(9):1008–1020, September 1991.

[RG2] T.J. Graettinger and B.H. Krogh. On the computation of reference signal constraints for guaranteed tracking performance. *Automatica*, 28(6):1125–1141, 1992.

[RG3] E.G. Gilbert, I. Kolmanovsky, and K.T. Tan. Discrete-time reference governors and the nonlinear control of systems with state and control constraints. *International Journal of Robust and Nonlinear Control*, 5(5):487–504, 1995.

[RG4] A. Bemporad, A. Casavola, and E. Mosca. Nonlinear control of constrained linear systems via predictive reference management. *IEEE Transactions on Automatic Control*, 42(3):340–349, March 1997.

[RG5] A. Bemporad. Reference governor for constrained nonlinear systems. *IEEE Transactions on Automatic Control*, 43(3):415–419, March 1998.

[RG6] J. McNamee and M. Pachter. The construction of the set of stable states for contrained systems with open-loop unstable plants. In *Proceedings of the American Control Conference*, pages 3364–3368, Philadelphia (PA), USA, June 1998.

[RG7] D. Angeli and E. Mosca. Command governors for constrained nonlinear systems. *IEEE Transactions on Automatic Control*, 44(4):816–820, April 1999.

[RG8] E.G. Gilbert and I. Kolmanovsky. Fast reference governors for systems with state and control constraints and disturbance inputs. *International Journal of Robust and Nonlinear Control*, 9(15):1117–1141, 1999.

[RG9] J. McNamee and M. Pachter. Efficient nonlinear reference governor algorithms for constrained tracking control systems. In *Proceedings of the American Control Conference*, pages 3549–3553, San Diego (CA), USA, June 1999.

[RG10] J.S. Shamma. Anti-windup via constrained regulation with observers. *Systems and Control Letters*, 40:1869–1883, 2000.

## $\mathcal{H}_\infty$ based anti-windup designs

[H1] S. Miyamoto and G. Vinnicombe. Robust control of plants with saturation nonlinearity based on coprime factor representation. In *Proceedings of the 35th Conference on Decision and Control*, pages 2838–2840, Kobe, Japan, December 1996.

[H2] P. Weston and I. Postlethwaite. Linear conditioning schemes for systems containing saturating actuators. In *Proceedings of the 4th Nonlinear Control Systems Design Symposium (NOLCOS)*, pages 702–707, Enschede, The Netherlands, July 1998.

[H3]  C. Edwards and I. Postlethwaite. An anti-windup scheme with closed-loop stability considerations. *Automatica*, 35(4):761–765, 1999.

[H4]  S. Crawshaw and G. Vinnicombe. Anti-windup synthesis for guaranteed $\mathcal{L}_2$ performance. In *Proceedings of the Conference on Decision and Control*, pages 1063–1068, Sidney, Australia, December 2000.

[H5]  P.F. Weston and I. Postlethwaite. Linear conditioning for systems containing saturating actuators. *Automatica*, 36(9):1347–1354, 2000.

[H6]  S. Crawshaw and G. Vinnicombe. Anti-windup for local stability of unstable plants. In *Proceedings of the American Control Conference*, pages 645–650, Anchorage (AK), USA, May 2002.

[H7]  S. Crawshaw and G. Vinnicombe. Anti-windup synthesis for guaranteed $l_2$ performance. In *Proceedings of the American Control Conference*, pages 657–661, Anchorage (AK), USA, May 2002.

[H8]  S. Crawshaw. Global and local analysis of coprime factor-based anti-windup for stable and unstable plants. In *Proceedings of the European Control Conference*, Cambridge, UK, September 2003.

[H9]  G. Herrmann, M.C. Turner, and I. Postlethwaite. Some new results on anti-windup-conditioning using the Weston-Postlethwaite approach. In *Proceedings of the 43rd Conference on Decision and Control*, pages 5047–5052, Atlantis (BA), USA, December 2004.

[H10]  M.C. Turner, G. Herrmann, and I. Postlethwaite. Further results on full-order anti-windup synthesis: Exploiting the stability multiplier. In *Proceedings of the IFAC NonLinear Control Systems Symposium*, pages 1385–1390, Stuttgart, Germany, September 2004.

[H11]  M.C. Turner, G. Herrmann, and I. Postlethwaite. Incorporating robustness requirements into antiwindup design. *IEEE Transactions on Automatic Control*, 52(10):1842–1855, 2007.

**Model recovery anti-windup designs**

[MR1]  J.K. Park and C.H. Choi. Dynamic compensation method for multivariable control systems with saturating actuators. *IEEE Transactions on Automatic Control*, 40(9):1635–1640, 1995.

[MR2]  J.K. Park and C.H. Choi. Dynamical anti-reset windup method for discrete-time saturating systems. *Automatica*, 33(6):1055–1072, 1997.

[MR3]  A.R. Teel and N. Kapoor. The $\mathcal{L}_2$ anti-windup problem: Its definition and solution. In *Proceedings of the European Control Conference*, Brussels, Belgium, July 1997.

[MR4]  A.R. Teel and N. Kapoor. Uniting local and global controllers. In *Proceedings of the European Control Conference*, Brussels, Belgium, July 1997.

[MR5]  A.R. Teel. Anti-windup for exponentially unstable linear systems. *International Journal of Robust and Nonlinear Control*, 9:701–716, 1999.

[MR6] C. Barbu, R. Reginatto, A.R. Teel, and L. Zaccarian. Anti-windup for exponentially unstable linear systems with inputs limited in magnitude and rate. In *Proceedings of the American Control Conference*, pages 1230–1234, Chicago (IL), USA, June 2000.

[MR7] L. Zaccarian and A.R. Teel. A benchmark example for anti-windup synthesis in active vibration isolation tasks and an $\mathcal{L}_2$ anti-windup solution. *European Journal of Control*, 6(5):405–420, 2000.

[MR8] J.K. Park, C.H. Choi, and H. Choo. Dynamic anti-windup method for a class of time-delay control systems with input saturation. *International Journal of Robust and Nonlinear Control*, 10:457–488, 2000.

[MR9] L. Zaccarian, A.R. Teel, and J. Marcinkowski. Anti-windup for an active vibration isolation device: Theory and experiments. In *Proceedings of the American Control Conference*, pages 3585–3589, Chicago (IL), USA, June 2000.

[MR10] L. Zaccarian and A.R. Teel. Anti-windup, bumpless transfer and reliable designs: A model-based approach. In *Proceedings of the American Control Conference*, pages 4902–4907, Arlington (VA), USA, June 2001.

[MR11] L. Zaccarian and A.R. Teel. Nonlinear $\mathcal{L}_2$ anti-windup design: An LMI-based approach. In *Proceedings of the Nonlinear Control Systems Design Symposium (NOLCOS)*, pages 1298–1303, Saint Petersburg, Russia, July 2001.

[MR12] C. Barbu, R. Reginatto, A.R. Teel, and L. Zaccarian. Anti-windup for exponentially unstable linear systems with rate and magnitude limits. In V. Kapila and K. Grigoriadis, editors, *Actuator Saturation Control*, chapter 1, pages 1–31. New York, Marcel Dekker, 2002.

[MR13] A. Bemporad, A.R. Teel, and L. Zaccarian. $\mathcal{L}_2$ anti-windup via receding horizon optimal control. In *Proceedings of the American Control Conference*, pages 639–644, Anchorage (AK), USA, May 2002.

[MR14] F. Morabito, A.R. Teel, and L. Zaccarian. Results on anti-windup design for Euler-Lagrange systems. In *Proceedings of the IEEE Conference on Robotics and Automation*, pages 3442–3447, Washington (DC), USA, May 2002.

[MR15] L. Zaccarian and A.R. Teel. A common framework for anti-windup, bumpless transfer and reliable designs. *Automatica*, 38(10):1735–1744, 2002.

[MR16] G. Grimm, A.R. Teel, and L. Zaccarian. The $l_2$ anti-windup problem for discrete-time linear systems: Definition and solutions. In *Proceedings of the American Control Conference*, pages 5329–5334, Denver (CO), USA, June 2003.

[MR17] A. Bemporad, A.R. Teel, and L. Zaccarian. Anti-windup synthesis via sampled-data piecewise affine optimal control. *Automatica*, 40(4):549–562, 2004.

[MR18] F. Morabito, S. Nicosia, A.R. Teel, and L. Zaccarian. Measuring and improving performance in anti-windup laws for robot manipulators. In B. Siciliano, A. De Luca, C. Melchiorri, and G. Casalino, editors, *Advances in Control of Articulated and Mobile Robots*, chapter 3, pages 61–85. New York, Springer Tracts in Advanced Robotics, 2004.

[MR19] F. Morabito, A.R. Teel, and L. Zaccarian. Nonlinear anti-windup applied to Euler-Lagrange systems. *IEEE Transactions on Robotics and Automation*, 20(3):526–537, 2004.

[MR20] D. Nešić, A.R. Teel, and L. Zaccarian. $\mathcal{L}_2$ anti-windup for linear dead-time systems. In *Proceedings of the American Control Conference*, pages 5280–5285, Boston (MA), USA, June 2004.

[MR21] L. Zaccarian and A.R. Teel. Nonlinear scheduled anti-windup design for linear systems. *IEEE Transactions on Automatic Control*, 49(11):2055–2061, 2004.

[MR22] L. Zaccarian, E. Weyer, A.R. Teel, Y. Li, and M. Cantoni. Anti-windup for marginally stable plants applied to open water channels. In *Proceedings of the Asian Control Conference*, pages 1702–1710, Melbourne (VIC), Australia, July 2004.

[MR23] S. Galeani, A.R. Teel, and L. Zaccarian. Output feedback compensators for weakened anti-windup of additively perturbed systems. In *Proceedings of the IFAC World Congress*, Prague, Czech Republic, July 2005.

[MR24] L. Zaccarian, D. Nešić, and A.R. Teel. $\mathcal{L}_2$ anti-windup for linear dead-time systems. *Systems and Control Letters*, 54(12):1205–1217, 2005.

[MR25] L. Zaccarian and A.R. Teel. The $\mathcal{L}_2$ ($l_2$) bumpless transfer problem: Its definition and solution. *Automatica*, 41(7):1273–1280, 2005.

[MR26] S. Galeani and A.R. Teel. On a performance-robustness trade-off intrinsic to the natural anti-windup problem. *Automatica*, 42(11):1849–1861, 2006.

[MR27] S. Galeani, S. Onori, A.R. Teel, and L. Zaccarian. Nonlinear $\mathcal{L}_2$ anti-windup for enlarged stability regions and regional performance. In *Proceedings of the Symposium on Nonlinear Control Systems (NOLCOS)*, pages 539–544, Pretoria, South Africa, August 2007.

[MR28] S. Galeani, S. Onori, A.R. Teel, and L. Zaccarian. Regional, semiglobal, global nonlinear anti-windup via switching design. In *Proceedings of the European Control Conference*, pages 5403–5410, Kos, Greece, July 2007.

[MR29] S. Galeani, S. Onori, and L. Zaccarian. Nonlinear scheduled control for linear systems subject to saturation with application to anti-windup control. In *Proceedings of the Conference on Decision and Control*, pages 1168–1173, New Orleans (LA), USA, December 2007.

[MR30] S. Galeani, A.R. Teel, and L. Zaccarian. Constructive nonlinear anti-windup design for exponentially unstable linear plants. *Systems and Control Letters*, 56(5):357–365, 2007.

[MR31] L. Pagnotta, L. Zaccarian, A. Constantinescu, and S. Galeani. Anti-windup applied to adaptive rejection of unknown narrow band disturbances. In *Proceedings of the European Control Conference*, pages 150–157, Kos, Greece, July 2007.

[MR32] G. Grimm, A.R. Teel, and L. Zaccarian. The $l_2$ anti-windup problem for discrete-time linear systems: definition and solutions. *Systems and Control Letters*, 57(4):356–364, 2008.

[MR33] F. Forni, S. Galeani, and L. Zaccarian. Model recovery anti-windup for plants with rate and magnitude saturation. In *Proceedings of the European Control Conference*, Budapest, Hungary, August 2009.

**Direct linear LMI-based anti-windup designs**

[MI1] V.M. Marcopoli and S.M. Phillips. Analysis and synthesis tools for a class of actuator-limited multivariable control systems: A linear matrix inequality approach. *International Journal of Robust and Nonlinear Control*, 6:1045–1063, 1996.

[MI2] M. Saeki and N. Wada. Design of anti-windup controller based on matrix inequalities. In *Proceedings of the 35th Conference on Decision and Control*, pages 261–262, Kobe, Japan, December 1996.

[MI3] M.V. Kothare and M. Morari. Multivariable anti-windup controller synthesis using multi-objective optimization. In *Proceedings of the American Control Conference*, pages 3093–3097, Albuquerque (NM), USA, June 1997.

[MI4] M.V. Kothare and M. Morari. Stability analysis of anti-windup control schemes: A review and some generalizations. In *Proceedings of the 4th European Control Conference*, Brussels, Belgium, July 1997.

[MI5] B.G. Romanchuk and M.C. Smith. Incremental gain analysis of piecewise linear systems and application to the antiwindup problem. *Automatica*, 35(7):1275–1283, 1999.

[MI6] E.F. Mulder and M.V. Kothare. Synthesis of stabilizing anti-windup controllers using piecewise quadratic Lyapunov functions. In *Proceedings of the American Control Conference*, pages 3239–3243, Chicago (IL), USA, June 2000.

[MI7] M. Saeki and N. Wada. Synthesis of a static anti-windup compensator via linear matrix inequalities. In *Proceedings of the 3rd IFAC Symposium on Robust Control Design*, Prague, Czech Republic, June 2000.

[MI8] G. Grimm, Jay Hatfield, I. Postlethwaite, A.R. Teel, M.C. Turner, and L. Zaccarian. Experimental results in optimal linear anti-windup compensation. In *Proceedings of the Conference on Decision and Control*, pages 2657–2662, Orlando (FL), USA, December 2001.

[MI9] G. Grimm, I. Postlethwaite, A.R. Teel, M.C. Turner, and L. Zaccarian. Case studies using linear matrix inequalities for optimal anti-windup synthesis. In

*Proceedings of the European Control Conference*, Porto, Portugal, September 2001.

[MI10] G. Grimm, I. Postlethwaite, A.R. Teel, M.C. Turner, and L. Zaccarian. Linear matrix inequalities for full and reduced order anti-windup synthesis. In *Proceedings of the American Control Conference*, pages 4134–4139, Arlington (VA), USA, June 2001.

[MI11] E.F. Mulder, M.V. Kothare, and M. Morari. Multivariable anti-windup controller synthesis using linear matrix inequalities. *Automatica*, 37(9):1407–1416, September 2001.

[MI12] Y.Y. Cao, Z. Lin, and D.G. Ward. An antiwindup approach to enlarging domain of attraction for linear systems subject to actuator saturation. *IEEE Transactions on Automatic Control*, 47(1):140–145, 2002.

[MI13] Y.Y. Cao, Z. Lin, and D.G. Ward. Antiwindup design for linear systems subject to input saturation. *Journal of Guidance Navigation and Control*, 25(3):455–463, 2002.

[MI14] G. Grimm, A.R. Teel, and L. Zaccarian. Results on linear LMI-based external anti-windup design. In *Proceedings of the Conference on Decision and Control*, pages 299–304, Las Vegas (NV), USA, December 2002.

[MI15] G. Grimm, A.R. Teel, and L. Zaccarian. Robust LMI-based linear anti-windup design: optimizing the unconstrained response recovery. In *Proceedings of the Conference on Decision and Control*, pages 293–298, Las Vegas (NV), USA, December 2002.

[MI16] E.F. Mulder and M.V. Kothare. Static anti-windup controller synthesis using simultaneous convex design. In *Proceedings of the American Control Conference*, pages 651–656, Anchorage (AK), USA, June 2002.

[MI17] J.M. Gomes da Silva Jr and S. Tarbouriech. Anti-windup design with guaranteed regions of stability: An LMI-based approach. In *Proceedings of the Conference on Decision and Control*, pages 4451–4456, Maui (HI), USA, December 2003.

[MI18] G. Grimm, J. Hatfield, I. Postlethwaite, A.R. Teel, M.C. Turner, and L. Zaccarian. Antiwindup for stable linear systems with input saturation: An LMI-based synthesis. *IEEE Transactions on Automatic Control*, 48(9):1509–1525, September 2003.

[MI19] G. Grimm, I. Postlethwaite, A.R. Teel, M.C. Turner, and L. Zaccarian. Case studies using LMIs in anti-windup synthesis for stable linear systems with input saturation. *European Journal of Control*, 9(5):459–469, 2003.

[MI20] G. Grimm, A.R. Teel, and L. Zaccarian. Establishing Lipschitz properties of multivariable algebraic loops with incremental sector nonlinearities. In *Proceedings of the Conference on Decision and Control*, Maui (HI), USA, December 2003.

[MI21] S. Tarbouriech, J.M. Gomes da Silva Jr., and G. Garcia. Delay-dependent anti-windup loops for enlarging the stability region of time delay systems with saturating inputs. *ASME Journal of Dynamic Systems, Measurement and Control*, 125:265–267, June 2003.

[MI22] G. Grimm, A.R. Teel, and L. Zaccarian. Linear LMI-based external anti-windup augmentation for stable linear systems. *Automatica*, 40(11):1987–1996, 2004.

[MI23] G. Grimm, A.R. Teel, and L. Zaccarian. Robust linear anti-windup synthesis for recovery of unconstrained performance. *International Journal of Robust and Nonlinear Control*, 14(13–15):1133–1168, 2004.

[MI24] A. Syaichu-Rohman and R.H. Middleton. Anti-windup schemes for discrete time systems: An LMI-based design. In *Proceedings of the Asian Control Conference*, pages 554–561, Melbourne (VIC), Australia, July 2004.

[MI25] M.C. Turner and I. Postlethwaite. A new perspective on static and low order anti-windup synthesis. *International Journal of Control*, 77(1):27–44, 2004.

[MI26] F. Wu and B. Lu. Anti-windup control design for exponentially unstable LTI systems with actuator saturation. *Systems and Control Letters*, 52(3–4):304–322, 2004.

[MI27] J.M. Gomes da Silva Jr and S. Tarbouriech. Anti-windup design with guaranteed regions of stability: An LMI-based approach. *IEEE Transactions on Automatic Control*, 50(1):106–111, 2005.

[MI28] S. Galeani, M. Massimetti, A.R. Teel, and L. Zaccarian. Reduced order linear anti-windup augmentation for stable linear systems. *International Journal of Systems Science*, 37(2):115–127, 2006.

[MI29] S. Galeani, S. Onori, A.R. Teel, and L. Zaccarian. Further results on static linear anti-windup design for control systems subject to magnitude and rate saturation. In *Proceedings of the Conference on Decision and Control*, pages 6373–6378, San Diego (CA), USA, December 2006.

[MI30] M. Massimetti, L. Zaccarian, T. Hu, and A.R. Teel. LMI-based linear anti-windup for discrete time linear control systems. In *Proceedings of the Conference on Decision and Control*, pages 6173–6178, San Diego (CA), USA, December 2006.

[MI31] J.M. Biannic, C. Roos, and S. Tarbouriech. A practial method for fixed-order anti-windup design. In *Proceedings of the 17th IFAC Symposium on Nonlinear Control Systems*, pages 527–532, Pretoria, South Africa, August 2007.

[MI32] S. Galeani, S. Onori, A.R. Teel, and L. Zaccarian. A magnitude and rate saturation model and its use in the solution of a static anti-windup problem. *Systems and Control Letters*, 57(1):1–9, 2008.

[MI33] T. Hu, A.R. Teel, and L. Zaccarian. Anti-windup synthesis for linear control systems with input saturation: Achieving regional, nonlinear performance. *Automatica*, 44(2):512–519, 2008.

[MI34] C. Roos and J.M. Biannic. A convex characterization of dynamically-constrained anti-windup controllers. *Automatica*, 44(9):2449–2452, 2008.

**Other schemes**

[OS1] S.F. Graebe and A.L.B. Ahlén. Dynamic transfer among alternative controllers and its relation to antiwindup controller design. *IEEE Transactions on Control Systems Technology*, 4(1):92–99, January 1996.

[OS2] T.A. Kendi and F.J. Doyle III. An anti-windup scheme for multivariable nonlinear systems. *Journal of Process Control*, 7(5):329–343, 1997.

[OS3] S. Valluri and M. Soroush. Input constraint handling and windup compensation in nonlinear control. In *Proceedings of the American Control Conference*, pages 1734–1738, Albuquerque (NM), USA, June 1997.

[OS4] N. Kapoor, A.R. Teel, and P. Daoutidis. An anti-windup design for linear systems with input saturation. *Automatica*, 34(5):559–574, 1998.

[OS5] W. Niu and M. Tomizuka. A robust anti-windup controller design for motion control systems with asymptotic tracking subject to actuator saturation. In *Proceedings of the 37th Conference on Decision and Control*, pages 915–920, Tampa (FL), USA, December 1998.

[OS6] Y. Peng, D. Vrančić, R. Hanus, and S.S.R. Weller. Anti-windup designs for multivariable controllers. *Automatica*, 34(12):1559–1565, 1998.

[OS7] N. Kapoor and P. Daoutidis. An observer-based anti-windup scheme for non-linear systems with input constraints. *International Journal of Control*, 72(1):18–29, 1999.

[OS8] Q. Hu and G.P. Rangaiah. Anti-windup schemes for uncertain nonlinear systems. *IEE Proceedings on Control Theory Applications*, 147(3):321–329, May 2000.

[OS9] J.A. De Dona, G.C. Goodwin, and M.M. Seron. Anti-windup and model predictive control: Reflections and connections. *European Journal of Control*, 6(5):467–477, 2002.

[OS10] A. Rantzer. A performance criterion for anti-windup compensators. *European Journal of Control*, 6(5):449–452, 2002.

[OS11] P. Hippe. Windup prevention for unstable systems. *Automatica*, 39(11):1967–1973, 2003.

[OS12] A.H. Glattfelder and W. Schaufelberger. *Control Systems with Input and Output Constraints*. London, UK, Springer-Verlag, 2003.

[OS13] P. Hippe. *Windup in Control: Its Effects and Their Prevention*. Surrey, UK, Springer-Verlag, 2006.

**Anti-windup applications**

[AP1]  N. Kapoor and A.R. Teel. A dynamic windup compensation scheme applied to a turbofan engine. In *Proceedings of the 36th Conference on Decision and Control*, pages 4689 – 4694, San Diego (CA), USA, December 1997.

[AP2]  A.R. Teel and J.B. Buffington. Anti-windup for an F-16's daisy chain control allocator. In *Proceedings of the AIAA GNC Conference*, pages 748–754, New Orleans (LA), USA, August 1997.

[AP3]  A.R. Teel, O.E. Kaiser, and R.M. Murray. Uniting local and global controllers for the Caltech ducted fan. In *Proceedings of the American Control Conference*, volume 3, pages 1539 –1543, Albuquerque (NM), USA, June 1997.

[AP4]  K.J. Park, H. Lim, T. Basar, and C.H. Choi. Anti-windup compensator for active queue management in TCP networks. *Control Engineering Practice*, 11(10):1127–1142, October 2003.

[AP5]  G. Herrmann, MC Turner, I. Postlethwaite, and G. Guo. Practical implementation of a novel anti-windup scheme in a HDD-dual-stage servo-system. *IEEE/ASME Transactions on Mechatronics*, 9(3):580–592, 2004.

[AP6]  C. Barbu, S. Galeani, A.R. Teel, and L. Zaccarian. Nonlinear anti-windup for manual flight control. *International Journal of Control*, 78(14):1111–1129, September 2005.

[AP7]  C. Roos and J.M. Biannic. Aircraft-on-ground lateral control by an adaptive LFT-based anti-windup approach. In *Proceedings of the IEEE Conference on Control Applications*, pages 2207–2212, Munich, Germany, October 2006.

[AP8]  A.R. Teel, L. Zaccarian, and J. Marcinkowski. An anti-windup strategy for active vibration isolation systems. *Control Engineering Practice*, 14(1):17–27, 2006.

[AP9]  C. Roos, J. Biannic, S. Tarbouriech, and C. Prieur.  On-ground aircraft control design using an LPV anti-windup approach. In D. Bates and M. Hagstrom, editors, *Nonlinear Analysis and Synthesis Techniques for Aircraft Control*, pages 117–145. Springer-Verlag, London, LNCIS, 2007.

[AP10]  L. Zaccarian, E. Weyer, A.R. Teel, Y. Li, and M. Cantoni. Anti-windup for marginally stable plants and its application to open water channel control systems. *Control Engineering Practice*, 15(2):261–272, 2007.

[AP11]  G. Herrmann, B. Hredzak, MC Turner, I. Postlethwaite, and G. Guo. Discrete robust anti-windup to improve a novel dual-stage large-span track-seek/following method. *IEEE Transactions on Control Systems Technology*, 16(6):1342–1351, 2008.

## 11.6  ADDITIONAL REFERENCES

Some of the results presented in this book are taken from references not directly related to anti-windup design but that become relevant in some anti-windup design contexts. These additional references are organized here in two groups, the first one comprising papers addressing nonlinear systems with saturation specifically. These are listed next.

[SAT1]  P. Kapasouris, M. Athans, and G. Stein. Design of feedback control systems for stable plants with saturating actuators. In *Proceedings of the Conference on Decision and Control*, pages 469–479, Austin (TX), USA, December 1988.

[SAT2]  A.R. Teel. Global stabilization and restricted tracking for multiple integrators with bounded controls. *Systems and Control Letters*, 18:165–171, 1992.

[SAT3]  H.J. Sussmann, E.D. Sontag, and Y. Yang. A general result on the stabilization of linear systems using bounded controls. *IEEE Transactions on Automatic Control*, 39(12):2411–2424, 1994.

[SAT4]  K.D. Hammett, W.C. Reigelsperger, and S.S. Banda. High angle of attack short period flight control design with thrust vectoring. In *Proceedings of the 1995 American Control Conference*, pages 170–174, Seattle (WA), USA, June 1995.

[SAT5]  A.R. Teel. Semi-global stabilizability of linear null controllable systems with input nonlinearities. *IEEE Transactions on Automatic Control*, 40(1):96–100, January 1995.

[SAT6]  J.M. Berg, K.D. Hammett, C.A. Schwartz, and S.S. Banda. An analysis of the destabilizing effect of daisy chained rate-limited actuators. *IEEE Transactions on Control Systems Technology*, 4(2):171–176, March 1996.

[SAT7]  A. Megretski. $\mathcal{L}_2$ BIBO output feedback stabilization with saturated control. In *Proceedings of the 13th Triennial IFAC World Congress*, pages 435–440, San Francisco (CA), USA, 1996.

[SAT8]  A.R. Teel. On $\mathcal{L}_2$ performance induced by feedbacks with muiltiple saturations. *ESAIM: Control, Optimization, and Calculus of Variations*, 1:225–240, September 1996 (http://www.emath.fr/cocv/).

[SAT9]  A.A. Stoorvogel, A. Saberi, and G. Shi. On achieving $\mathcal{L}_p$ ($l_p$) performance with global internal stability for linear plants with saturating actuators. In *Proceedings of the 38th Conference on Decision and Control*, pages 2762–2767, Phoenix (AZ), USA, December 1999.

[SAT10]  A. Bemporad, M. Morari, V. Dua, and E.N. Pistikopoulos. The explicit linear quadratic regulator for constrained systems. *Automatica*, 38(1):3–20, 2002.

[SAT11]  T. Hu, Z. Lin, and B.M. Chen. An analysis and design method for linear systems subject to actuator saturation and disturbance. *Automatica*, 38(2):351–359, 2002.

[SAT12]  A. Syaichu-Rohman and R.H. Middleton. On the robustness of multivariable algebraic loops with sector nonlinearities. In *Proceedings of the Conference on Decision and Control*, pages 1054–1059, Las Vegas (NV), USA, December 2002.

[SAT13] A. Syaichu-Rohman, RH Middleton, and MM Seron. A multivariable nonlinear algebraic loop as a QP with application to MPC. In *Proceedings of the European Control Conference*, 2003.

[SAT14] H. Fang, Z. Lin, and T. Hu. Analysis and control design of linear systems in the presence of actuator saturation and $\mathcal{L}_2$ disturbances. *Automatica*, 40(7):1229–1238, 2004.

[SAT15] T. Hu, A.R. Teel, and L. Zaccarian. Stability and performance for saturated systems via quadratic and non-quadratic Lyapunov functions. *IEEE Transactions on Automatic Control*, 51(11):1770–1786, November 2006.

The second group involves more general references, most of them focusing in various ways on nonlinear systems or on computational tools or on specific application studies cited in the examples discussed in the book.

[G1] M. Grant and S. Boyd. CVX: MATLAB software for disciplined convex programming. (http://stanford. edu/ boyd/cvx).

[G2] B. Armstrong, O. Khatib, and J. Burdick. The explicit dynamic model and inertial parameters of the PUMA 560 arm. In *Proceedings of the IEEE Conference on Robotics and Automation*, pages 510–518, San Francisco (CA), USA, April 1986.

[G3] A.S. Morse, D.Q. Mayne, and G.C. Goodwin. Applications of hysteresis switching in parameter adaptive control. *IEEE Transactions on Automatic Control*, 37(9):1343–1354, September 1992.

[G4] P.I. Corke and B. Armstrong-Hélouvry. A search for consensus among model parameters reported fot the PUMA 560 robot. In *Proceedings of the IEEE International Conference on Robotics and Automation*, 1608–1613, San Diego (CA), USA, May 1994.

[G5] P. Gahinet and P. Apkarian. A linear matrix inequality approach to $\mathcal{H}_\infty$ control. *International Journal of Robust and Nonlinear Control*, 4:421–448, 1994.

[G6] T. Iwasaki and R.E. Skelton. All controllers for the general $\mathcal{H}_\infty$ control problem: LMI existence conditions and state space formulas. *Automatica*, 30(8):1307–17, August 1994.

[G7] G. Mester. Adaptive force and position control of rigid-link flexible-joint SCARA robots. In *Proceedings of the IEEE Industrial Electronics Conference*, volume 3, pages 1639–1644, Bologna, Italy, September 1994.

[G8] P. Gahinet. Explicit controller formulas for LMI-based $\mathcal{H}_\infty$ synthesis. *Automatica*, 32(7):1007–1014, July 1996.

[G9] M. Jankovic, R. Sepulchre, and PV Kokotovic. Constructive Lyapunov stabilization of nonlinear cascade systems. *IEEE Transactions on Automatic Control*, 41(12):1723–1735, 1996.

[G10] F. Mazenc and L. Praly. Adding integrations, saturated controls, and stabilization of feedforward systems. *IEEE Transactions on Automatic Control*, 41(11):1559–1578, November 1996.

[G11] S. Majhi and D.P. Atherton. Modified Smith predictor and controller for processes with time delay. *IEE Proceedings on Control Theory Applications*, 146(5):359–366, September 1999.

[G12] A.R. Teel. Asymptotic convergence from $\mathcal{L}_p$ stability. *IEEE Transactions on Automatic Control*, 44(11):2169–2170, November 1999.

[G13] D.Q. Mayne, J.B. Rawlings, C.V. Rao, and P.O.M. Scokaert. Constrained model predictive control: Stability and optimality. *Automatica*, 36(6):789–814, 2000.

[G14] W. Desch, H. Logemann, EP Ryan, and ED Sontag. Meagre functions and asymptotic behaviour of dynamical systems. *Nonlinear Analysis*, 44(8):1087–1109, 2001.

[G15] J. Lofberg. YALMIP: A toolbox for modeling and optimization in MAT-LAB. In *Proceedings of the CACSD Conference*, Taipei, Taiwan, 2004.

[G16] Y. Wang and S. Boyd. Fast model predictive control using online optimization. In *Proceedings of the IFAC World Congress*, pages 6974–6997, 2008.

[G17] S. Boyd, L. El Ghaoui, E. Feron, and V. Balakrishnan. *Linear Matrix Inequalities in System and Control Theory*. Philadelphia (PA), USA, Society for Industrial and Applied Mathematics, 1994.

[G18] F.H. Clarke. *Optimization and Nonsmooth Analysis*. Philadelphia (PA), USA, Society for Industrial and Applied Mathematics, 1990.

[G19] P. Gahinet, A. Nemirovski, A.J. Laub, and M. Chilali. *LMI Control Toolbox*. The MathWorks Inc., 1995.

[G20] H.K. Khalil. *Nonlinear Systems*, 3rd ed. Englewood Cliffs (NJ), USA, Prentice Hall, 2002.

[G21] M. Morari and E. Zafiriou. *Robust Process Control*. Englewood Cliffs (NJ), USA, Prentice Hall, 1989.

# Index

active constraint, 214
algebraic loop, 86, 213
  strongly well-posed, *see* strong well-posedness
  well-posed, 37
algorithms, summary, xi–xiv
angle of attack, 229
anti-windup, 3
  augmentation, 32–33
  continuous-time vs. sampled data, 34
  direct linear, *see* direct linear anti-windup
  external vs. full authority, 34
  filter, 32
  history, 269
  linear vs. nonlinear, 34
  LMI-based synthesis, 70–72
  model recovery, *see* model recovery anti-windup
  objectives
    global vs. regional, 27–29
    qualitative, 26–31
  quantitative objectives, 40–47
  response, *see* response, anti-windup
  static vs. dynamic, 34
approximate dynamics, 165
arc length, 187
averaged reference, 210

band-limited white noise, 242
bandwidth, 187
basin of attraction, 202, 215, 222
bumpless transfer, 235–240
bypass duct area, 180

Caltech ducted fan, *see* example, Caltech ducted
  fan
center of gravity, 259, 265
centrifugal, 246, 255
Cholesky factorization, 213
circular trajectory, 261
closed loop
  anti-windup, 32–33
  compact, 78, 111
  external, 112
  full-authority, 112
  representation, 78–80, 110–113
    external, 80
    full-authority, 79

saturated, 26
  unconstrained, 23–24
command governor, 22
compact closed loop, *see* closed loop, compact
computed torque controller, *see* controller, computed torque
constrained quadratic
  optimization, 208
  programming, 211
controller
  computed torque, 16, 250, 253
  loop shaping, 20, 102
  LQG, 10, 14, 90, 239
  LQR, 228
  PI, 6
  PID, 4, 16, 88, 250, 266
  two degrees of freedom, 13
  unconstrained, 23
convergence to zero, 40
convex
  combination, 205
  solution set, 52
core engine pressure ratio, 180
Coriolis, 246, 255
cubic interpolation, 260
cvx, 57

daisy chain, 229
damped mass-spring, *see* example, damped mass-
  spring
dead-time plant, 232
deadzone, 48, 59
decentralized
  constraint, 209
  saturation, 25
delay, 232
Denavit-Hartenberg, 265
detectable pair, 196, 209
direct linear anti-windup, 75–154
  dynamic, 109–154
  external, 92–97, 130–141
  full-authority, 81–92, 98–107, 114–130, 143–
    152
  global, 81–97, 114–141
  plant-order, 114–122, 130–135, 143–146
  reduced-order, 122–130, 135–141, 146–152
  regional, 98–107, 141–152